AUSTRALIA

Phillip Knightley is the author of many books. He is best known for *The First Casualty*, *The Second Oldest Profession*, *Philby, KGB Masterspy*, *The Secret Lives of Lawrence of Arabia* and *A Hack's Progress*. He divides his time between Britain, Australia and India.

ALSO BY PHILLIP KNIGHTLEY

The First Casualty
The Second Oldest Profession
Philby, KGB Masterspy
An Affair of State
The Rise and Fall of the House of Vestey
A Hack's Progress
The Death of Venice
Suffer the Children
The Secret Lives of Lawrence of Arabia

Phillip Knightley

AUSTRALIA

A Biography of a Nation

Published by Vintage 2001

2 4 6 8 10 9 7 5 3

First published in Great Britain by
Jonathan Cape 2000

Vintage
Random House, 20 Vauxhall Bridge Road,
London SW1V 2SA

Random House Australia (Pty) Limited
20 Alfred Street, Milsons Point, Sydney,
New South Wales 2061, Australia

Random House New Zealand Limited
18 Poland Road, Glenfield, Auckland 10, New Zealand

Random House (Pty) Limited
Endulini, 5A Jubilee Road, Parktown 2193, South Africa

The Random House Group Limited Reg. No. 954009
www.randomhouse.co.uk

A CIP catalogue record for this book
is available from the British Library

ISBN 0 09 977291 4

Papers used by Random House are natural, recyclable products made from wood grown in sustainable forests. The manufacturing processes conform to the environmental regulations of the country of origin

Printed and bound in Great Britain by
Bookmarque Ltd, Croydon, Surrey

To my extended family, in particular Maria Tarifa.

CONTENTS

Terra australis incognita, a Latin term meaning "the unknown south land". Ancient geographers had always suspected its existence . . . The application of the anglicised version, *Australia*, to the whole continent, belonged to its circumnavigator, Matthew Flinders, in 1805.

– *The Oxford Companion to Australian History*

CHAPTER ONE

A NEW DAY DOWN UNDER

'The Customs Officer looked at our four children and our huge pile of luggage and asked us if that was all there was. We assured him it was and he said, 'Welcome to Australia' and waved us through. We were still stunned by this lack of formality when we came out into the arrivals hall. And there was Gough Whitlam, the Prime Minister of Australia, and his wife standing at the front of a large crowd and they all burst out singing *Waltzing Matilda*. What a welcome, we thought. What a country.'

– Mary Gates, immigrant, 1973

The new day's sun, rising out of the Pacific Ocean, first touches Australia at Mount Warning on the coast of New South Wales. The morning star dims, the Southern Cross fades and the sun's first pale rays light up the swirling mists that cling to the mountain's twin peaks. As the sun grows stronger, the mist slowly lifts revealing as far as the eye can see the deep greens of a subtropical rainforest – huge trees, ferns and vines, lichens and mosses, streams and pools.

Aboriginals call Mount Warning *Wollumbin*, meaning 'fighting chief of the mountains', because when surrounded by thunder and flickering lightning it appears as if the gods are fighting there. It is an awesome, dramatic and sacred place, the remnant of a volcanic cone which burst forth from the earth millions of years ago while Australia was separating from Gondwana, the great southern continent which once included Antarctica, South America, Africa and India.

Captain James Cook, of His Britannic Majesty's Royal Navy, the man who discovered for Britain more of the earth's surface than anyone else, sailed north past Mount Warning in HMS *Endeavour* in 1770 on his first southern voyage. On 15 May, Cook's scientist, Joseph Banks, recorded: 'At sunset a remarkably peakd hill was in sight, 5 or 6 leagues off in the country, which about it was well wooded and lookd beautiful as well as fertile.' The next day *Endeavour* struck reefs, so Cook named the easily visible twin peaks Mount Warning as a guide to other navigators he knew would pass that way.

Cook did not, of course, 'discover' Mount Warning. The Aboriginals had arrived at least 50,000 years earlier, probably from the Indonesian archipelago, and their descendants had slowly explored the whole continent, giving a name to every creek and headland, every valley and mountain. Cook may not have been even the first white man to see the east coast.

The Great South Land had been an idea that haunted the imagination of the seventeenth century when brave men in little ships set sail from Europe in search of the wealth of the East. There are indications – nothing stronger – that the Portuguese, who owned a colony on the island of Timor only three hundred miles north-west of Australia, may have visited not only the nearest Australian mainland in the north but also the east coast two hundred years before Cook. Not all the mysteries of this vast, untamed land have yet been solved.

Down on the coast at Byron Bay, almost in the shadow of Mount Warning, and the most easterly point of the Australian coastline, the sunrise wakes up no one. A couple of surfers are watching the waves from the beach and are speaking in tongues about left- and right-hand breaks, but they have been up all night. It will be another hour or two before cafes see barefoot men, with sunbleached dreadlocks and wearing sarongs knotted at the waist, sip coffee with blonde women in bright kaftans, arms jingling with Indian silver bracelets, earth mothers who are skilled at keeping one eye on their men and the other on berry-brown children with painted faces who have just spotted a whale spouting out at sea.*

This is the home of Aussie hippydom, a throwback to the seventies era of serious surfers, of ease, peace, relaxation and splifs, of dropouts and greenies. This may be the new Millennium, but Byron Bay, nestling in its own little corner of the world with its white sandy beaches washed by warm winter currents, has yet to notice it. Its hippies have just grown older. And since they were always rebels, now that they feel the hand of time they are protesting against the conventional way of death.

'When we were younger we went to twenty-first birthday parties, then it was weddings, then it was christenings and now it's funerals,' says Amare Pearl, a herbalist. 'The way I see it is that we were part of the sixties baby-boomer generation and we've tried to do everything differently. Now we're older we're having a look at illness and death and we're going to do that our own way as well.'

The main activist in the home funeral movement is Zenith Virago, a Byron Bay paralegal, who runs the natural death centre and has plans to market recycled cardboard coffins that will sell at $100 each. It was from her that I heard

* All the events described in this chapter happened – but not necessarily on the same day.

the Nick Shand story. Shand, the genial editor of the local newspaper, the *Byron Shire Echo*, made a running joke out of a new course at the Adult Community Education Centre. There, listed among snake identification, patchwork quilting and creative writing was a workshop for 'custom coffin making'. Shand was still joking about it when in 1996 he had an accident and death caught him totally unprepared. The home funeral movement swung into action. His friends dug the grave in Shand's garden behind the multi-occupancy house where he had lived. His family washed and dressed the body and a local carpenter made the coffin.

On the day of the funeral, Shand's open coffin was loaded on the back of a truck and driven in a slow procession to a community hall for a memorial meeting attended by more than a thousand people. Then Shand was laid to rest beneath a mound of flowers and wine bottles. Everyone agreed that it was the best funeral they had ever attended and enrolments for the coffin-making course actually went up. 'It heightened people's realization that you never know who'll be next and that you'd better live it up while you can.'

But until it happens Byron Bay's hippies fight to keep the old lifestyle. They have seen off McDonald's and Pizza Hut. They stopped Club Med dead in the Australian courts. But can they defeat fame? Australian movie magnate John Cornell bought and refurbished the Beach Hotel. His business partner Paul Hogan, of *Crocodile Dundee* fame, moved into the area with actress Linda Kozlowski and built an Italianate mansion at Possum Creek. Olivia Newton-John was seen in town and Kylie Minogue was photographed topless on the beach.

This worries Rusty Miller, the 1965 United States surfing champion who arrived in Byron Bay in 1970 hunting the world's best wave. 'California was getting very crowded and surfers don't like crowds. I loved what I found here and I stayed. I was the original eco-tourist.' Now he is married and has a small daughter, teaches surfing and is a passionate environmentalist.

'Byron Bay has swung back and forth between boom and bust for a hundred and fifty years. First there was timber and centuries-old rainforests were felled and exported. Then there was dairy farming on the cleared land, then whaling, then sandmining. Now we have tourism which is potentially the most sustainable industry. But we are learning that this too, is finite, if we get it wrong.'

Byron Bay does strange things to people. Paul McCarthy arrived from Sydney in 1973 for the surf, stayed on and is now a living oxymoron – a real estate agent who is against development. 'The Club Med people tried to tell us that they were doing us a favour by bringing money into town. We don't need those kind of favours. In future people will travel hundreds of miles to come to Byron Bay just because there's no Club Med, no McDonald's, no

Pizza Hut. We love this place. It's about lifestyle. You can be who you want to be. No one gives a rat's arse whether you wear a thong or a three-piece suit in the street. Byron Bay has a soul and a heart that beats, and it's transparent.'

Gervase Griffiths used to be a journalist in Sydney but gave it up in protest against the soullessness of it all. He and his brother bought a run-down dairy farm in 1972 and have transformed it into the Tree Tops Environment Centre with five timber lodges, a restaurant and an art gallery. All waste products are recycled into the forest as compost and the aim is total self-sufficiency in water, electricity and home-grown food.

Griffiths says, 'I'm an ordinary Australian doing an ordinary job. I've had a good upbringing and a balanced set of values and I apply those to create a balanced lifestyle for my family and others. Our evolution has accelerated to the stage where we have reached a critical point very quickly – it's less than five minutes to midnight. People keep talking about what we should be doing but where are the examples? I believe we are putting thinking into practice here at Treetops. We're the custodians of this property for a short period of time, during which I hope we will add something. How many people can say that?'

Four minutes later than Mount Warning, the sun creeps over the ocean opposite Sydney Heads, picks up a few surfers at Bondi Beach, and then shines straight down the Harbour. The light touches the edges of a scattering of clouds and the reflections of their outlines on the water casts strange golden patterns. Seen from the Harbour Bridge, the sails of the Opera House, backlit by the sun, look like giant shark fins.

It is only just after dawn but this is a working harbour and two squat tugs move past the Botanic Gardens, leaving on their port side an early ferry from Manly with a cargo of commuters – heading for Circular Quay and the central business district – and sweep into Woolloomooloo Bay where a Royal Australian Navy destroyer is singling up its mooring lines. There are already swimmers doing laps in the Andrew (Boy) Charlton Pool and the ubiquitous joggers puff their way along Mrs Macquarie's Walk as it winds its way around the harbour foreshore.

As the sun strengthens, the water sparkles and defines the harbour's many coves and estuaries reaching like green fingers into the red-roofed inner suburbs. In one of these, Pyrmont, the fish market has already been busy for hours. Hedonistic Sydney is the seafood capital of the world and no other city offers such a range of marine life. More than 150 types of fish and crustacean – from albacore to yellow-tail, through Balmain bugs, carp, catfish, goatfish, groper, pigfish, pipis, salmon, shark, snapper, eels, wahoo, pomfret, seventeen varieties of crab and eighteen varieties of prawns – are spread over white marble slabs in colourful confusion.

★ ★ ★

Margaret Simpson is shopping for a dinner party she plans for this evening. She is in the butcher's shop at the Castle Mall Shopping Centre, one of the largest in Sydney. This is not a corner shop butcher because it is no longer enough in Australia in 2000 to display your lamb chops, sirloin steaks, and neatly-plucked chickens, and trim a bit of the fat off the cutlets if the customer requires it. Today's customer, man or woman, wants a dish that they will just have to heat. Not a supermarket TV dinner, but a fresh, appetizing meal with all the hard work done by the butcher.

Margaret Simpson runs her eye along the display, a tribute both to the range of food available in Australia and to the contribution made by its multicultural society to the way Australians eat – marinated chicken stir-fry, Thai chicken cutlets, satay chicken kebabs, barbecue chicken, marinated chicken wings, tandoori chicken cutlets, chicken schnitzel, garlic chicken steaks, chicken pastry wheels, sun-dried tomato chicken breast, chicken and corn rolls, chicken strudel, honey soy wings, chicken cacciatore, chicken enchilada, chicken mignon and herbs, chicken lasagna, Italian roll-ups, chicken pastry with mango and cheese, lamb coconut curry, steak diane, honey soy spare ribs, satay beef, marinated pork ribs, barbecue pork ribs.

She makes her choice, collects her car, and heads off to work, reflecting that Australians did not always eat as well as this.

It is now 8 a.m. and before his first meeting of the day Peter Ryan, the Commissioner of the New South Wales Police Service, is sitting alone at his desk in police headquarters in College Street, Sydney, wondering what to do about the case of Constable Kim Hollingsworth – policewoman, undercover agent for the Police Royal Commission, nightclub stripper, and prostitute.

Her story has gripped Australia more than Ryan's other problems: eight murders in one week, the suicides of 11 police officers facing exposure for corruption, a power struggle with the Police Minister, hostility from the police union and resentment at Ryan's linking of organized crime to ethnic groups in Sydney.

Hollingsworth once earned $400 an hour as a prostitute in some of Sydney's best-known legal brothels, including the famous 'Touch of Class', which accepts all credit cards as well as personal cheques and will run a monthly account by arrangement. As well as this work in the sex industry, Hollingsworth has claimed she did regular strip performances at police parties in Sydney and provided 'discount sex' for police officers afterwards.

Her life story, as she told it to a New South Wales Industrial Commission hearing her complaint that she had been wrongfully dismissed from the Police Service, gripped everyone who heard it. The daughter of a police officer, she had run away from home at 20 and had drifted into prostitution and striptease. But all the while her ambition was to join the police force and after an interview she was accepted for training at the New South Wales Police Academy at Goulburn, outside Sydney.

There she was recognized by a young detective who remembered seeing her at a police strip night. She thought her police career had come to an end but it turned out that all he wanted her to do was act as a part-time madam at a brothel he was planning to open. Outraged by this, Hollingsworth went to the Royal Commission investigating police corruption and offered to go undercover.

The Commission set up a 'sting' operation, concealing a video camera in Hollingsworth's flat and recruiting a number of civilians to play the part of 'crooked coppers' and pimps so as to give the 'sting' the right atmosphere – Inspector Gerald (really a tow-truck driver) and Eddy from the Black Garter escort agency (really a maintenance man). But all they caught on video was the young detective accepting a $100 bribe.

By now Hollingsworth's role in the sting had come to the attention of the police and after checking her background, they sacked her, claiming that she had concealed her career in the sex industry on her application form and had described her only previous jobs as student, model and shop assistant. (Hollingsworth claimed there was insufficient room on the form to include all her sex work.) Worried about her safety now that her role in the sting operation was known to the Police Service, Hollingsworth called on the Royal Commission to honour its promise and enter her into the witness protection programme.

She had understood that this would involve a new identity, financial support and a new life in another Australian city. Instead, she told the Industrial Commission, 'They gave me a one-way plane ticket to Adelaide and a loan of 600 dollars and suggested that I go back to work in a brothel so as to repay the loan.'

The Industrial Commissioner thought that this was all a bit too much. He found that Hollingsworth had been a victim of 'a fundamental denial of natural justice and a procedural unfairness' and ordered that the police either give her back her badge, gun and uniform or pay her compensation. Commissioner Ryan, who had been brought in from Britain because the New South Wales government thought that only an outsider would have a chance of stamping out a culture of corruption in the New South Wales police service, now had to decide what to do about Hollingsworth – reinstate, compensate or appeal.

Could the service really afford to see this woman back in uniform? True, she had given evidence that her work in brothels was not really prostitution because she had engaged only in what she called 'trick sex', somehow avoiding penetration while keeping her clients satisfied. But against this had to be weighed stories from the Industrial Commission hearing that Hollingsworth had told the court reporters during tea breaks that she had had cosmetic surgery to enlarge her bust and had found the the implants enabled her to move each breast independently. When the press corps expressed its scepticism, Hollingsworth had offered to demonstrate its truth.

Ryan's decision was further complicated by newspaper reports that the police case before the Industrial Commission had not been helped by the police lawyer, Terry Anderson. Mr Anderson, who had been appearing daily wearing a dark, three-piece suit, arrived one morning at the hearing in a pink dress, carrying a handbag, and with a frizzy ginger hair-do. When the hearing convened, Anderson announced that he had had a sex-change operation and should hence be referred to as Teresa.

Peter Ryan's instinct was to appeal. He would discuss it with the lawyers later.

In Waverley, an inner eastern suburb of Sydney, John Shaw, the editor of *Europe Business Review*, is working on an article for his magazine. He taps out: 'This week, the University of New South Wales had to cancel a Sydney seminar on Australia's largest economic partner because of lack of interest.' He gets in a few more facts – this partner has held that position for the past seven years; Australia's first White Paper on Foreign and Trade Policy devotes only one of its ninety pages to this partner.

Now, good journalist that he is, Shaw is ready to spring his surprise on his readers. 'No, we're not talking about Japan, the Association of South-East Asian Nations, or the United States – it's the European Union, the world's largest trade group, comprising 15 nations and 350 million mainly affluent customers.'

The theme of the article is dear to Shaw's heart – Australia may be physically in south-east Asia, it may have a substantial minority of Asian-Australians among its population, and the 'Asia First' seminar circuit among diplomats and academics may be thriving, but Australia cannot ignore its historic, cultural and economic links with Europe.

The 'Asia First' supporters say that Shaw's attitude is only to be expected – he is an Anglo-Australian, a '£10 Pom', part of the wave of British migrants who poured into Australia in the early postwar years. Shaw says that this is true, but points out that he married into a slice of early Australian history – his wife, Liz, is a descendant of Tommy Marx. 'Tommy rescued the legendary Peter Lalor, leader of the 1854 Eureka Stockade republicans,' Shaw says. 'Lalor was shot by the troops and had an arm amputated. Marx got him away by horse and cart to Geelong to escape the police. Lalor hid there until charges were dropped and then lived on to become Speaker of the Victorian Parliament. That's a rare brush with fame for an immigrant boy.'

His article finished, Shaw turns to a project his sister-in-law Jocelyn has been working on. Tidying up some family papers, she came across carbon copies of a badly typed diary the 20-year-old John Shaw had kept during the family's voyage to Australia in 1951. She re-typed it and has circulated it to the nine children and six grandchildren descended from the Shaw family who took the plunge, left England forever, and made a new life in Australia.

The diary documents a way of travel between the Mother Country and Australia that has gone forever, a leisurely journey by sea that induced a sense of both loss and rebirth, that emphasized the enormous distance between the two countries, but which, at the same time, recalled the glories of a great maritime Empire then slipping reluctantly into the twilight.

The decision to migrate to Australia was taken by Shaw's mother, Clarice, from a mixture of motives. The end of the war had been followed by a series of appalling English winters. At the same time press photographs from the 1946–7 MCC tour of Australia showed Australian crowds in summer dresses and shirt sleeves watching cricket played in bright sunshine. National Service looked as if it was going to go on forever and Clarice, as the mother of four sons whom she wanted to do well in life, was worried about the loss of two years out of their careers. The international situation was grim – the Korean War had broken out and the Cold War was intensifying. But the scales were tipped by letters from Clarice's uncle in Geelong in Victoria. Uncle Jack had migrated to Australia during the Depression and had done well there in the fruit trade. 'Great opportunities here for everyone; what are you waiting for?' was the tenor of his letters.

The Shaw family, all six of them, sailed from Tilbury on the old P & O liner *Ranchi*, on Wednesday, 19 September 1951. John Shaw began his diary that very day.

'Shortly before dinner which was at 8 p.m. saw the brilliant lights of Margate, to the background of close harmony singing provided by a group of Scots fellows in the saloon.'

20 September: 'Wakened by steward at 6.30 a.m. with tea. Ship's announcements enabled us to pick out the salvage ship and marker buoys above the sunken *Affray*, the submarine which went down with all hands April last. The Tannoy system was used for records for an hour or so, 'Rose, Rose' and 'Autumn Leaves' among the selection. England and Australia seem a long way off, both in distance and in thoughts. With such a variety of distraction on board, the fact that most people are going to a strange land to make and shape a life from scratch is pushed into the background for the time-being.'

21 September: 'Real Biscay swell moving the boat about alarmingly.

Couldn't face sausages, tomatoes and fried potatoes. The coloured waiter who we have Christened (secretly, of course) Ali Baba, grinned sardonically when I left the table having eaten so little. Cinema shows start this evening. Michael Wilding in *Into the Blue*, appropriately enough. The crack Argentinian liner *President Peron* passed going full steam ahead on her maiden voyage, a beautifully-built streamlined ship in light blue, dark blue and white.'

22 September: 'Went along to the immigration library with Mum and got some literature describing Geelong. Saw SS *Strathdene* mid-morning. She was heading home from Australia. Had a chat with Ronnie James, the British ex-lightweight champion who pointed out to me Frank Banner, the famous Australian professional runner who is returning home after running in the United States.'

23 September: 'Really warm for the first time since leaving England. The sea has assumed a blue one never sees, quite startling to the eyes after the grey English Channel. Managed to get Don Bradman's *Farewell to Cricket* from the ship's library. Most interesting, particularly his article on the bodyline controversy of the Larwood and Jardine era.'

8 October: '*Ranchi* is steaming south down the Gulf of Suez and going faster than ever before on this trip. Nothing noteworthy occurred during our enforced four-day stay there [for engine repairs] except when a refuse boat alongside caught fire and some stupid woman ran along the main deck shouting, 'Save my children, the ship's on fire.' The flames were quickly put out without much damage to the craft but with much damage to the rubbish it carried. It was probably a typical Gyppo trick to avoid the labour of unloading it ashore.'

10 October: 'One of the Australian Army types boasted to his companions of his drinking prowess. Said he had knocked back twelve Worthingtons during an evening. He added that this required the strictest training, and wasn't something that could be achieved at once. SS *Chusan*, the P & O's crack liner, passed us homeward bound after taking the English cricketers to Bombay. She is a very fast ship being newly-built and passed us at about 20 knots.'

16 October: 'A tote on the *Ranchi* Derby meeting with races such as the Aden Amble, Colombo Cup and Ceuta Stakes. Decided with dice. We managed to strike it lucky and had two payouts of two and eightpence and two and eleven pence.'

17 October: 'Three RAF Mosquito fighters skimming the waves wingtip to wingtip roared past on the port side as they headed north-west at an enviable speed. Dad, who hates deck chairs, thought up this parody of Kipling

> The pinch of the fingers for laughter,
> The bark of the shins for mirth.

You're nearer to death in a deck chair,
Than anywhere else on earth.'

18 October: 'Awoke to find *Ranchi* in the centre of Colombo Harbour with half a hundred merchant ships of every nation lying around us. Berthed immediately alongside was the Orient Line's *Otranto*, another ship bound for Melbourne and Sydney with emigrants. Ashore the shops display all the articles England makes for 'export only'. The best Lancashire cloth, and fine Worcester and Spode china are the best examples of how England has successfully built up the thriving export business of the postwar years. British cars hold their own, numerically at least, with those from the USA in this city. We landed at the same time as a party of troops who had come ashore from the *Empire Windrush*, berthed next to the *Ranchi*, and bound for Singapore.'

20 October: 'Crossing of the Equator ceremony with King Neptune. Punishments were in the hands of a fierce-looking barber who lathered the victims with cold porridge, shaved them with a large wooden razor and then hurled them without ceremony into the pool. This last stage of the long journey covers the 3,000 odd miles to Fremantle in Western Australia.'

The diary ends at this point as the young John Shaw became engrossed in his approaching new life in Australia. (He went on to become a senior correspondent for *Time* magazine in Saigon, Moscow and many other cities.) Jocelyn concludes, 'As I type this script, written by the 20-year-old John, all the boys except John are grandfathers. Dad died at 72, living long enough to see the next generation start to flourish. Mum is still alive at 86 [she died in 1999] and lives in Geelong, as does Clive. John lives in Sydney, David and Derek in Melbourne. The Shaw family's total contribution so far to the population of Australia is *15*.'

Moi Moi Treborlang is a Sydney publisher who works from her flat in Kings Cross, the city's cosmopolitan centre. A member of Sydney's old Chinese community, Moi married Robert Treborlang, a Jerusalem-born Australian writer she met when they were both at Sydney University. Her family story shows the continuing link between southern China and Australia and how one of the newest countries in the world offered the people of one of the oldest the chance of a better life.

Moi Moi's grandfather, King Nam Lo, came to Sydney from Canton in 1871 on the SS *Brisbane*, when he was only 16. 'Or when he was about 16,' Moi Moi says. Few of the records relating to Chinese immigrants of that period were accurate. Australian officials had difficulty with Chinese names and since they did not think it mattered much, often wrote down any name that occurred to them. Father Cumines, a priest on the *Brisbane*, advised King Nam Lo to forget his Chinese name, which delineated the Cantonese village in which he was born, and adopt a Western name. 'Use my name,' he said. 'We'll call you Young Cumines.' As for age, the Chinese took on the age they thought people would want them to be and the Australian official seemed happiest with 16.

Young worked and saved until he had accumulated enough money to move into a small house at 85 George Street in the Rocks area of Sydney, then a tough seamen's quarter near where the southern pylon of the Harbour Bridge now stands. Bean curd had something to do with his early success. He had a relative in Melbourne's Chinese area, Little Bourke Street, who knew how to coagulate soy beans to make tofu at a time when it was highly prized by all the Chinese workers and immigrants. Young became his distributor.

Young was naturalized in 1882. Documents show him as living at 85 George Street, and running a general store and an export-import business. What they do not show is that he had leased from the government a four-storey building on the other side of the road, where the Sydney international shipping terminal now stands. Here Young opened a boarding house for seamen and for Chinese immigrants on their way to the South Pacific islands. This meshed nicely with his shop because the boarders would often leave money with him as credit against orders they would send back to Sydney for delicacies like Arnott's biscuits. When Young died his family found in the back of the shop an old steel moneybox full of envelopes containing various amounts of cash and meticulously-kept records of who it belonged to – 'Joong Hing Loong – £4 2s. 6d'.

Young married three times. None of the family knows anything about the first wife, one with bound feet, but the second marriage to one Fanny Fong, who was part Portuguese, produced Moi Moi's father. 'The thing to remember about my Dad was that he and my mother brought up seventeen children,' Moi Moi says. 'They had five boys and two girls of their own. And Dad's brother died leaving ten kids under 16. Dad and Mum brought up the lot of us. There was never any thought of putting some in care or anything like that. They were family and the family would look after them.

'Mum was born in Sydney, the daughter of a neighbour, and Dad used to nurse her as a baby. He always said that he knew then he would marry her because she was the prettiest baby he had ever seen. But then her mother died and her father took her back to China to be brought up by her father's first wife – he felt he couldn't bring the first wife here because she had bound feet.

'Dad didn't meet her again until she was 15 and he was on a visit to China. They decided then they would get married as soon as she was 17. They did, but when Mum came to Sydney she cried every night for the first year. She had had a house full of servants in China. She had never washed clothes or cooked anything in her life. Now in Sydney she had to do all this and learn English too.

'My mother had had a classical education at home in China,' Moi Moi remembers. 'She believed in a good education for girls and sent me to Dover Heights school. I was the only Chinese girl in a school dominated by Jews. All my friends were Jewish. When I began to read the English classics I found that my mother had read them all in Chinese. The family believed in the advantages of a good education and the next generation are now lawyers, dentists, psychologists and engineers.'

As the Cumines family expanded, there was an easy interchange of some members between Australia, China and Hong Kong. And although they were educated in Australia and considered themselves Australians, some Chinese beliefs persisted. 'My father's sister Alice had two moles – one above her lip and one on her right foot. A Chinese fortune teller saw them and told her that she would never want for anything in her life.

'Well, in the thirties she married Duckie Louey and they started a bus company in Hong Kong with a couple of trucks with benches in the back. They lost the lot when the Japanese came but when the war ended they commandeered everything back again, the business took off, and by the late seventies it was the largest private bus company in the world. Although she was a great personality in Hong Kong, when Alice lived in Sydney few Australians ever heard of her. Only the younger generation raised in Australia accept society publicity. Aunt Alice would never be in a social column.'

Like many Chinese-Australians, Young Cumines's descendants had long lives. Moi's father kept the shop in George Street going until the late 1950s when he sold it and retired. He died in 1995 at 88. Her mother died in 1997 aged 86.

Some Australians criticise Chinese-Australians, claiming that they regard Australia only as a place to make money and that they fail one of the major tests of allegiance to the country in that they all want to be buried back in China to be near their ancestors. When the Chinese-Australians I spoke with passed this test, I added another – the Norman Tebbit test. (Tebbit, a British Conservative politician, complained that when, say, an Indian cricket team visited England, the local Indian immigrant community did not support England but the visiting team.)

'Suppose a Chinese basketball team came to Sydney and played against Australia,' I asked the owner of a Chinese restaurant in Castle Hill. 'Who would you cheer for – the Chinese team or the Australian one?' His face twisted in the agony of the decision. Finally, with the acumen and tact that has made the Chinese in Australia one of the most successful communities, he smiled and said, 'Whichever one was winning.'

* * *

Bruce Gates is waiting on the fairway of the 14th hole of the Yeppoon Golf Course, near Rockhampton, Queensland, for a big red kangaroo to move off the green. A kangaroo family has adopted the club as its home territory and the club members have had to learn to put up with them until they decide to move on. In the old days the greenkeeper would have shot the lot but these days he recognizes that members are eco-conscious and might object.

True, Australia slaughters five million kangaroos per year for their skins and meat but this barely keeps the kangaroo population under control. In 1999 it numbered between 20 and 25 million, beating the human population by between 2 and 7 million. This is because the kangaroo has adapted to Australian conditions with an ease that has defeated humans. It needs little water. In temperatures where humans need ten litres a day, the kangaroo can get by on two litres a week. (They can even survive on saltwater if pushed.) Their breeding pattern is astounding. The female begins her reproductive cycle at 15 months of age which, since she has a life expectancy of 20 years, is as if the human female began having babies at the age of five.

Her gestation period is only 33 days, after which the newly-born kangaroo climbs from the uterus into the mother's pouch and attaches itself to a teat. The mother then mates again but keeps the new embryo 'in suspension' in her uterus until the first 'joey' starts leaving the pouch for brief excursions. When it leaves the pouch permanently after 235 days, or dies for any reason, the mother then activates the dormant embryo and it is born about 24 hours later. So a female kangaroo has three offspring on the go at any one time – one in the uterus, one in the pouch and one at her side.

Gates, as an immigrant interested in Australian wildlife, knows all this, but it does not ease his mixture of irritation and amusement that the red kangaroo on the green ignores his shouts and the waving of his club. If the kangaroo does not move soon he will play his shot and risk approaching the green anyway.

Gates, like a lot of migrants attracted to Australia, is a risk-taker by nature. Born in Quebec City in 1936, he married young and in 1961 quit his job to go around the world on a shoestring. Both Bruce and his wife, Mary, were impressed with Australia and New Zealand but had no intention of leaving Canada until one summer's evening in 1972. As Gates recalled in 1999: 'We were staying at Lac St Joseph just north of Quebec City. Northern lights brightened the sky. It was mid-July, the temperature was just seven degrees above freezing and we could see our breath. We were shivering with the cold in midsummer. We decided there and then to leave Canada.'

Their first preference was New Zealand. They paid the fares for themselves and their four children and settled in Auckland, only to decide that the weather was still not warm enough. Only Queensland would melt the ice of Canada out of their bones. They flew to Sydney. 'We were the last to clear customs,'

Mary remembers. 'The customs officer looked at our four children and our huge pile of luggage and asked us if that was all there was. We assured him it was and he said, "Welcome to Australia" and waved us through.

'We were still stunned by this lack of formality when we came out into the arrivals hall. And there was Gough Whitlam, the Prime Minister of Australia, and his wife standing at the front of a large crowd and they all burst out singing *Waltzing Matilda*. What a welcome, we thought. What a country. Of course we learnt later that they were filming a sequel to the *Adventures of Bazza McKenzie*.'

The Gateses decided to drive north up the coastline until they found a town where they felt warm and then settle there. Yeppoon was it. Right on the Tropic of Capricorn, a sleepy central Queensland town with no traffic lights, no stop signs, no parking meters, no roundabouts, a town with magnificent beaches with hardly a soul on them, a town where everyone knew everyone, the children walked around shoeless and the Gates's youngest daughter, Kara, would later ride her horse bareback to school. It was sunny most days of the year and the temperature hovered around 25°C.

Gates and his wife walked into the office of a local real estate agent, looked at some houses for sale, found one they liked and then made an offer to the owner, giving him until 4 p.m. the same day to make up his mind. He decided that this was a mad foreign family and sold. Except for some overseas postings, Bruce and Mary Gates have lived in Yeppoon ever since. 'We quickly developed a sense of belonging to Australia and took out Australian citizenship as soon as we were allowed to do so,' Mary Gates remembers. 'Everyday life was so full of good experiences. Take the basset hound picnic. We had a basset hound and we were members of the basset hound club which organized a basset hound picnic. The hounds arrived dressed as babies in prams, as policemen, lifesavers. There were races, obedience tests, team games. It was a day filled with gaiety, fun and laughter. It typified one of the things we like about Australians – their willingness to enjoy life, to have a good time, their sense of the ridiculous.'

Bruce Gates helped found the Cattlemen's Union of Australia as a way to try to get a better deal from the big British companies which bought Australian cattle. He went all over Queensland in light aircraft to meet cattlemen but it took him some time to become accustomed to the nonchalant attitude that cattlemen have to flying their little planes over such vast distances, often at night, with none of the facilities of a modern airport.

'Flying with the president of the Cattlemen's Union, Graham McCamley, from Longreach to his property near Marlborough we arrived in the dark, so Graham radioed his wife, Shirley, to turn on the lights for landing,' Gates recalled. 'The other passenger looked at me aghast and said, "Last time I flew in here with Graham he radioed Shirley to turn on the lights and she turned on the lights on the tennis court. I hope she gets it right this time." No lights came on, and then I saw two cars, lights on, heading for the grass runway. One parked at the beginning of the runway, the other at the end, their headlamps at high beam pointing at each other. We came down and landed between them. That's Australian bush flying for you.'

The first rays of the sun are just creeping through the slats of the venetian blinds in the intensive care ward of the district hospital in Mount Isa, western Queensland. Bill Harcourt, a Sydney investment adviser who comes from an old Anglo-Australian family, is examining the bandage on his left ring finger, nearly severed in a boating accident four days earlier, to see if any blood has seeped through. Although it is only dawn he has given up trying to get back to sleep – the two other patients sharing the ward with him had crept out last night, going down the fire escape to dodge the matron, and had returned loudly drunk at 2 a.m.

True, their injuries were not too serious – the older one, Darren, had a knife wound, the younger one, Cameron, a paralysed arm after being hit by a bull. But Bill, 66, felt a bit of a wimp when he declined to join them on their nocturnal hunt for some grog. Mount Isa is a hard-drinking, hard-living copper mining town, and city investment advisers who prefer to lie in a hospital bed rather than drink with mates do not make a good impression.

Harcourt and his wife, Karen, had been on a fishing holiday at Sweers Island in the Gulf of Carpentaria. This is not the sort of holiday you would ever find in the Australian Tourist Board brochures. In fact, Australians do not want tourists to know about it because they want to keep it for themselves. It is the ultimate outback fishing experience and only people absolutely mad about fishing are welcome there. Harcourt says, 'You fish all day every day – Spanish mackerel, barracouta, coral trout, queenfish, rock cod as big as a man. It's a fisherman's idea of heaven.'

To get there you fly into Burketown on the mainland on a commercial airline and then by light aircraft to a small strip on Sweers Island. The island is nothing to look at – rocky, flat and covered with scrub. Accommodation is at the 'Sweers Hilton', a series of reinforced aluminium cabins and steel shipping containers bolted down on to large slabs of concrete so that they will not blow away during the cyclone season. Twenty-four guests sleep at the 'Hilton' and eat their meals in a steel aircraft hangar. They range from city judges to cattle station managers, from investment advisers to farmers. Often all they have in common is a love of fishing.

The resort is run by two couples, originally from Mount Isa, who are mad about fishing themselves, and they are as representative of modern multicultural Australia as you are likely to find – Ray Atherinos, a Greek-Australian who is married to Salme, a Finnish-Australian; and Tex Battle, an Australian-Australian, who is married to Lynn, an Irish-Australian. They provide all meals but alcohol is extra and some guests prefer to bring their own.

When Harcourt was there he wondered whether the other guests would look askance because he had brought a whole case of chardonnay. He need not have worried. Another guest, 'Bull', a Queensland farmer who was 6ft 8 and weighed 22 stone, drank enormous amounts of beer during the day, but in the evening turned to his 'special mix' – Bailey's Irish Cream, raw eggs, condensed milk, cream of coconut and whisky. On good nights he would sit out in front of his container under the stars, tell stories and finish a gallon.

Everyone at the resort had their own boat, twin hull aluminium with a powerful engine. You could fish where you liked – everyone had their favourite spot – but you had to be back for lunch if you said you were going to be, and you had to be back by 7 p.m. come what may. Harcourt realized how seriously Ray and his partners took this safety rule when he and Karen arrived back for lunch one day ten minutes late. 'Ray was furious,' Harcourt remembered. 'He said he was about to go out looking for us.'

On the day of the accident, Harcourt and Karen were fishing off Bentinck Island, an Aboriginal settlement, well-known in the area because the elders have banned alcohol on the island. Harcourt had just thrown the anchor overboard when he spotted a tangle in the chain and tried to clear it while the anchor was still on its way down. The chain caught his finger and mangled it, leaving the end hanging loose and bleeding heavily.

The boat was equipped with a small box labelled: 'Do not open except in a life-threatening situation.' (They learnt later that opening it activates a satellite SOS signal that could bring aircraft and ships rushing to the area.) They decided that Harcourt's accident was not yet life-threatening and Karen managed to get him back to Sweers Island.

There Ray telephoned the Flying Doctor Service at Mount Isa, 1000 kilometres south. This service was established in 1928 at Cloncurry, Queensland, with one flimsy aircraft and a primitive pedal wireless set, to provide a speedy medical service for isolated areas. Now, over the telephone, it was agreed that a doctor would fly to Burketown and meet Harcourt who would be flown there in a four-seater, single-engine aircraft belonging to one of the other guests. Ray was reassuring. 'There's a great doctor at Isa,' he said. 'A coony-looking little fella, Indian I reckon, but he's the best in the game. Lots of crushed limbs from the mine and stabbings and knifings in Isa, so he's had plenty of experience. You'll be in good hands.'

Harcourt and Karen, still in their swimming costumes, were hurried on board the plane, where the pilot hardly added to their confidence by saying,

'Now let's see. Burketown. Burketown. I've flown into it lots of times and I can never find the bloody place.'

But he spotted tin roofs flashing in the sun and landed. They were met at the airport by an ambulance and Bush Matron Bev, famous Australia-wide for having operated on a man who had been thrown from his horse and was dying from pressure on his brain. Bush Matron Bev drilled through his skull with a handbrace and an awl, relieved the pressure, and saved his life.

Bush Matron Bev gave Harcourt a shot of Pethidine to relieve the pain and forty minutes later a flying doctor swept out of the sky in a twin-engine turbo-prop Beechcraft plane with a full operating theatre on board. He decided that Harcourt could wait until they flew back to Mount Isa. Harcourt was on the operating table there an hour later – more quickly, he worked out, than if his accident had happened in Sydney.

Meanwhile Karen, still in her bikini, was being looked after by the hospital social worker, who fitted her out in clothes donated to the hospital, loaned her $50 to buy toothpaste, toothbrush and shampoo, booked her into a motel, and warned her not to walk the short distance from the motel to the hospital after 7.30 a.m. because of the danger of heatstroke.

She also briefed Karen on relations with the local Aboriginal community. They were reasonably good. The owner of a local airline, a pilot originally from Melbourne, had fostered a young Aboriginal and had sent him to school in Melbourne. He had also sponsored another Aboriginal to do pilot training. However, there were certain social rules to observe. If she found herself in a public place where there were Aboriginal men, she should not make eye contact and if she sat down she should sit with her back to them.

During his four days in hospital, Harcourt has learnt from his fellow patients that the matron not only runs the hospital but behaves like a mother to younger inmates. Cameron, the teenager whose arm had been paralysed by the charging bull, told the matron that his boss at the cattle station had sacked him immediately after the accident. He would now have to find another job as soon as possible because he had no money. 'I'll soon fix that,' the matron said. She rang the station manager who, when he heard it was the matron on the line, quickly agreed that Cameron should have three weeks' compensation.

Harcourt's finger is as good as it ever was. His medical bills came to – zero. The Flying Doctor Service is funded by donations from all over Australia. The Mount Isa District Hospital is a public hospital funded by the Queensland state government. The Australian system of public health care according to need, no matter who you are or your ability to pay, had worked. And the tyranny of distance that could have made this system useless in a country of such vast space, had been conquered by modern technology.

★ ★ ★

There is a place called Cameron Corner in outback Australia where the borders of South Australia, New South Wales and Queensland all meet. Country people sometimes drive their visiting city friends to places like Cameron Corner, stop the car, get out, wait for the noise of the engine cooling to end and then ask them, 'What can you hear?' When the puzzled city friends says, 'Nothing,' the country people say, 'Yeah. The sound of absolute silence, mate. When did you last hear that?' Then they say, 'Look over there.' And after a while the city friends realize what they are seeing on the distant horizon is the curvature of the earth, again something you do not get to see in the city.

Not far away from Cameron Corner is a 2,600 square kilometre – kilometre, not acres – sheep station, 'Wertaloona'. Its co-owner, Peggy Kileen, is up early today to drive out to meet boundary rider Brian Lock, who has come to repair a huge hole in the dingo fence that passes through Peggy's property.

The dingo fence is 5,614 kilometres of continuous wire netting, two metres high, which runs from Ceduna in South Australia, near the uppermost point of the Great Australian Bight, across into New South Wales and finally on to Dalby, near Toowoomba in Queensland. It is the world's longest fence, two and a half times longer than the Great Wall of China, can be seen from outer space and costs $3 million a year to maintain.

It shuts off the lower right-hand third of Australia, where the sheep live, from the upper left-hand two-thirds, where the country's wild dog, the dingo, lives. Dingoes, who are a relative of the Indian wolf, are thought to have arrived in Australia some 4,500 years ago as companions or trade items for south-east Asian seafarers, and turned out to be remarkably suited to life in Australia. Sleek, swift, smart and able to survive long periods without water, they multiplied rapidly, and after colonization began preying on sheep, a $5 billion a year industry. The dingo fence is the industry's answer.

It needs 52 boundary riders to maintain it, each with an allocated section. It is lonely work. Australia is roughly the size of the United States but has only one-fifteenth of the population, so hundreds of thousands of square miles show no sign of human habitation. Brian Lock looks after 280 kilometres of the dingo fence including the 90 kilometres which runs through 'Wertaloona' grazing land. Today he is repairing a gap about a metre wide, caused by a kangaroo. 'This'll give you an idea of how strong roos are,' Lock says. 'This is tough wire netting and the roo has ripped through it like tissue paper.' He quickly knits the two edges together again and then buries the lower edge about half a metre into the ground to prevent the dingoes from burrowing their way under the fence.

Peggy Kileen thanks him and says goodbye. Lock has more fence to check and the earlier in the day he can do it, the easier it is – in the early afternoon, temperatures reach 53°C in the sun and 43°C in the shade. 'We've lost about

2000 sheep and I'd say 90 per cent of those were from dingo attacks,' Peggy says. 'It's heartbreaking to think about the sheep that the dogs kill. Lots of the time the dogs just maim the sheep and it dies slowly. It makes you very angry and you get that way that you hate them and want to shoot them because what they do to the sheep is cruel. If the maulings and killings continue, we might have to go like the neighbours and run all cattle.'

It is after breakfast at Kakadu National Park in the middle of the Northern Territory and the tourists are getting out of the bus where the road dips into a river ford. Daryl, the driver and guide, is warning his charges not to go too close to the water's edge. 'There was this local fella fishing with his son. Some of his mates arrived on the other side and he decided to wade over and join them. A crocodile had passed by a little while earlier and everyone told him not to risk it. He thought he knew better. Got about halfway when the croc hit him. Terrible sight.'

Some of the tourists think Daryl is exaggerating. Others who have seen the local Darwin publication, *The Territorian*, are not so sure. It carried a picture of an Aboriginal, Johnny Banjo, his head swathed in bandages. Johnny, it seemed, had bent down to take a drink from the Daly River and a big, black crocodile rose from the depths, clamped his jaws around Johnny's head and pulled him under. Johnny desperately gouged his fingers into the crocodile's eyes and when it relaxed its grip he dragged himself to the river bank and got away.

'Yeah,' says Daryl when reminded of the story. 'He was lucky. No one knows how many Abos the crocs eat. The Abos come down to the river bank for a party and a bit of a drink. They fall asleep. Next morning Charlie's not there. His mates think he's gone walkabout. Couple of weeks later no one's seen Charlie and only then do they get around to realizing that a croc must've come up on the bank during the night and got him. No chance of running away from a croc, you know. They can bring down a galloping horse.'

The tour includes a ride in a boat to see crocodiles close up. Presumably that is safe? 'Big boat is OK,' says Daryl. 'But there was this woman in a canoe going along the river bird-watching. She was a university lecturer, an environmentalist, so she thought she knew what she was doing. Then she paddled into this old croc's territory and he didn't like it. He nudged the canoe a couple of times and she got worried he was going to tip her out, so she grabbed some overhanging branches and climbed into a tree. But crocs can jump, so he came right up the tree after her and got her around the waist and brought her down into the water.

'He was giving her the old death-roll – they hold you in their jaws and then roll their body right over so as to break your spine – when she managed to grab an underwater branch. He let her go and then grabbed her a second time. But it couldn't have been a good grip because he let her go a second time. This time when she got away she managed to claw her way up the river bank. He started to come after her again, then for some reason let her go. She crawled a couple of miles and the rangers found her, half her insides hanging out. She's lucky to be alive. I'll say this for her. She didn't blame the croc. "It was my fault," she said. "I was in his territory."'

In preparation for the trip on Alligator River, Daryl hands out some brochures about crocodiles. They have been around for 200 million years, can grow to seven metres long, weigh more than a tonne and have no natural predator except man. Related to dinosaurs, the crocodile evolved to outlast them all. It can live in salt, brackish and fresh water and drink all three. It can see equally well by day or night and has a transparent membrane, like a pair of swimming goggles, to protect its eyes underwater. It needs to eat only once a week and can go for several months without eating at all. It has a continuous supply of teeth – if one breaks there is another underneath. If it needs to sink quickly in an emergency, it can move its large organs backwards in its body cavity, like a submarine shifting ballast. Once on the bottom it can reduce its heart rate to two or three beats a minute and remain submerged for up to an hour without breathing.

The female crocodile lays a clutch of about 50 eggs in the wet season and leaves it to the weather to decide what the sex of the baby crocodile will be when it hatches. At 32°C the indications are that it will be a good season with plenty of food so the baby crocodiles will all be males. Below 32°C suggests that food will be scarce and that the crocodile community will need lots of breeding mothers, so the eggs decide to come out female. Either way, if there looks like being a danger of over-population, the males are programmed to eat some of their young.

Crocodiles live mainly on fish but may take large land animals, like people. They are smart and patient. They try to identify habits in their prey, like returning to the same spot each day to drink or swim or fish. Then they lie in ambush and wait for the right moment to strike. The crocodile has only a couple of muscles to open his mouth but 40 to close it, so once he has his jaws around his prey, the chances of escape are slim. He prefers his meat well hung, so after killing his prey he hides it under a river bank or in a mangrove swamp, and returns for it after a week or so. No other crocodile – or animal – dares touch it.

As Daryl told us next morning, 'There was this American woman on a yacht trip. She'd seen *Crocodile Dundee* but it made no difference. Went for a swim and a croc got her. The captain called Broome and they sent a couple of croc experts who worked out where this big croc might have hid the body. Sure enough they found it in the mangroves, put it in a body bag to take back to Broome and put the body bag in the stern of their big launch. They're anchored waiting for dawn and suddenly the croc comes up over the transome like the shark in that movie *Jaws* and tries to get the girl's body back. The croc is thinking that body's mine and no one's gonna take it. The skipper started the engine and they got out of there quick smart, I can tell you.'

In 1972, the Federal government imposed a ban on hunting crocodiles because of fears they might become extinct. Even a crocodile known to have killed a human was not shot. Instead it was trapped in a steel cage baited with pig meat, a noose slipped around its jaws, and blindfolded and docile, it was transported to a crocodile farm where it was used for breeding young crocodiles for the crocodile skin industry – $20 million a year from France and Japan.

But in 25 years the crocodile population expanded from about 5,000 to 70,000 and as they lost their fear of man and more tourists came to see them, attacks became more frequent. So in 1999 there was a legal hunt to reduce the numbers. But, as Daryl said, the crocodile will go on being – as it has for millions of years – the toughest, most inspiring, most dangerous animal in Australia. 'If you don't believe it,' he says. 'Then when you're in Broome, go and see "Old Three Legs". Big croc. About 18 feet long. Something bit his leg off so he couldn't compete with the other crocs in the Ord River. He just moved on to a cattle station and took to eating a big bull a week, running them down on his three legs.'

Karl Goeschka is asleep on the back seat of his hired four-wheel drive Toyota in the South Australian desert near Halligan Bay, close to Lake Eyre North. It is only 8 a.m. but already the temperature is 35°C. In a few hours' time it will hit 60°C just as it has for the past eight days. When he wakes, Goeschka, an Austrian tourist, will drink a little water from a 400 litre container then go back to the car or sit under a small, corrugated iron lean-to, and wait to be rescued.

His girlfriend of 11 years, Caroline Grossmueller, had gone for help six days earlier after their Toyota campervan had become bogged down in loose sand. After spending two days trying to get it out they had abandoned further attempts and assessed their situation. They were 80 kilometres from William Creek, their last stop and the nearest place to find help. They had no radio. They had left a note at William Creek Hotel setting out their travel plans, but how long would it be before they were missed? They decided there were two courses of action – they could wait for someone else to pass their way, which, in such an isolated area could be days or even weeks; or they could try to walk back to William Creek.

Goeschka doubted if he was strong enough to walk that far in such fierce heat, especially as he was not accustomed to it. Grossmueller, a fit, well-built young woman, felt that she could do it. She took a two-litre plastic bottle, filled it with water, and put it with some food into a small backpack. She said she thought it would take her three days at the most to get to William Creek. Goeschka should wait by the Toyota until she returned with help.

That was six days ago. Goeschka had slowly slipped into a near trance-like state, sleeping a lot, waking only to drink a little, scan the horizon, and then return to the campervan or the lean-to. But today, as he lies down, there is the rumble of an approaching vehicle and over the sand dunes comes a big South Australian police truck.

Goeschka springs to his feet and moments later is shaking hands with Senior Constable Paul Liersch and Mal Anderson, the publican from the William Creek Hotel. His face is full of relief. He has been saved. Liersch chats with the Austrian for a few minutes and establishes that he is lucid and in reasonable physical shape before using his satellite telephone to call his police station and report. Then he asks Goeschka to sit in the back seat of the police truck with him and this tough outback policeman tackles the task he has been dreading.

Caroline Grossmueller had kept her promise to her partner and brought him help. But she did not make it herself. She had covered 40 kilometres and still had water and food when she collapsed from heat exhaustion and died on the track. Two German tourists had found the remains of her body, and a note written in blue ink pinned to an outback sign. They had raised the alarm.

'Goeschka started punching the seat when Paul told him,' Anderson remembers. 'Then I think Paul got him to lie down inside the truck. I don't know. I couldn't see. I didn't want to see. Then after a while Paul asked me to sit with the Austrian. The poor fellow kept trying to explain to me why he couldn't talk his girlfriend into them both going or both staying back with the Toyota. He wasn't the stronger of the two. She was. Then he talked about their life together over the previous 11 years and how much he loved her. I got the feeling that she'd come from a hard background and that once he'd met her things were going good – until now.'

The little Cessna gains height to get over the escarpment and bounces as it hits the updraft from the valley below. Fortunately, it is still three hours to breakfast and the shock of the sight in front of us immediately takes away any nausea: the lost world of the Bungle Bungle. There is nothing else like it in Australia – not Ayers Rock, not the Barrier Reef, not Kakadu National Park. There is nothing else quite like it anywhere, which is why it is the eighth natural wonder of the world.

The Bungle Bungle is a magic place, a lunar landscape, an aberration in time, a space trip into the world as it was 350 million years ago. All the way to the horizon stretches an extraordinary vision of sheer rock walls, high ridges, deep gorges, huge chasms and thousands of giant stone beehives banded with orange, black and green. The early morning sun creates strange shadows, and the total lack of any evidence of human beings gives you the eerie impression that you have somehow strayed out of our solar system, perhaps to the Planet of the Bees.

Until a few years ago this magnificent natural wonder, covering some 270 square miles of the long-forgotten Kimberley region of Western Australia, was known only to local Aboriginals and the occasional stockman. Then an enterprising television reporter making a documentary on the cattle industry stumbled on the Bungle Bungle, was bowled over by its beauty and shot some film. When it dawned on the West Australian government that it had a new tourist attraction on its hands, it declared the area to be a national park, gave it its original Aboriginal name, Purnululu, and carried out studies on how best to protect it.

The Bungle Bungle is fragile not only because of its age but because it consists largely of sandstone with stripes of silica (orange), moss (green), and lichen (black). No one knows why it is striped like this or why it has worn the way it has, and too large an influx of tourists and their infrastructure might cause the whole thing to crumble away before scientists can find out. Fortunately it is still very hard to get there overland and, anyway, the best views are to be had by flying over it.

The nearest commercial airport is at Kununurra and if you like isolated airports, then this one is for you. For a start, the co-pilot is also the baggage handler and the booking clerk. There are no taxis. There is a taxi phone but no one answers. However, this is hospitable, outback Australia and the Hertz girl will always give you a lift into town, or if you have a reservation at the Overland Motel, then the manager will probably come to pick you up.

Flying over the Bungle Bungle also gives you a chance on the way to see the Argyle diamond mine, the biggest producer of diamonds in the world. Machines that look huge even from the air slice away at the ochre-coloured hillsides. The men who operate these earthmovers live in Perth and are flown in to work twelve-hour-a-day shifts for two weeks and are then flown back to Perth for two weeks' leave. The earthmovers and trucks are air-conditioned and equipped with stereo sound, and the recreational facilities at the mine village are so lavish that it is known as the 'Argyle Club Med'.

The Kimberley region is a land of such excesses that the occasional Texan who visits it is rendered speechless. To begin with, this long-forgotten corner of Australia is three times larger than England. This is not the dry heart of Australia so familiar from photographs and television. This is a mixture of ferrous red earth, of a million termite mounds, gum trees, coolabahs and boabs, of lush cattle country and irrigated farming land, with monsoon rains in the summer and a winter of rainless, balmy days when the sun shines out of a cloudless, deep-blue sky and at night the stars go on forever.

The people match it. What could a Texan say, for example, about Michael Durack, the pioneer who founded the local cattle industry by driving 7,500 head of cattle right across Australia from Queensland, a journey that took him two years, and who then settled on seven million acres? Or the fact that the Argyle Dam, which irrigates the area, is nine times the size of Sydney Harbour? In fact what can any visitor say about the sheer size of Australia, except that if you imposed a map of Australia on a map of Europe, it would stretch from Oslo to Cairo, from Madrid to Moscow.

Even phlegmatic British visitors tend to lose their cool here. Lord McAlpine of West Green, a devout lover of Western Australia who restored an old pearling master's house in Broome for his Australian residence, said: 'Why go to Ayers Rock when you can go to the Bungle Bungle? Why go to the Great Barrier Reef when the coral reef at Exmouth is every bit as good but is only one hundred metres offshore, the water is unpolluted and the deep sea fishing has never been exploited?'

Euan Ferguson, of the *Observer*, wrote: 'If you ever get the opportunity to go to the Kimberley, you should sell various unwanted members of your family to take it. The experience will live forever.'

Peter Leitans, Superintendent of Services Delivery for the giant Hamersley mining group, shakes his head over the problems of running the communications system for a company that has mines scattered all over the north-west of Western Australia. Hamersley's use systems that are at the leading edge of new technology and are meant to be reliable under all circumstances. But since this is the unpredictable Australian outback, faults occur that are not easily traced.

Leitans has on his desk a couple of cases that he wants to use to show how they provided the inspiration for a solution to one of his longest-running problems – intermittent faults between microwave dishes at some mine sites and Perth headquarters. He fingers one file. 'The communication tower on Mount Nameless had a fault that occurred at the same time every day – sundown – and when we couldn't find it ourselves we brought in Telstra [Australia's largest telecommunications service].

'They couldn't find it either, but did work out that the problem disappeared of its own accord during the day. They bandied about theories of heating and cooling, expansion and contraction of metal, and in the end sent someone up the tower to have a close look. He found a nest of wasps who hung around during the night interfering with the circuitry, but who vacated the premises during the day. If we had kept the tower under observation we would have noticed the wasps and that should have been our clue.'

So when the microwave problem popped up, the first thing Leitans did was to get someone to watch the dishes. 'After a long period of observation we got the answer,' Leitans says. 'We didn't believe it at first, but in the end we had to.' Every morning, almost to a timetable, a flock of cockatoos would fly noisily in and perch at the back of the microwave dishes. Tentatively at first and then with increasing and obvious enjoyment, squawks and screams, they would peck away at the back of the dishes.

'They were getting a tingle on the beak from the low voltage electricity,' Leitans says. 'They liked the buzz so much that they invited their friends and more and more cockatoos came to try it and so our problems grew. They got hooked on the buzz and it became like a recreational drug for them. Once we'd scared them off, our problems went away.'

Hannah McGlade is a trainee law lecturer at Murdoch University in Perth, one of only two Aboriginal law lecturers in Australia. It is only 9 a.m. but she is already wilting slightly, both with the West Australian summer heat and her crowded schedule. She was up at 6.30 a.m., got her three-year-old daughter ready for kindergarten, squeezed in a little kung fu practice for a class this evening, worked on a lecture she is preparing, did some study for the Masters degree she is doing in international law, and is now looking at a case she is considering taking to the Australian Human Rights and Equal Opportunities Commission.

The file is headed 'Uncle Ben Taylor'. Uncle Ben is a respected elder member of the Aboriginal Roman Catholic laity and lives in an outer suburb of Perth. When he heard that Pauline Hanson, leader of the One Nation Party, was going to speak in a local hall during the 1998 election campaign, he went along to hear her. Afterwards he got involved with some white Australians in a political argument. Someone, worried that it might escalate into violence, called the police.

Uncle Ben has a heart condition, and when he gets excited – as he was that night – he sometimes froths at the mouth. The police officers took one look at him and charged him with spitting at them. In court he tried to explain but the police put a series of officers in the witness box to back up the spitting charge. Hannah says, 'The magistrate was in an awkward position. He was not impressed with the police case but . . . So he convicted Uncle Ben and then ordered that no conviction be recorded. I think he's got a good chance before the Commission.'

Hannah is a living rebuttal to those Australians who say that Aboriginals have no ambition and are low achievers. Her story also shows that white Australia is beginning to provide opportunities for Aboriginals who are prepared to take advantage of them.

Hannah's mother is a member of the Nyungar tribe, which lives between Perth and the south-western town of Esperance. She was a teacher and an activist for Aboriginal rights but fell in love with a traditional Aboriginal man when she was teaching him in a Perth jail. When they married he wanted to return to the outback, so they moved to the Kimberley. There, Hannah was lucky to escape being forcibly taken from her mother as a 'half-caste child' and put into care. Her mother lied and managed to convince the authorities that Hannah was half Chinese and not covered by the State's assimilation policies.

Her mother encouraged Hannah's education, and radicalized her by taking her to demonstrations and rallies as a child. At 17 Hannah had finished a year studying journalism and was working in an Aboriginal radio station. 'Then I decided that there was little hope of changing things through journalism or arts and that there'd be a better chance with law, so I switched. The Australian National University in Canberra was making a special effort to attract Aboriginal law students so I went there when I was 20 and got my degree. Then I came back to Murdoch University because it was offering cadetships for Aboriginals to train as law lecturers.'

Hannah stood for Parliament in 1998 as a Democrat – 'They are strongest against racism and are the best supporters of Aboriginal land rights' – but lost, even though she doubled the Democrat vote. She has plans to be a top civil rights lawyer, to set up a museum of remembrance on Rottnest Island, off Perth, a former jail for Aboriginals where many are buried, and to write a book about Yagan, an Aboriginal leader murdered by a teenage boy in the early years of the nineteenth century whose head was sent to Britain as a trophy.

This is the inspiring professional side of the Hannah McGlade story. The personal one is less encouraging. 'At the ANU in Canberra I was invisible. In Perth I have only two white friends because I have been let down so often that I stopped making the effort. I like Australia – the land, the climate, the freedom, the multiculturalism. There's been a lot of progress but there's still a long way to go. I've got an Aboriginal girl as a client who was chatting with a white Australian friend in the school playground when another girl came up, an English migrant, and started chanting, 'White, white, join the National Front tonight. Black, black, get back to your habitat.'

It is afternoon tea-time and Dame Rachel Cleland, 93 years of age and a leading activist in the West Australian anti-logging campaign, is entertaining some union leaders from the timber industry in her apartment overlooking the Swan River, Perth. The union leaders, big muscular men, have been blockading Parliament House for days with their huge trucks and getting nasty with the local 'greenies'. Now they perch on the edge of pastel-silk upholstered sofas surrounded by antiques and expensive paintings, sip tea from fine China cups and listen to Dame Rachel, a powerbroker in the Liberal (conservative) Party, tell them that they cannot go on cutting down old-growth forests.

'I've enjoyed meeting you all,' says Dame Rachel. 'And I appreciate how polite you've been. But I have to tell you that nearly 90 per cent of Western Australians are now opposed to logging. It doesn't even make commercial sense – government royalties are far too low. So naturally your employers are going for their lives clearing *our* forests.

'What I want you to do is go away and put your thinking caps on and come back with a solution that will satisfy everybody. If I agree with it, then I will work to have it recognized.' And with that, they nod, get up and file obediently out.

Dame Rachel is living proof that a country where the maxim used to be: 'If it moves, shoot it; if it grows, cut it down; if it's valuable, dig it up; if it flows, dam it,' has become acutely conscious in recent years of the need to preserve the environmnent. There have always been Australians aware of how they were damaging their country. In 1890, the Californian reformer Henry George was met by enormous crowds in Sydney when he came to talk about 'Humankind's Relations with Nature', part of a Pacific exchange that had enriched both Australia and California in many ways, including architecture, irrigation and horticulture – Australia sent the gum tree, California the pine.

But commercial and industrial development took precedence after the Second World War and it was not until the 1960s and 1970s that concern for the environment became popular again when a series of successful 'green bans' showed what a combination of angry, mostly middle-class citizens, and trade union power could achieve.

Environmentalists stopped the demolition of the historic Rocks area of Sydney, and of the Regent Theatre in Melbourne; the clearing of Kelly's Bush in Hunters Hill, a suburb of Sydney; the damming of Lake Pedder and the Franklin River in Tasmania, and sandmining on Fraser Island in Queensland.

And in many cases, to the alarm of conservative politicians, protests against environmentally damaging projects were led not by students, trade union leaders, hippies, 'full-time greenies and weirdos', but by their own party members. Dame Rachel's husband, Sir Donald Cleland, was a founder of the Liberal Party and its first national director, and Dame Rachel remains a powerbroker in the party to this day.

'As a child I played in the great Jarrah forest by the Swan River, with its orchids and gorgeous birds like the red-tailed black cockatoo. Today only one tree remains of that forest, surrounded by mansions and exotic gardens. I spent a lot of time out of Australia accompanying my husband. When I got back to

Perth in 1979 I was outraged to discovered that most of the old-growth forests would be gone by early next century if logging continued at the same rate.'

She got together a group of experts and developed and published a new forest policy, with logging based on tree plantations. But the timber industry has a turnover of $850 million a year and powerful friends in government.

'The best hope seemed to me to be with the tourist industry, already worth $500 million a year. So I sent out letters to three hundred tour operators and visited many myself. There was a very good response. Now the pressure is on the Liberal government in Western Australia. Do they want to remain in government, or risk losing their most faithful supporters?'

Back on the east coast, summertime extends the light until 9 p.m. Tony Healey is sitting at a picnic table alongside the swimming pool in the backyard of his house in Sandringham, a seaside suburb of Melbourne. He is writing a postcard to a cousin in London. Healey is not a rich man, but like many Australians he manages to live like one.

He was born in an old workers' terraced cottage in North Melbourne into a family originally from Ireland. His father was a glazier; his mother did clerical jobs at the Melbourne Markets to supplement the family income. Healey went to a Christian Brothers school, left at 13, discovered he had a flair for publicising good causes and now runs his own small public relations company.

His mother lives in the country not far from Melbourne and today he has taken his partner, Tanya, and their children to see her and celebrate Christmas. Healey pours himself a cup of coffee, takes up his pen and writes:

> Melbourne. Xmas Day, 1997
> 9pm, 31 degrees, light sea breeze.
>
> Have just returned from Christmas dinner at Mum's in the Yarra Valley. Started with drinks locally. Champagne, of course. Visited Froggatts for some antipasto – olives, artichokes, parmesan and more champagne. Gifts were opened under a large oak tree and after eggnog we dined on grilled prawns, grilled lobster, abalone and heaps of the obligatory oysters. Washed down with a fine Australian chardonnay. Traditional turkey and vegetables followed, then summer pudding with a classical botrytis rot dessert wine to finish.
>
> Back home in Nelson Street we all had a quick dip in the pool, some sparkling burgundy with grilled baby calamari, Moroccan lime chicken, cos lettuce with fresh beetroot and a Caesar dressing.
>
> Anyway, with the sun just setting and before a poolside coffee and

Armagnac, thought it timely to be in touch. Hope you enjoyed your Christmas.

Sleep warm,

Tony.

CHAPTER 2

BEGINNINGS

Could the dregs of Britain's dungeons, its repressed and criminalized working class and the disaffected and deprived peasants of England's oldest colony, Ireland, if sent against their will to an empty land at the other end of the world, remake themselves and create a respectable outpost of British civilization in the Antipodes?

I grew up in Australia believing many of the myths about it, including the principal one about its history – that it was a country in which nothing had ever happened, a dull, utterly uninteresting place inhabited by aggressively familiar people disappointed that they were forever condemned to be once-removed Englishmen.

Writing this book I was surprised to discover another country, one that seethed with passion, one full of love, hatred, conflict, terrorism, kidnapping, atrocities, murder, secret armies and a class and religious struggle, that was frequently on the brink of rebellion and civil war. This was not a country of Poms Down Under, suppressing their longing for Britain's 'green and shaded lanes', miserable that 'Home' was so far away. It was a young country – Australia is only one hundred years old; before 1901 it was a series of colonies linked to Britain through the Colonial Office in London. Here the universal themes of promise and disappointment, success and failure, were acted out – but in a peculiarly Australian manner. Does the land make the people?

The twentieth century will be remembered for the failure of a great social experiment, Communism. Its collapse in 1989 was such a triumph for the West that some claimed it heralded the end of history – from now on there was no path for people to follow other than that leading to the democratic free market.

But while Communism was foundering in the northern hemisphere, in the south there was a different social experiment going on, yet another attempt by mankind to find an answer to the age-old question – how should we order our affairs so as to get the best out of our brief life on this planet? In Australia, the world's newest country, an eclectic mix of people from the Old World were trying to create a different society, one best described by Ryszard Kapuściński, the Polish author of *Imperium*, a history of the rise and fall of Soviet Communism. Kapuściński told me that as he travelled around the old Soviet Empire doing his research for *Imperium*, person after person would say something on these lines:

'Ryszard, we made the right decision to get rid of Communism; it wasn't working. But my God, this capitalism is hard. I don't know if we can take it. And aren't we entitled to an alternative to the American model? You've been all over the world. Isn't there a country somewhere that has found a middle way – where market forces rule but where the government looks after the kids and the old and the sick and the poor? Somewhere where the bosses give the workers a reasonable deal? Somewhere where people help each other instead of just looking after themselves?'

And Kapuściński told them: 'Yes. It's called Australia.' Only to see them look bewildered. He said he was amazed so few Europeans knew anything at all about Australia, or had any idea that despite problems, the experiment of the middle way was still going on. Some thought Australia was just an extension of Britain in the South Pacific. Others that it was a little United States. Those who had sought enlightenment from British friends often ran into British prejudices, like that expressed by journalist Jeremy Clarkson: 'With the south of France, God can say, "I did OK there." But we must never let Him forget Australia, a vast and useless desert full of spiders that'll kill you, and men in shorts.'

Kapuściński, who has been to Australia, did his best. No, Australia was not another United States and never would be. The Pilgrim Fathers were people of strong religious beliefs who were convinced they were following Divine will in founding America for God's elect. Australia, on the other hand, was founded by criminals and their jailers.

The Pilgrims stepped gracefully ashore at Boston on to a seaside rock still revered to this day. The first British settlers landed in Sydney Cove, where the criminals persuaded the jailers to dole out some rum and Australia was born in an orgy of drunkenness, violence and sexual debauchery. The paths the two nations have taken have been different ever since.

Kapuściński's interrogators naturally wanted to know how – after such an inauspicious start – Australia had become a country that Kapuściński now so much admired. 'Ah,' Kapuściński would say, 'That's a fascinating story.'

We know why the British wanted to found a colony on the east coast of Australia, but determining what sort of colony they intended it to be is more difficult. Britain in the closing years of the eighteenth century was on the brink of revolt by the working classes. The prisons, the hulks and the poorhouses were full. The government of William Pitt had looked at several places in Africa as possible sites for overseas prisons but had rejected them.

By 1786 the situation was so desperate that a decision had to be made quickly, and by January the following year the first convicts bound for Botany Bay had been loaded on to ships of the First Fleet and were ready to sail. On the surface then, the settlement of New South Wales on the eastern coast of Australia was to be a penal colony pure and simple, an enormous jail to confine Britain's criminal classes – the world's first concentration camp, according to author Robert Hughes in *The Fatal Shore*.

Yet there must have been more to it than this. In deciding whether to go ahead with Botany Bay, the British government had heard evidence that the new colony might also produce tea, coffee, tobacco, silk and spices – another India. And if the climate was inhospitable for white men to undertake this work, then perhaps Chinese slave labour could be introduced. Historian Alan Atkinson, author of *The Europeans in Australia*, believes that the British wanted to create a community of peasant proprietors. 'At Botany Bay, so it was thought, convicts would be peasants in a country of their own.'

Logic supports this view. If Botany Bay was to be a Gulag pure and simple, then Britain would have chosen its worst convicts to send there – men and women serving life sentences. Instead, many convicts in the early transports had only seven-year sentences, some even less. Next, if Botany Bay was to be a Gulag why go to such pains to make certain that the convicts arrived in reasonable health? Surely the more that died on the long voyage the better. Yet we know from *The Great Shame*, by Tom Keneally, not a writer to favour the British view, that the surgeon-general on each convict transport was in trouble with his superiors if too many of his charges died en route to Australia.

So it came as a surprise to me, having, like a lot of my generation, been brought up on stories of the brutality of the British jailers, to learn that convicts on the better transport ships were issued with sturdy clothing and a pair of shoes (the first time many had ever worn them). Laundry was done on deck and bedding aired regularly. In balmy climes the convicts were encouraged to sing and dance. They were served one substantial meal a day (which included meat and sometimes a glass or two of wine) and two snacks. There was a daily dose of lemon juice to avoid scurvy. To alleviate boredom some transport ships had lectures given by the ship's officers or better-educated convicts. The ships themselves were carefully inspected and only those classed as A1 or A2 were chosen. As a result, the convict voyage to Australia was distinctly safer than emigrant voyages, either to Australia or to America. True, being transported to Australia was not a P & O cruise, but it was not as bad as some anti-British propaganda has made out.

When they arrived, the convicts found not a Gulag but a colony where English law applied. Henry and Susannah Kable, convicts on the First Fleet, could not locate their luggage when they landed so they sued the captain of their transport ship for losing it and were awarded £15 damages by the judge-advocate. Most convicts found that the authorities were more interested in integrating them into the workforce than punishing them for their misdeeds. In economic terms they were viewed as near ideal migrants. They were young, healthy and skilled. Many were literate and numerate. But their biggest attraction was that they had to do whatever the state told them – they could be directed into the public or private sector – and they did not have to be paid wages, just basic provisions.

What this all adds up to is a major economic and sociological experiment which has not yet run its course. Could the dregs of Britain's dungeons, its repressed and criminalized working class and the disaffected and deprived peasants of England's oldest colony, Ireland, if sent against their will to an empty land at the other end of the world, remake themselves and create a respectable outpost of British civilization in the Antipodes? (The land, of course, was not empty and we will come to that.)

Well, they did. Over the next hundred years other settlements sprang up in Australia, each with its own laws, defence forces, railway systems and local loyalties. But, in general, they all prospered, so that by the beginning of the twentieth century the descendants of those first convicts and jailers were enjoying one of the highest living standards in the whole world. The conclusion has to be that, despite its many horrors, transportation to Botany Bay was one of the more humane prison experiences of the eighteenth century.

Transportation ended in 1868. Many of the convicts had indeed stayed on, and instead of a nation of peasant proprietors, each scratching a living from a European-sized plot, enormous areas of land were given over to growing wheat and raising sheep and cattle. The discovery of gold, silver, copper and a plentiful supply of coal, timber and whale oil convinced many that Australia could become a new America.

Charles Darwin, passing through Sydney in 1836 on HMS *Beagle*, said that Australia was *the* country to grow rich in. 'There are now men living who came out as convicts . . . who are said to possess without doubt an income of from 12 to 15,000 pounds per annum [well over £100,000 in today's money].' Land deals could provide spectacular returns. In 1795 the government deeded 55 acres of Sydney waterfront land – later to become the whole industrial suburb of Pyrmont – to a foot soldier called Thomas Jones for . . . nothing!

A year later Jones sold it to another soldier for £10. He kept it for a couple of years and then swapped it with John Macarthur for a gallon of rum. By 1836 nearby land was selling at auction for £10,000 an acre.

A bountiful country then, of great potential, admittedly given to capitalism's cycle of boom and bust, but one where everyone was optimistic about the inevitability of progress and cherished the absolute conviction that their children would be better off than they were. But what sort of people were they – rich ex-convicts, rich ex-soldiers, extremely rich graziers packing off huge wool bales to the hungry mills of the Motherland?

It depends on whose history you read. One school has it that Australia, like its founders, remained British to the core, and its history is the story of national growth fostered by British genius, enterprise and respect for an ordered and hierarchical society. The country's inhabitants, even those born there, were 'Anglo-Australians,' linked forever by language, culture, tradition and kin. 'Australia', said one Prime Minister, Billy Hughes, himself born in London, 'is as much a part of England as Middlesex.' Another Prime Minister, Robert Menzies, born in Melbourne, an Anglo-Australian by temperament, said he was proud of being 'British to the bootstraps'.

The other school has it that a distinctive character developed, that someone calling himself an Australian, an 'Aussie', grew out of the antagonistic factions of, on the one hand, the Irish Catholics who made up a major proportion of the convicts, and, on the other, the Protestant English, Scots and Welsh, who made up a major proportion of the jailers.

These Australians wanted to forge a country which was not a poor copy of Britain or, indeed, any part of the Old World. They regarded Europe as a cemetery, populated by the most cultured people in the world, who every now and then went mad and did terrible things to each other. They wanted to leave all the old cultural and religious prejudices behind and create a new and better society. But how could they when the governing classes deliberately modelled Australia's education system on that of Britain, so as to instill a loyal spirit in the minds of each new generation? Australian school children learnt the names of English monarchs, the geography of the United Kingdom and the history of the British Empire. It was very effective. As late as 1998 the Ipswich Rotary Club saw nothing strange in singing 'Land of Hope and Glory' as a tribute to Britain on the anniversary of the outbreak of the Second World War.

Was it any wonder that 100 years after European settlement Australia had no national song, no heroes, little literature and no national days? And as late as 1930 when an Australian Prime Minister, James Scullin, of Irish descent, put forward a fellow Australian, Sir Isaac Isaacs, of Jewish descent, to be the next Governor-General (the monarch's representative in Australia and until then always an Englishman), Scullin had to sail to London and endure a bitter confrontation with the King before he would agree.

The argument between these two schools of history over what Australia is rumbles on. Crucial to its outcome is whether it can be established that there is an Australian character distinctly different from his or her forebears, no longer an Antipodean Englishman, an 'Aussiefied-Pom', a Colonial or a second-hand European, but *Homo Australiensis*. If so, who is he or she, how did they develop and in what way are they different?

I had just started work on this book when I saw Andreas Whittam Smith, the former editor-in-chief of the British Independent group of newspapers in London and told him during our conversation what I was doing. Whittam Smith, who had visited Australia many times, said he would be interested to see whether I would be able to answer the question which had intrigued him above all others – why Australians had turned out to be so different from Americans.

'After all,' Whittam Smith said, 'both were originally British colonies, both English-speaking, both frontier communities and both faced hostile indigenous people – Native Indians and Aborigines. Yet your American has developed a culture of rugged individualism, each man for himself, scorn for the weak, winner take all, while Australians have an ethos of mutual help and social obligation – your great belief in mateship.'

Some Australians will tell you that there is a whole book to be written about mateship. D. H. Lawrence devotes pages of *Kangaroo* to the topic and was obviously fascinated by it. The Australian author Henry Lawson saw mateship as a central prop of Australian culture and wrote about it often, stressing its importance in Australian life. The Australian National Dictionary defines it as: 'The bond between equal partners or close friends; comradeship; comradeship as an ideal.'

Australian men call each other 'mate' all the time but this is more form than substance, and anyway 'mate' is different from 'mateship'. A new generation of books about the Australian environment offers the theory that the harsh and unforgiving nature of the country imposed certain characteristics upon first the animals and plants, next the Aboriginals, and finally the white man, and that these characteristics form the foundations of mateship.

The Australian ecosystem is so fragile, the theory goes, that animals and even plants manage to survive not by Darwin's triumph of the fittest, by nature 'red in tooth and claw', but by co-operation. Thus we find Australian plants that recycle nutrients for each other in highly efficient ways, and birds and animals that help each other for the good of the group. Take the kookaburra, one of the best-known Australian native birds, whose call resembles raucous laughter.

The young male does not leave home as soon as he is old enough but hangs around to help his parents find food for his little brothers and sisters, and to do his share of incubating new eggs, sometimes foregoing for ever the chance of finding a mate and setting up his own home.

'Pioneers had to be tough and self-reliant, and they had little time for recreation,' writes Tim Flannery in *The Future Eaters*. 'But most importantly they had to stick together, for hard times could see all perish if the community was not cohesive. Thus, mateship was esteemed above almost any other value.' In *Among the Barbarians*, Paul Sheehan gives examples of how mateship saved lives in Japanese prisoner-of-war camps in the Second World War. 'In the camps the Australians discarded their differences and became a tribe, a tribe which was always the most successful group. The core of this success was an ethos of mateship and egalitarianism which not only survived the ultimate dehumanizing duress of the death camps, but shone through as the dominant Australian characteristic.'

Gavan Daws in *Prisoners of the Japanese* compares the behaviour of each nationality in the camps. The British hung on to their class structure to the point of death. The Americans were the capitalists, the gangsters, even charging interest on borrowed rice. The Australians kept trying to create tiny welfare states. 'Within little tribes of Australian enlisted men, rice went back and forth all the time, but this was not trading in commodity futures [like the Americans]; it was sharing, it was Australian tribalism.'

When the Japanese freighter *Rokyu Maru* carrying 649 Australian and 599 British prisoners of war was torpedoed by American submarines in September 1944, a group of English survivors who had been drifting for days sighted a huge pontoon of rafts carefully linked together. Daws quotes one of the Englishmen who spotted the rafts: 'They seemed organized compared with the disintegrated rabble we had become during those last two days. They may have been drifting aimlessly as we had; but at least they had all drifted together. A lot of them still wore those familiar slouch hats – we had caught up with the Aussie contingent. One of them looped us with a dangling end of rope. 'Cheers, mate,' came the friendly voice. 'This is no place to be on your lonesome.'

But – and there is no sense in denying it – Australian mateship is mainly for men. It was – and is – difficult to be mates with a woman. Here is a little experiment you can try if, when visiting Australia, you have been invited to dinner and will be leaving the country shortly. Ask the women around the dinner table one by one who their best mate is. Most, if not all, will name their husbands. Then ask the men one by one who *their* best mate is. Most will name a male friend, often from their schooldays. The expression on the face of a woman who has just proudly named her husband as her best mate only a few seconds later to hear him, completely unabashed, name his golfing partner, has to be seen to be believed.

This resentment may explain why some modern, liberated Australian women see mateship as essentially corrupt. Columnist Susan Mitchell points out that in golf there is an expression, a 'gimme', meaning a short putt that should be simple to hole out. Mates give it to each other. 'Women, on the other hand, make their opponents play the shot . . . That's why women make better leaders and managers, because no one gets "gimmes".'

Historically it was difficult to be mates with someone who was not white because a fierce racism in those early years generally excluded from mateship anyone whose skin colour, appearance, costume or behaviour was different. The Chinese, Aboriginals, Arabs, Indians, Jews, Africans were all considered fair targets for discrimination of all types. The Australian Labor Party, dedicated to the brotherhood of man, coupled this noble sentiment with a declaration of its belief in 'racial purity'.

The influential magazine the *Bulletin* had as its slogan 'Australia for the White Man' on its masthead until 1961 when the incoming editor, Donald Horne, author of *The Lucky Country*, removed it. (It had been invented by Tom Courtney, the associate editor of the *Daily Telegraph*.) Early *Bulletin* cartoonists created a series of national stereotypes to make racist points – Chunder Loo, the turbaned Indian; Sissy Ah Wong and Billy Ah Song, the slippery Chinese; Aaron Shemabonee Isaacs, the long-nosed Jew.

A classic example of such racist cartoons is a Norman Lindsay cartoon from 1916 headed 'The Insult'. It shows two Chinese children, one eating a piece of cake. Sissy Ah Wong: 'Give us a bit.' Billy Ah Song: 'Will I?' Sissy Ah Wong: 'Go on, yer dirty Jew.'

Women, even though they could not be mates, were to be treated decently but quickly put down if they showed intellectual aspirations. In a leading article headed 'The Great Woman Question', the *Bulletin* offered the view of most nineteenth-century Australian men: 'The equality argument is an absurd one. Women are far from that stage of progressive rationalism when they can take their stand on the same platform as men. Women, as a totality, are far inferior to men, as a totality.

'Women's enfranchisement is a Tory proposal. The Tories champion the alleged cause of women because the women of today are, as a rule, Tories; almost every woman is a queen-worshipper, a prince-worshipper, a parson-worshipper, a lord-worshipper.' In the same issue was this item of news: 'A man fishing in the Murray River at Echuca last week caught a dead woman. He might have done worse. He might have caught a live one.'

Mateship carried personal obligations beyond trust, reliability and a willingness to share. Because life in early outback Australia, the home of the 'bush barbarians', was so tough, acts of sustained heroism were sometimes necessary to survive. But no man can be a hero to his mate; it would upset the egalitarian nature of the relationship. So Australians developed an attitude which combined bombast about Australia as a whole – 'God's own country . . . the best in the world . . .' with wry understatement of personal achievement, a determination to show that nothing really matters, so why care too much about it.

D. H. Lawrence noticed it. The principal character in *Kangaroo*, Richard Somers, an Englishman who has come to Australia 'to start a new life and flutter with a new hope', is asked for his opinion on Australians. He replies: 'It seems to me they think it manly, the only manliness, not to care, not to think, not to attend to life at all but just tramp blankly on from moment to moment, and over the edge of death without caring a straw. The final manliness.'

Lawrence and other writers thought that the only thing that could shake Australians out of this attitude would be bloodshed. 'It always seems to me,' said Somers, 'that somebody will have to water Australia with their blood before it's a real man's country. The soil, the very plants seem to be waiting for it.' Henry Lawson, too, thought violence might be necessary to establish an Australian identity.

> We'll make the tyrants feel the sting
> Of those that they would throttle
> They needn't say the fault is ours
> If blood should stain the wattle.

But Australians were well aware of how the Civil War had torn America apart – Australian newspapers were still carrying articles about its causes and lessons 30 years after it had ended – and shrank from that path, although, as we shall see, at times they came much closer to civil war than many today are prepared to admit.

The reasoning of idealistic Australians in the 1800s was simple – Australia is a country of limitless potential. If we are genuine in our wish to create a new society, to leave behind the class divisions of the Old World, to avoid servility and poverty, then we should divide everything in a fair and reasonable manner. So 'fair and reasonable' became the touchstone of the Australian way of life – 'Give us a fair go, mate . . .' and 'Fair crack of the whip, sport' – and it has been repeated incessantly in pleas and judicial decisions, trade union conferences and parliamentary debates. It appeals to a vital human need.

Matt Ridley in *Origins of Virtue* explains how citizens of early hunter-gatherer societies evolved a strong sense of fairness so as to ensure that all of them would commit themselves to communal enterprises. This placed an obligation on the powerful to treat their peers and their underlings decently, and a duty on those underlings and peers to demand that decent treatment. Our sense of fairness is therefore inherited, a natural human emotion that many feel is threatened by *laissez-faire* capitalism. This sense of fairness spilled over into the Australian business world, to the amazement of Americans. 'Australians start with the notion that if you are making a profit, you must be screwing somebody,' said American executive Bob Joss in 1999. 'In Australia, profit is often a dirty word.'

Other Australian characteristics flow from their commitment to a fair and reasonable social system. Australians hate officiousness and authority – especially when embodied in military officers and policemen – because, in their experience both are unlikely to give you a fair go. They are suspicious about discipline and have a great wariness about elites – except, perhaps, in sport – because those imposing the discipline are unlikely to be reasonable and elites are reluctant to extend a fair go beyond their own members.

Lawrence saw all this as a recipe for anarchy. He has Somers conclude: 'In Australia nobody is supposed to rule, and nobody does rule . . . The proletariat appoints men to administer the law, not to rule. These ministers are not really responsible, any more than the housemaid is responsible. The proletariat is all the time responsible, the only source of authority. The will of the people.' Somers accepts that this is real democracy but he does not like it. He is an Englishman with an Englishman's instinct for authority and in Australia authority was a dead letter.

It is difficult to exercise authority in an egalitarian society. Somers/Lawrence finds himself longing for the certainty of India, for the responsibility of command, the pleasure of obedience. In Australian society, long before Manning Clark gave the young historian Geoffrey Searle the following advice on how to behave in Britain: 'Call no biped lord or sir and touch your hat to no man,' Australians had extended the original idea of egalitarianism to 'No one can tell me what to do.'

As Lawrence writes: 'There was no giving of orders here; or if orders were given, they would not be received as such. A man in one position might make a suggestion to a man in another position and this latter might or might not accept the suggestion according to his disposition.' How could a country run like this? Was all that stood between Australia and anarchy the shadow of British authority reaching out across thousands of miles of sea – Britain, Empire, the King, the Governor-General?

But the country *did* run and, in a manner unprecedented anywhere in the world, passed law after law to improve the welfare of its citizens. Votes for women (18 years ahead of the United States, 16 years ahead of Britain, and 70 years ahead of Switzerland); the secret ballot; free and compulsory education for all children; old age and invalid pensions; safety at work, fixed working hours, minimum wages, and a legal arbitration system to settle disputes between bosses and workers and even – as early as 1908 (and of some personal interest) a pension of £1 week for distressed authors. The benefits were quickly reflected in the wellbeing of white Australians. (Aboriginals were another matter.) The country soon had one of the lowest infant mortality rates in the world and it would have been hard to name any other country that looked after its aged as well – both indicators of a decent, caring society. In 1999, despite economic rationalism, Australia was still spending 54 cents of every government dollar on health and welfare, a greater proportion than the United States spent on defence in the Second World War.

As befitted a new country, Australians approached their social legislation with a fresh eye. Take elections. Gaining the right to vote had been too hard a battle down the years to be treated lightly, so Australians, contrary to their normal views on being ordered around, decided that its citizens should be compelled to exercise their franchise. You can still spoil your ballot paper and thus vote for no one, but by law, on pain of a fine if you do not, you have to go to the polling station, or send in your postal ballot. As a result, the turnout in Australian elections always approaches 100 per cent, which makes it difficult to claim that the result does not reflect the will of the people.

In other ways, too, Australia runs a more democratic democracy than, say, that paragon of the free world, the United States. To start with, most Americans do not vote at all. They stay away from the polls in their millions. Next, although we read a lot about political funding scandals, most Americans (93 per cent) have never given a single cent to any political party. Yet only a very rich man or a professional politician with access to millions of dollars stands a chance of running for President – the four-yearly national election campaign lasts more than a year and the cost runs to billions of dollars. In contrast, Australian Prime Ministers have included an engine driver (Ben Chifley), a trade union official (Bob Hawke), and a one-time bottlewasher (Billy Hughes). The Australian election campaign every three years lasts about a month and costs less than a single Senate race in California.

Finally, most American citizens have never met their elected official at any level – not even the local councillor. In Australia, television has lately taken away some of the personal contact between candidate and voter, but before that every candidate was expected to turn up in a hall in their constituency or in the street if they were brave enough, and submit themselves to questions – and the odd rotten tomato or two – from voters. And in small-town Australia many are on handshaking terms with the mayor and most of the councillors.

* * *

This has been the optimistic view of the Australian experiment, the one Kapuścińzki himself holds and spreads. There is another view. This is that the Australian middle way has failed, that the country is a white island in an Asian sea, uncertain where to look for comfort. Nanny Britain turned her face to Europe and abandoned Australia. Nanny America has withdrawn from the South Pacific and could not care less about Australia. Australia today stands or falls on its own.

The divisions between the Australians and the Anglo-Australians or Austral-Brits have not been resolved – witness the debate over whether the country should become a republic, whether the Head of State should be – as former Prime Minister Paul Keating so aptly put it – 'one of us', or remain the Queen. The debate is bitter and divisive because it runs the risk of reviving old sectarian enmity between Irish Catholics (republicans) and Anglo-Australians that disfigured Australian life for so many years and which had apparently died.

The battle for 'a fair go' has been fought out between capital and labour with many a long and bloody strike. In the nineteenth century, the bosses had the upper hand but the unions became very powerful in the twentieth century until a new factor appeared – free market reform and international competition. The idealistic values so important to Australians – egalitarianism, mateship, solidarity and a fair go – went head-to-head with the new world order, and in 1998 Australians witnessed on the waterfront the sort of violent clashes between capital and labour they thought had gone forever. The war goes on.

Early Australians who so strongly preached mateship and the brotherhood of man did not extend this to the Aboriginals, who turned out to have occupied this 'empty' land for over 50,000 years before the white man arrived and did not take kindly to being dispossessed. Aboriginals held their own in the first campaigns, but the arrival of the repeating rifle made their cause a hopeless one and they were eventually defeated, although clashes continued until the 1920s. The peace terms were harsh because 'Aborigines were not really human', and included reparations paid to the whites in the form of Aboriginal children.

But a more enlightened Australia agreed in the 1990s to renegotiate the peace terms and, after admitting that Australia had not really been empty when the white man first arrived, went on to sign what was, in effect, the Mabo Peace Treaty. This came into effect on 22 December 1993, sponsored by Prime Minister Paul Keating, an act that on its own assured him a place in Australian history.

The country still struggles to overcome its early racism. In 1973 the government of the day announced that to be an Australian no longer meant you had to be white. This was a courageous step because it was saying to the four generations of Australians since Federation, 'We are now going to change direction.' White Australia was over and in the eighties and nineties Australia opened its doors to so many Asians that some demographic projections in 1999 showed that within 25 years one in five of the population would be Asian-Australian. A social and cultural revolution took place, a new Australia emerged, one praised by President Clinton as the only country in the world that had managed to get multiculturalism to work.

Not everyone thought that this was a good thing and the cry of 'Populate or perish' was replaced with 'Stop immigration now', with newspapers full of articles along the lines of 'The New Ghettos of Sydney'. Pauline Hanson, a Queensland politician, built a whole political party – admittedly one with a short effective lifespan – on a platform of stopping Asian immigration and ending positive discrimination for Aboriginals.

The lucky Australians in high-leisure cities like Sydney and Melbourne continued to enjoy a hedonistic lifestyle the rest of the world only dreamed about, ignoring two scientific developments that may shatter Australia's sense of freshness and hope. The first was the identification by scientists in the 1980s of the 'El Niño Southern Oscillation' (ENSO), a cycle that powers Australia's weather and makes it the only continent on earth where the overwhelming influence on the climate is a *non-annual* climatic change.

The European farmer knows that no matter how harsh the winter, spring will come, the snow will melt, summer will ripen the crops and there will be an autumn harvest. ENSO, with an unpredictable cycle ranging from two to eight years, ensures that the Australian farmer can enjoy no such certainty. All he can expect is that sooner or later he will have to cope with floods and bushfires and droughts that can last for years.

The other event was that the end of the Cold War released satellite imaging from its spy-in-the-sky role and it became available for important peacetime uses. A satellite 400 miles in space and using various imaging processes can conduct mining surveys, determine water depth and wave patterns, examine crops and vegetation, and report on their health. It can see things the human eye cannot and no corner of the planet is beyond its reach.

In the early 1990s, satellite imaging of Australia showed that far from being a rich and bountiful land, it is a continent with a thin skin of soil which is getting thinner all the time. Even areas of deep soil and good rainfall have poor tree growth because the soil is deficient in nutrients. Erosion and salinization are common. This led some scientists to conclude that at nearly twenty million the country has already reached its optimum population and that immigration should cease immediately.

So has Australia already run its course, reached the end of its history only two centuries after the first penal settlement and only a hundred years after it became a nation? Is the dream still alive or did it go spectacularly wrong somewhere? David Foster, the scientist-turned- novelist, says, 'Our record here is undistinguished and may be short-lived. I see Australia as a potential tragedy because it's a civilization that has gone into premature decline.'

Or, put with typical Australian bluntness by the poet Laurie Duggan:

> I like the way we've
> been able to fuck things here
> as good as anywhere else
> in only half the time.

CHAPTER 3

FEDERATION

The British working in India had 61 levels of status, something that the Indians recognized and understood because they had an aristocratic hierarchy of their own. Why did Britain accord such respect to this Indian aristocracy and so little to Australia? Was it because Australia had no aristocracy at all?

The last topic occupying the attention of most Australians as the twentieth century loomed was the impending union of the six states – federation. They had something far more important to worry about – would they live to see Christmas?

Bubonic plague was raging throughout the country. It had started in Adelaide, South Australia, the previous year and the city's harbour had been declared 'an infected place'. Even though there was as yet no Federal government, the Premiers of all the colonies met frequently to discuss matters of national interest, and had ordered a major clean-up at ports across Australia. For a while it had appeared that the disease had been contained.

But then on 19 January 1900, Arthur Payne, of Dawes Point, Sydney, was diagnosed as having the disease. Payne was a carter at the Sydney wharves and it was a reasonable bet that the plague which laid Payne low had been brought to Sydney by rats on a ship from Adelaide. But, in keeping with the anti-foreigner sentiment of the time, many said that rats off a ship from French New Caledonia were to blame. Payne and his family were put into quarantine and police supervised a fumigation of his house which caused something of a stir among his neighbours. 'There is no need to panic,' said Dr Ashburton Thompson, the President of the Board of Health. 'Every precaution is being taken. I know what the plague is and I'm not going to let it get a foothold in Sydney.'

He failed. By August, 103 citizens, including Payne, had died from it. Payne's case inaugurated a ten-year epidemic in every state except Tasmania which, for once, blessed its isolation. Sydney, population 50,000, was hit worst of all, probably because its public hygiene was the worst. A vast clean-up campaign saw 750 tons of debris removed from inner city houses and dumped at sea on the first day. An immense collection of old timber, bagging, bedding, hen coops and fowl houses was burnt on the street. Rotten wooden floors and planking were taken up and tons of filth and dead rats removed.

The campaign was conducted like a military operation. Constables and contractors would without warning surround and barricade an area. No one was allowed to leave no matter what their business and families deprived of an income faced starvation. 'Some of the poor fellows made a little catching rats, for which the Health Department paid a penny a scalp,' recalled Billy Hughes, a British immigrant later famous as a politician. 'The rat has an almost even-money chance. If he bites or scratches you, you will probably die. If you kill the rat and get bitten by one of the ten million fleas which live on him, then all your worries in this world are over. It's a great game.' Still, this appealed to the Australian gambling instinct and there was no shortage of volunteers.

When the rubbish and filth had been removed, houses, yards and even streets were disinfected and whitewashed – one woman complained that the workers had even whitewashed her piano. Early in April, forty policemen quarantined the Chinese quarter, the notorious Wexford Street. 'The Orientals, frenzied at the outrage, indulged in the wildest profanity,' said a contemporary report. 'Special constables were ordered to accompany the health inspectors to protect them while discharging their tasks from the attacks of the offended Chinks. The premises, which frequently consisted of gambling hells and opium dens, were generally in a revolting condition.'

These days Australia is so frequently linked with health and vigour that it is hard to believe that earlier Australians lived in a country ravaged by disease. An outbreak of typhoid fever in Melbourne in 1889 killed more than 400 and incapacitated thousands. 'Typhoid fever . . . is with us, of us, among us, upon us,' said the President of the Inter-Colonial Medical Congress of Australasia, Mr T. N. Fitzgerald. 'It defies legislation, administration. It laughs at Boards of Health and triumphs ruthlessly and always.' It appeared he was right – just when a nationwide rat extermination programme had reduced bubonic plague cases to a trickle, typhoid fever came rushing back and killed 630 in one year alone – 1905.

It is obvious now that poor sanitation was largely to blame. London had had a sewerage system since the late nineteenth century, but turn-of-the-century Australian cities relied on the 'nightcart' to remove their effluent. The nightcart travelled up and down city and suburban streets collecting pans full of sewage and replacing them with fresh empty ones. Nightcarts were still operating in some otherwise highly developed Australian suburbs until the 1940s.

A Melbourne bank manager, J. H. Barrows, described what this was like back in the early days. 'How many a hot, sultry summer night have I thrown open my window panting for heaven's pure air, and I have swallowed – a nightcart. Or hastening to some other window, hoping there to find consolation, I have found a stench. Our gutters and drains, our river and nightcart are a discredit to our boasted civilization.'

In Queensland, dengue fever was endemic. At one stage one in five of the population was diagnosed as suffering from it. In Sydney's eastern suburbs, the posh areas of Double Bay and Vaucluse had the highest average of typhoid cases in the city because of 'the large number of goats' which went scavenging through the streets and lanes. But, to many Australians, the suggestion that it was their own unhygienic environment that caused plagues and fever was unacceptable. They knew the real source of the problem – the Chinese and 'the Queen's Nigger Empire'.

Why were those early Australians so racist? 'Australia for the white man' . . . 'Rule Britannia/Britannia rules the waves/No more Chinamen in New South Wales.' There are the obvious reasons. Some Australians feared that the Colonial Office would bring indentured labour to Australia from other parts of the empire, as it had done in Fiji and the African colonies, and destroy 'the working man's paradise'. There was – and remains – the fear that the many millions of Asians to the north of Australia would cast longing eyes on the one apparently rich but sparsely populated outpost of European white civilization in the area. But there has to be more to it than that.

My own theory is that jealousy played a major role. Australians saw themselves as the youngest country in the empire and therefore deserving most attention from the Mother Country. They also felt they most resembled Mother – they spoke English without an American accent; they did not have large French- or Dutch-speaking groups in the country, like Canada and South Africa; they played cricket; named their suburbs after English places – Brighton, Sandringham, Ramsgate, Windsor; cherished British-made goods; and fawned on the British royal family.

Why then was India the favourite child, the jewel in the crown? Why did India have a Viceroy and the Australian colonies only governors? Why did the pick of Britain's civil servants want to serve in India in large numbers while Australia got only a few second-raters? The British working in India had 61 levels of status, something that the Indians recognized and understood because they had an aristocratic hierarchy of their own. Why did Britain accord such respect to this Indian aristocracy and so little to Australia? Was it because Australia had no aristocracy at all?

The *Bulletin* often referred to 'the Queen's Nigger Empire'. This expressed a resentment that most of the Queen's subjects were coloured and yet some of them were treated by the British establishment as if they were the equal of white Australians. So Australia's racism came not from working-class British immigrants – a common accusation – but from older-generation Australians who resented the attention Britain paid to its coloured dominions.

In support of this theory I draw attention to two laws which Australians rushed to introduce once they had become members of a united country and no longer subject to the restraining hand of the Colonial Office. The first was the Immigration Restriction Act, 1901, the basis of what was commonly called the White Australia Policy, and the second was the Quarantine Act, 1908. The purpose of both was to keep Australia white and 'pure' by keeping out all Asians.

Until the Liberal (conservative) government of Harold Holt began to dismantle it in the 1960s, the Immigration Restriction Act became notorious. It had a beautifully simple 'catch-all' clause that brooked no argument or legal appeal: 'Any person who when asked to do so by an officer fails to write out at dictation and sign in the presence of the officer a passage of fifty words in length in a European language dictated by the officer is a prohibited immigrant.' In practice the immigration officer chose a language he expected the applicant not to know. If, perchance he did, the officer moved on to another language, choosing languages more and more obscure until the applicant inevitably failed.

(One of my first reports as a journalist on the *Herald*, Melbourne, in the late 1940s was a court case concerning two poorly-educated Fiji Indians, Mohammed Razak and Mohammed Sadiq, who had jumped ship hoping to be allowed to stay in Australia. They were convicted and deported under the Immigration Restriction Act for failing to write a single word of a dictation test in – Italian.)

But the effectiveness of the other law, the Quarantine Act, in keeping Australia white was not properly assessed until Dr Alison Bashford of the University of Sydney wrote 'Quarantine and the Imagining of the Australian Nation' in 1997. Dr Bashford revealed that quarantine was one way in which the newly federated Australia was encouraged to see itself as an integral whole, as a new island-nation.

This island-nation had the advantage of being a long way from the Old World and therefore new, pure, healthy and white. But it was also located in Asia which was seen as dirty, diseased and 'coloured'. Dr Bashford writes,

'Quarantine was the one strategy by which the new government could imagine itself at war with its region – thus defining the new Australia geographically, politically and racially against Asia – without actually being at war.' So the threat of invasion by disease made the external frontiers of the nation meaningful in a way which created an identity for Australia.

'There were two dominant images [in Australia] of Chinese men,' writes Dr Bashford, 'first, as a sexual and moral threat to white women and thus a threat to national integrity; and second, as purveyors of disease.' All passengers and crews of ships were checked at the quarantine line for signs of smallpox and plague. But if they were Asians, they were already 'foreign bodies' to the Australian nation imagined as white and pure, and irrespective of whether they were actually diseased or not, they were in most cases excluded under one Act or the other.

The quarantine authorities made no secret of this. Australia's foremost public health bureaucrat of the twentieth century, Dr J. H. L. Cumpston, wrote that one of the objects of quarantine was 'the strict prohibition against the entrance into our country of certain races of aliens whose uncleanly customs and absolute lack of sanitary conscience form a standing menace to the health of any community.' No one doubted that by 'alien', Dr Cumpston meant 'Asian'. He would have justified these views by pointing to the great plagues in China which at alarmingly short intervals killed millions.

Australians also justified their racism and racist legislation on economic grounds. They were trying to create a new society, one with high wages and a short working week, and everyone knew that Asians would work long hours for low pay. The only way to protect Australians' standard of living was to exclude all Asians as a threat to the Australian working man and his unions. (In 1997, old-time union stalwarts, who had long since abandoned their prejudices about Asians and 'coloured labour', nevertheless could not help pointing out that the ranks of strike breakers during a bitter waterside workers' dispute were thick with Pacific Islanders.)

Today, when governments and individuals devote so much effort and ingenuity to making life as long and risk-free as possible, the brief, brutal and often dangerous lives of Australians in the early days of the century come as something of a surprise. Average life expectancy for a man was only 55 years and 2 months. A woman could expect to live for 58 years and 10 months – that is if she did not die earlier in childbirth. Large families were common. Many a woman had a toddler at her skirt, another at her breast and a third in her womb. No wonder a Melbourne nurse reported that it was, in her experience, common practice for mothers to administer morphia – freely available then – to their babies to give themselves 'a free half day'.

The loss of a child from sickness or accident was not unusual. The day the first electric trams ran in George Street, Sydney, was not the happy occasion it might have been. A three-year-old girl, Rosalind Cortese, wandered onto the tracks near Circular Quay, went under the wheels of an approaching tram and was cut to pieces. The contemporary reports are very matter of fact – no apportioning of blame, no calls for better parental supervision of children, nor for their education about the dangers of traffic, no demands for better safety devices on trams – just an acceptance that in Australia life itself is a risk and children sometimes get killed. Perhaps this was a British-abroad, stiff-upper-lip attitude – Ranee Margaret, the wife of Charles Brooke, the second 'White Rajah of Sarawak', wrote in her autobiography about a voyage to Britain in 1873: 'The three children we were taking home with us died within six days of one another and were buried in the Red Sea.'

Travelling in Australia was a major risk in itself. Founderings, wrecks, rammings, collisions, groundings and ships that simply vanished went on year after year, culminating in 1911 with the worst disaster in Australia's maritime history, when the SS *Yongala* went down off Cape Bowling Green, south of Townsville, Queensland, with all its 142 passengers and crew.

What was going on? The Australian coast is a treacherous one and the weather often extreme. But there were other, preventable factors. Before Federation, the colonies lacked a national co-ordinated policy on navigation in dangerous waters. After the loss of the *Yongala* a captain said that the area between Mackay and Townsville was still not sufficiently lit and making a passage through the area at night was a risky venture, no matter what the weather or the experience of the crew.

And although the Victorian Marine Court decided that the foundering of the SS *Glenelz* in 1900 with the loss of 31 lives was 'the act of God', seamen who knew the ship claimed that her plates were only 1/16 of an inch thick in places and that they had been covered with a thick coat of paint to help hold them together. The *Bulletin* commented in typical style, 'The verdict of the Court exonerates the owners and the surveyors and all the live people, and makes things smooth all round: and if the dead people have any objections to make, the Court thinks they ought to get up and make them and not skulk in an un-British fashion at the sea bottom and think evil of their betters.'

Train travel appeared to be just as risky. In April 1908 two trains packed with holiday-makers collided in Victoria killing 44 and injuring 412. Some of the carriages caught fire and a number of trapped, injured people were burnt alive. Some of these train crashes had a Mack Sennett or a Charlie Chaplin quality to them which, if it were not for the loss of life, would have been not only funny but a healthy corrective for the commonly held belief that Australians were natural masters of machinery.

In 1887 a special passenger train carrying 400 holiday-makers set out from Sydney for a day's excursion to the Hawkesbury River. At Beecroft the engineer realized the engine was not powerful enough to pull all nine carriages up the slope to Hornsby, so he uncoupled two carriages and left them behind, telling their passengers to pass the time in community singing until he returned. He then took the remaining seven carriages to Hornsby, uncoupled them, steamed back to Beecroft, collected the other two, returned to Hornsby, joined up the whole train, and set off downhill to Peats Ferry, the train's destination. But it apparently did not occur to the driver that the engine which was insufficiently powerful to pull the nine carriages up the hill might not have the braking capacity to stop them going downhill. The train relentlessly gathered speed, shot out of the terminal at Peats Ferry Station doing 70 mph, went over the bank and into the river, killing the driver. The carriages concertinaed and four passengers were killed.

So it was risky to travel by ship or by train, but perhaps the most dangerous thing to do in early Australia was to go to work, especially in the railways – 'The engine bars are splashed and starr'd/They've killed a shunter in the yard' – and the mines. Explosions, accidents, rockfalls, flooding and fires killed workers in large numbers in coal mines, lead mines, copper mines, gold and silver mines.

In 1902, an underground explosion killed 96 workers at Kembla, New South Wales, when coal gas was ignited by a miner's flare lamp. Safety lamps had been thought unnecessary because no one suggested that gas was present in the pit. But the press, rather than questioning mining safety, seemed much more interested in miners' acts of heroic 'mateship'. When floodwaters in a gold mine near Coolgardie in Western Australia trapped miner Modesto Vareschetti, an Italian immigrant, his mate, Frank Hughes, saved his life. Vareschetti managed to find an air pocket on a narrow ledge above the floodwater and spent nine days and nights waiting for the water to recede so that he could get to safety. Every day Hughes donned diving gear and swam down through the floodwaters to bring his mate food, light and comfort. On the tenth day he was able to lead Vareschetti out of the mine thus winning, said the *Sydney Morning Herald*, 'a chorus of congratulations and rejoicing'. It was a story that touched more than one chord. An Australian miner with 'incredible courage' had saved an immigrant whose survival was remarkable – 'considering his Italian temperament!'

And always present, harsh and unforgiving, was Australia itself – drought, floods, heatwaves, hurricanes, cyclones and bushfires. The turn of the century found the country gripped by a great drought that had begun in the middle 1890s. When Henry Lawson toured some of the drought-stricken areas, he found 'nothing but dust and sand and blazing heat for hundreds of miles'.

Victoria had had its driest year on record and bush fires raged in many areas. South Australian wheat yields had suffered heavy decline, the Kimberley area of Western Australia had experienced its worst drought in sixteen years, and sheep numbers in the Northern Territories had dropped by more than a half since 1897. (There were still more than 100 million sheep and less than 4 million people.) Farmers gave in, handed their property over to the banks or agents, and headed for Sydney.

How did they cope with this tough life, these Federation Australians? In photographs they mostly appear well fed, decently clad and content. (Thirty years earlier, according to an index developed in 1997 and projected back to 1870, the Australian colonies led the world in the well-being of its white citizens.) The women wore dresses with exaggerated shoulders, corsets that pinched in the waist, ankle-length skirts, gloves and hats. They put on hats for picnics, hats for playing tennis and hats for the beach. Hats – black toppers, black bowlers, or flat boaters, were also *de rigueur* for men. They wore dark, double-breasted suits, shirts with starched fronts and cutaway collars and waistcoats crossed with watch chains. Beards or moustaches were almost universal.

In general, they ate well – they became the biggest meat eaters in the world – if monotonously. 'We eat meat and we drink tea,' said Dr P. E. Muskett. 'Meat eating in Australia is almost a religion.' Even the less well-off managed to eat meat three times a day, partly because of a custom that allowed a housewife to go to her butcher and exchange yesterday's unused meat for today's fresher cuts. The butcher then sold on the returned meat at a cheaper price – a practice not made illegal until 1906. Even the down-and-out could afford a saveloy, a large bright-red sausage made from slaughterhouse scraps that sold for a penny. There were fancy restaurants and oyster bars in the capital cities and unusual dishes for those who wished to try them, like black swan casserole, roast bush turkey and koala and kangaroo steaks. Typical meal times had been established by an etiquette book published in 1834 as eight, twelve and six o'clock and they remain little changed today.

These early Australians lived in city and suburban houses designed for Britain – narrow terraced affairs with high ceilings, dark corridors, steep pitched roofs and solid brick walls. But this was changing. A cavity wall as insulation against Australian heat had become a fairly standard building practice by the late 1890s, and soon after Federation Australian architects were complaining that many features of British house design had no place in Australia. 'Steep pitched roofs are used here and are a great mistake,' said Florence Parsons, the country's first woman architect. 'There is no snow or hail to throw off, so they are quite unnecessary. I usually design with the Australian climate in mind – bedrooms facing East to harness the morning's sunshine and sitting rooms facing north and south to capture the breeze.'

For the city dweller there was plenty of entertainment to occupy the leisure hours. 'Every Australian worships the Goddess of Sport,' said one English author, and he was right. 'The Melbourne Cup is the Australian National Day,' said the American author Mark Twain, visiting Australia on a lecture tour. 'It would be difficult to overstate its importance.' Cricket, Rugby Union, Australian Rules football, golf, hockey, tennis and boxing were all booming. And so that recently arrived migrants, nostalgic for ice-skating, a recreation popular in the Old World, could enjoy it in tropical Australia, in 1903 an enterprising businessman opened the world's first indoor artificial ice rink in Adelaide.

Cinemas were just over the horizon. In September 1900 a crowd of 4,000 saw a show at the Melbourne Town Hall produced by the Salvation Army. Using techniques that foreshadowed the worldwide success that the American journalist Lowell Thomas achieved twenty years later with his *Lawrence of Arabia*, the Melbourne show mixed lecture, lantern slides and thirteen special-shot motion picture segments. Five years later, John and Charles Tait made *The Story of the Kelly Gang*, about the infamous Australian bushranger, Ned Kelly, and claimed it to be the world's first full-length feature film. Films about bushrangers then became so popular that the New South Wales government decided that they were glorifying crime and promptly banned them.

Plays and musicals from London's West End and New York's Broadway did well at Sydney and Melbourne theatres, despite a tendency for the theatres themselves to burn down – the Tivoli in Sydney and Her Majesty's in Melbourne both going up in flames within three years. The quality of the productions was a matter of opinion, most locals enjoying everything. But visiting critics like Lewis Wingfield decided: 'The drama in the Antipodes? There is none. Both plays and players are beneath criticism.'

The motor car had arrived and the role that it would play in Australian life became apparent quite early. A syndicate of Melbourne businessmen launched a simple car, the Pioneer, in 1897, and three years later two of them drove it from Bathurst in New South Wales to Melbourne – a distance of 500 miles in ten days – to demonstrate its reliability. The *Age* newspaper rightly saw what an important development this was, predicting that the rise of the 'horseless carriage' or 'motor buggy' would spell great industrial and social change for people all over the world. The only mistake the *Age* made was to say that the streets would become cleaner and quieter. 'The clatter of ten thousand shod hoofs' and the 'eternal rattle of iron wheels' giving way to 'the soft murmur of the pneumatic tyres of the motor car'. (Although those who worry about the pollution of the atmosphere caused by cars should reflect on the air pollution which would have resulted from the enormous number of horses that would be around today if the automobile had not been invented.)

Australians were also getting ready to fly way back in 1894. Lawrence Hargrave took to the skies at Stanwell Park, south of Sydney, in a large box-kite, hovering for several minutes at the end of a 16 foot long wire. Hargrave was taken seriously in the international scientific community and the famous German scientist Otto Lilienthal said of him, 'If there is one man more than any other who deserves to succeed in flying through the air, that man is Lawrence Hargrave of Sydney, New South Wales.' Others were less impressed. 'Men will never fly,' said the then Chancellor of Sydney University, Sir Normand MacLaurin. Men ignored him and kept trying.

In 1910, Fred Constance, a mechanic from Adelaide, took off from a field at Bolivar, South Australia, in a French-built Blériot monoplane and flew for three miles. He was in the air for five minutes and twenty-five seconds and so was able to claim the first powered aeroplane flight in Australia, just pipping the American escapologist, Houdini, who flew for two miles at Diggers Rest, outside Melbourne, twenty-four hours later.

But for a true Australian story that expressed that 'give it a go' aspect of the national character we turn to a Melbourne electrical engineer, John Duigan. Duigan had followed all the earlier attempts to fly with close interest. He had never seen an actual aeroplane but he had read articles about them in magazines, had studied photographs, and had bought and read a book called *Natural and Artificial Flight*. Although for many this might seem a rather thin background for building an aeroplane and risking your life by flying it, Duigan went confidently ahead. Working in his spare time in a shed at a sheep station outside Melbourne, he built a wooden biplane 35 feet long with a wingspan of 24 feet 6 inches, and then persuaded a Melbourne engineer to construct a 20 horsepower petrol engine to power it. To even his own amazement, Duigan took off at his first attempt and flew 24 feet before prudence made him land again.

Now there was no stopping these early aviators. The following year William Hart became the first Australian to receive a pilot's licence after passing tests conducted under the rules of the Royal Aero Club of Great Britain. Then he took part in an air race from Botany Bay to Parramatta, beating American ace W. B. 'Wizard' Stone, who got lost and had to land at Belmore. Hart took 25 minutes and 53 seconds to cover 14½ miles and at one stage touched 70mph. Greater days for Australian aviation were still to come, but it had been a splendid, inspiring start.

One of the reasons that Australians showed so little interest in Federation was that politicians had been talking about it but doing nothing for far too long. It had been first suggested in 1847 but every proposal, every discussion seemed to get sidetracked. When the six state Premiers got together someone would invariably raise the question of Federation, the others would express enthusiasm, but all that would happen was that yet another committee would be established to consider the proposal in detail.

Over the years 'historic' conference followed 'historic' conference, banquet followed banquet, stirring speech followed stirring speech – and nothing, but nothing, followed year after year. The truth was that each colony was keen to protect its own interests and there were sharp political differences – New South Wales was freemarket, Victoria protectionist, New South Wales had supported the North in the American Civil War, Victoria the South. The railway tracks were different gauges, duties were sometimes levied on goods moving from one state to another. There were quarantine restrictions between states, some of which remain to this day, and different public holidays – still not sorted out.

There were also petty jealousies. In 1878, New South Wales announced that it was considering changing its name to Australia. It claimed the right to do so because not only was it the first Australian colony – a point it still emphasizes in its official publicity – but more native Australians had been born there than in all the other colonies put together. Victorians could scarcely contain their anger. If New South Wales did this, said one MP, then would the Victorian Premier rename his state Australasia? 'No,' said the Premier 'because then New South Wales might well call itself The Southern Hemisphere.'

South Australia considered itself a cut above the other states because it had never been a convict colony and refused to take part in celebrations for the New South Wales centennial, claiming that South Australians had no reason to celebrate the establishment of 'distant penal settlements'. Western Australia resented the fact that the other states considered it so far away that it could be safely ignored. (Perth, the capital of Western Australia, remains the most isolated large city in the world.) Queensland had delusions of expansion and wanted to annexe eastern New Guinea – and perhaps the New Hebrides, Samoa and all the islands lying north and north-east of New Guinea, an area in total larger than France or Germany – before the Chinese could do so.

(The Earl of Derby made a caustic note on this request, 'I asked them whether they did not want another planet all to themselves and they seemed to think it would be a desirable arrangement, if only feasible. The magnitude of their ideas is appalling to the British mind . . . It is certainly hard for four millions of English settlers to have only a country as big as Europe to fill up.')

The New South Wales Premier, Sir Henry Parkes, 'the Father of Federation', refused to give up. Parkes, originally from Warwickshire, England, came to Australia in 1839 and after a shaky career as a customs officer, ivory turner, businessman and newspaper proprietor, entered politics in 1854 and became an expert at cobbling together unlikely coalitions over which he presided as Premier – when not struggling to avoid bankruptcy. Although his critics mocked him for sentimental exaggeration when he talked of Australia's future, he nevertheless had an appealing turn of phrase – he wanted, he once said, 'to open the batting for the innings of the people'.

What he needed to do, he decided, was to inspire ordinary Australians with his vision. The venue he chose was yet another official banquet at Tenterfield, a small farming town on the New South Wales northern tablelands. It may not have been a calculated decision on Parkes's part – there is much anecdotal evidence that the Father of Federation had a lot to drink that night. He made a powerful historical case for Federation, comparing Australia with the USA. The population of Australia was now about the same as when America formed the United States 'and surely what the Americans have done by war, Australia can bring about in peace,' Parkes said.

But his strongest point was defence, and here Britain dealt him the 'Chinese menace' card to play. The British army had sent a professional soldier, Major General Sir Bevan Edwards, to Australia earlier in the year to examine the defence forces of the colonies. Parkes now revealed that General Edwards had concluded that the only way to prevent a Chinese invasion of Australia would be by merging the forces of all six colonies. Federation would ensure that this happened.

March and April 1891 saw the longest, most lavish, most impressive, most important meeting on Federation yet. Held in Sydney, it lasted six weeks and included not just the usual banquet, but dinners, balls and picnics. The convention chairman was, of course, Sir Henry Parkes, who at a 'gala banquet' held in the new Centennial Hall on 2 March made a speech which included the phrase that was to signify the essence of Federation – 'one people, one destiny'. A draft constitution for the United States of Australia was adopted and sent back to the individual parliaments for approval – and yet another four years ticked away. Sir Henry Parkes died, aged 80, in 1896, his dream of one people, one destiny, unfulfilled.

But in March the following year it looked at last as if Federation was imminent. Delegates elected by popular vote assembled in Parliament House, Adelaide, to hammer out a compromise on disagreements over the Draft Constitution. The central question, that age-old one, was where should power reside – with the people, or with some elite minority group?

As Alfred Deakin, a delegate from Victoria and future Prime Minister of Australia, said, 'The convention was divided between those who have faith in the wisdom of the people and those who believed in the right of the few to rule.' Edmund Barton, who was to become the country's first Prime Minister, finally convinced the delegates that their duty to save Federation meant that they had to choose the middle way. The constitution presented to a referendum in each of the colonies in 1898 offered a Federation with a Governor-General, appointed by the Queen, who would have vested in him all the powers of the Crown.

Yet there were still obstacles. In the first referendum ever held on a countrywide basis, New South Wales failed to reach the minimum number of Yes votes required. The Melbourne *Age* accused New South Wales of striking a blow below the belt at the federal cause. 'It was fondly hoped that 3 June would mark the beginning of a great Anglo-Saxon Commonwealth beneath Austral skies,' the *Age* said. 'We hoped that we should enjoy a constitution, not wrung like Magna Carta from a trembling monarch by the "mailed fists" of feudal lords, but drawn up by the representatives of free men to meet the needs of free men. And it is now evident that these aspirations would have been fairly satisfied had it not been for the treachery of the free-trade Premier of New South Wales.'

After negotiations with New South Wales, the supporters of Federation tried again in June the following year and this time there was a decisive Yes vote in New South Wales. Victoria, South Australia and Tasmania followed suit with majorities. In Queensland the Yes vote just made it, but Western Australia hesitated until July 1900, then signed up. It had been touch and go and a hundred years later Geoffrey Bolton, emeritus professor of history at Edith Cowan and Murdoch universities in Perth, paid tribute to the West Australian leaders: 'Despite serious and legitimate concerns about the economic effects of Federation on the West, they continued with the negotiations on the basis of good faith in the integrity and goodwill of their fellow delegates . . . and eventually, because of that good faith, it was possible for Western Australia to enter the Commonwealth, and for Australia to begin as it has continued: a nation for a continent.'

The Federation was complete. It had taken 23 years since it had first been suggested and thirteen years of active negotiation to achieve 'one people, one flag, one destiny'. But in many ways it was a botched job. The state leaders had been unable to jettison their colonial mentality, unable to make a clean break with the past, unable to think beyond a federation of colonies. Instead of a lean, efficient system of government suitable for a vibrant new nation, Australia ended up with fifteen houses of Parliament, three tiers of government (federal, state and local), a Governor-General, six state Governors, overlapping bureaucracies and a tangle of disparate laws, taxes, railway gauges, land titles and education systems. Federation even failed to remove state rivalry, and nearly a hundred years later states still competed with each other for foreign trade contracts and maintained their own representatives abroad. Writer Rodney Hall says, 'One can only imagine what hard-headed negotiators, let's say the French or the Koreans, think when, planning to invest in industry, they find themselves offered the opportunity of playing Tasmania off against Queensland, or Victorians against South Australians in a bidding game for supplies of cheap electricity or coal or whatever.'

One of the few good decisions – and this was the result of New South Wales versus Victoria rivalry rather than a conscious act – was eventually to choose Canberra as the site for the capital of Australia rather than Sydney or Melbourne. The philosopher Josiah Tucker, writing in the eighteenth century, warned against basing the centre of government in an already large city. 'All overgrown cities . . . ought not to be encouraged by new privileges to grow still more dangerous; for they are and ever were, the seats of faction and sedition, and the nurseries of anarchy and confusion.'

The trouble with Federation was that at no stage had this genuinely historic struggle really succeeded in capturing the imagination of the people. There had been no widespread determination among ordinary Australians to make their own nation and forge their own history, their own destiny, independent of Britain. In fact, in the long run-up to Federation, anyone who suggested publicly that Britain was not the greatest country in the world ruled by the wisest, most gracious Queen ever to ascend the throne, and that Australia was privileged to be part of her Empire, literally risked his life.

In June 1887, when gangs of trade unionists, republicans and some recently arrived labourers from Britain broke up a meeting at Sydney Town Hall called to celebrate Queen Victoria's Silver Jubilee, drowning out 'Three cheers for the Queen' with shouts of 'Three cheers for liberty', the organizers felt compelled to call another meeting to show that loyal citizens would not tolerate such behaviour.

The second meeting was held in the Exhibition Building. Police and squads of ushers recruited from the Protestant Orange Lodges, Sydney University, the Lancers, the Volunteer Naval Artillery and Newtown Football Club, kept order. Speaker after speaker put motion after motion expressing the meeting's devoted loyalty to Her Majesty the Queen, interrupted only for the crowd to sing 'Rule Britannia' at the tops of their voices. The meeting ended with a rousing rendition of 'God Save the Queen'. The organizers were well pleased – they had demonstrated their loyalty and had seen off the Irish, the Catholics and the rabble-rousing republicans.

The people's poet Henry Lawson was outraged and responded in verse with the 'Song of the Republic'.

> Sons of the South, aroused at last!
> Sons of the South are few!
> But your ranks grow longer and deeper fast
> And ye shall swell to an army vast
> And free from the wrongs of the North and past
> The land that belongs to you.

But the sons of the South were indeed few, and when the Mother Country called in her moment of need, their ranks thinned as Australians acknowledged the ties of blood, language and common culture. The moment of need was the Boer War, Britain's struggle to bring to heel the independent Boer states of South Africa. When Britain declared war on 5 October 1899 all the Australian colonies immediately offered to send troops. Anyone who expressed any doubt about the justice of the British cause was shouted down and a man who tried to argue with a band of 'patriots' was beaten to death. When bewildered South Africans confronted the first Australian troops to arrive in Cape Town and asked them why they had come, the troops had no answer other than patriotic platitudes. Even when the British army executed two members of the Australian contingent, Peter Handcock and Harry 'Breaker' Morant, an English immigrant, for shooting civilians and prisoners of war, there were few who protested that, irrespective of the men's guilt, they should have been tried by the Australian military authorities, not the British. The *Bulletin* concentrated instead on British perfidy, claiming that the two men were only following British orders to execute Boer prisoners of war. Not only was there no evidence that such orders existed, but the irregular unit to which Morant belonged, the Bushveldt Carbineers, was trying to snuff out the last of Boer resistance, and would not have been adverse to the shooting of civilians and prisoners of war to achieve this aim. Thanks first to the *Bulletin* and then to a film about him, Morant went on to become a hero in Australia and a potent anti-British symbol.

The Boer War was hardly a glamorous one for Australia. Three hundred Australian troops mutinied because an English officer told them that they were 'a lot of wasters and white-livered curs' and the Australian government had to intervene to get them pardoned instead of hanged. Of the 16,000 troops who went to fight in South Africa, more died from disease (267) than from enemy action (251). But as far as the Australian public was concerned Australia's participation had been a triumph, and there was a widespread feeling, which was to last until 1915, that Australia was invincible. 'There has been nothing so glorious in the history of Australia as your volunteering to go to South Africa to assist the British troops in their troubles there,' said the New South Wales Premier, William Lyne, welcoming home the first troops to return. 'We will shortly have a federation of the Colonies and this is a forerunner of it when we find Australian contingents going to the front, fighting together, shoulder to shoulder, heart to heart.'

Tuesday, 1 January 1901 was a perfect Sydney summer's day. Country towns in New South Wales were near-deserted and thousands of people poured into Sydney to celebrate what they had been told was the greatest moment in their history – the inauguration of the Commonwealth of Australia. A five-mile long procession made its way in brilliant sunshine from the Sydney Domain to Centennial Park.

In the main it was a workers' procession – labourers, miners, seamen, loggers, shearers, slaughtermen – all the trades' unions, but at Centennial Park there was a reminder of the other side of the occasion, the British one. The Governor-General elect, the Englishman Lord Hopetoun, a tall commanding figure in full court dress of black and gold, a former Lord-in-Waiting to Queen Victoria, listened to the reading of the Queen's Proclamation:

> We do hereby declare that on and after the first day of January, one thousand nineteen hundred and one, the people of New South Wales, Victoria, South Australia, Queensland, Tasmania and Western Australia shall be united in a Federal Commonwealth of Australia.

(Hopetoun did not last long as Governor-General; he resigned in 1902, complaining that the pay and expenses were not enough, and returned forthwith to England.)

The band played 'God Save the Queen', guns boomed, the crowd roared and the colourful uniforms and dark faces of an Indian regiment reminded everyone that Australia was still part of a British Empire that circled the globe. There too, as a further reminder of who headed this empire, were the Household Cavalry, the Highlanders, the Guards and the Royal Engineers.

The *Bulletin* needed no such reminding, and in a dissenting voice in its next issue called on all true Australians to proclaim 'Australia for Australians' and turn their backs on 'Queen Victoria's nigger Empire'. But this was a minority view. Most Australians, apathetic though they had been in the past, saw Federation as a landmark in their history – at last a continent for a nation, a nation for a continent, one people, one destiny. Only a few pessimists stopped to ask: But are we really a nation? If so, where are we going? And, most important of all, who are we? Our founding fathers, working on the nation's Constitution, laboured long over the definition of an Australian citizen. Then, typically, they gave up and left it out.

CHAPTER 4

THE WAR YEARS: GALLIPOLI

The Anzacs were young martyrs who died not for Britain and empire, but for Australia. Fourteen years after Federation they went to Gallipoli as troops from six separate states. But because of those who gave their lives in a hopeless cause, the survivors came back as Australians.

The First World War began with the promise of splendour, honour and glory. It ended in unparalled ferocity, an orgy of blood-letting and death, the effects of which still linger today – one writer says he traces English melancholy to the battlefields of the Somme.

What was it all about? What was the point? Had the world gone mad? The idea that ten million people might have died for nothing appears incredible. Yet the very first sentence in war historian John Keegan's latest book, *The First World War*, reads: 'The First World War was a tragic and unnecessary conflict.'

Other historians agree with him. Niall Ferguson in *The Pity of War* asks why the British did not just let the Germans slug it out with the Russians and French and so spare not just British lives but the future horrors of Bolshevism and Nazism and the incursion of America into Europe's affairs. As late as July 1914, the Chancellor of the Exchequer, Lloyd George, assured the House of Commons that relations with Germany were better than they had been for years. The Germans posed no real security threat to the British Empire.

Although Germany had a considerable Pacific empire in 1914 – Samoa, German New Guinea, the Bismarck Archipelago and the Marshall and Caroline Islands – some Australians would have agreed with this assessment. The *Age* newspaper in Melbourne decided that although Europe was turning into an armed camp, there was no reason for Australia to be alarmed. Others objected. If Germany and Britain were to go to war, they argued, and Germany won, how long would it leave Australia alone?

The debate had no time to mature. When Britain declared war on Germany on 4 August 1914 – the only Allied power to do so rather than the other way round – Australia immediately joined her even though few people or leaders thought of the war as an *Australian* one. But then, the Indians did not think of it as an *Indian* war, either. The very notion of Empire, duty and honour was such a powerful one that a wounded Indian soldier, recuperating in Brighton, could write to a friend in the Punjab: 'I have taken up service for my King, and whether I am spared to serve him faithfully, or whether I shall die on the battlefield, I shall go straight to heaven. We must be true to our salt.'

The Australian Prime Minister, Joseph Cook, expressed it differently but the sentiments were the same. He told the nation, 'Our duty is quite clear – namely to gird up our loins and remember that we are Britons . . . If the old country is at war, so are we.' The leader of the Opposition, Andrew Fisher, soon to become Prime Minister, agreed: 'We Australians will help defend the mother country to our last man and our last shilling.' Before the war was over it had almost come to that.

At first a patriotic fervour swept Australia. German clubs were attacked and Chinese areas smashed and looted – for the simple reason that Chinese were obvious foreigners. Henry Lawson changed sides and announced that his heart was with Britain. The few dissenting voices were howled down. The Australian Miners' Association could declare with prescience that the war would become 'a display of civilized savagery'. But street mobs roaring 'Rule Britannia' drowned out all other views. In fact there was more opposition to the war in Britain than in Australia. There the Free Churches passed an anti-war resolution, the professors and Fellows of Cambridge signed a petition in favour of neutrality, the *Manchester Guardian* carried a full-page advertisement announcing the formation of a league to stop the war, and there was a huge anti-war demonstration in Trafalgar Square.

The Australian government promised Britain that it would provide 20,000 soldiers and recruiting booths were soon overwhelmed. One reason for this, in a period of high unemployment, was the pay – six shillings a day, the average wage of an Australian worker. But most recruits denied that money was the main attraction and gave the usual jingoist reasons for volunteering: 'Music, drums, glory, the need to wipe out an infamous nation, to be part of the first Australian army.'

But this first Australian army was not going to war under a real Australian. Its general officer in command was William Throsby Bridges, the founder of the Duntroon Military College. Bridges was born in Scotland of an Australian mother, and educated in England and Canada. He had fought for the British army in the Boer War and although he believed in the 'imperial' approach to defence rather than an Australian one, he did fight to keep the Australian Imperial Force (AIF) – a name he had conceived himself – as one single national unit rather than allowing the British command to divide it among British units, as had been the practice with colonial troops in the past.

Having displayed their patriotism at the outbreak of war, Australians now began to denigrate their own volunteers. Charles Bean, who was to become the official war correspondent and later the official historian, quotes the manager of a Sydney newspaper as saying, 'They'll keep the trained British regular army for the front. The nearest these will get to it will be the line of communications.'

He had a point. By September, the behaviour of newly recruited soldiers on leave in Sydney and Melbourne had disgusted civilians. Fights broke out when the soldiers were ridiculed as 'six bob a day tourists', and when a newspaper revealed that recruits at one camp had been disciplined after sex orgies with a teenage girl, police had to intervene to stop the soldiers from invading the newspaper building to wreck it. These Australians were ready for war.

From Europe the news was uniformly bad. Germany was carrying all before it. The French were retreating towards Paris, Belgium had been occupied and the British army had been defeated at Mons in its first encounter with the Germans. Not that you would have learnt any of this from reading the Australian press. The long voyage which mails had to travel combined with the very meagre cable services meant that Australians knew little of that first disastrous winter of the war.

They did know that Australian troops were gathering in the great harbour of Albany, Western Australia, and on 1 November the *Orvieto* with General Bridges, his staff and 1,000 Victorians led the first contingent of the AIF on its way to Europe. Not far behind it came 10,500 Australian troops and 2,000 New Zealanders under the command of Colonel John Monash, a citizen soldier. Monash, a Melbourne engineer with a distinguished career, excluded from membership of the Melbourne Club, the temple of the Victorian establishment, because he was a Jew, was to find that the war was his path to glory.

To their disappointment, the soldiers of the AIF did not go straight into action against the Germans. Britain, struggling to provide winter camps for its own troops and the Canadians who had already arrived, agreed with the Australian government that mud, slush and an English winter might dent the Australians' morale. So the Secretary of State for War ordered the AIF to disembark in Egypt.

There they were placed under the command of Major-General William Birdwood, an English gentleman-soldier. He saw no sense in being other than what birth and upbringing had made him, so he played the part to the hilt. The Australians loved it. If they had to have an English officer to command them, then why not one from Central Casting, and in no time he was known throughout the Australian ranks as 'good old Birdie'. Other British officers followed his lead. One who wore a rimless monocle was confronted on the parade ground by lines of Australians with pennies in their eyes. He threw his monocle into the air, caught it in his eye, then turned to the parade and roared, 'All right, you bastards, let's see you do that.'

As for Egypt itself, the Australians hated it. They enjoyed the weather but found the country 'a land of sin, sand, shit and syphilis', and the Egyptians 'mercenary, shifty, and dirty', proof if the Australians needed it that the White Australia Policy was the right one. Stories went around of how Egyptian pedlars freshened up grapes by dipping them in urine and then selling them at exorbitant prices. The Australians reacted by bargaining with Egyptians with fists and sticks: 'The stick and plenty of it is the only language the Gyppoes understand.'

This maxim did not apply to Egyptian women and, along with their beer, the Australians took their 'horizontal refreshment' when they were on leave in Cairo's brothels. As a result, they went down with venereal disease at a rate that so shocked King George V – historians do not say how he learnt this – that he wrote to his representative in Australia, the Governor-General, to say how sorry he was and wondered whether excessively high rates of pay might have something to do with it.

In Cairo, the Australian soldiers did not give a damn what the King thought and on 1 April, the day all leave was stopped, an indication of embarkation for action at some unknown destination, a mob of them joined up with some New Zealand troops in a raid on the brothel area. Egyptian police who tried to stop them from burning houses found themselves involved in a full-scale riot with the Australians shouting, 'Finish the bastards off'. When the British Military Police proved no match for the Australians, the Lancashire Territorials had to be ordered out to restore order with fixed bayonets, an act that only confirmed the Australian view that you could not trust a Pommy soldier not to turn on you if his officers ordered him to do so.

The unknown destination turned out to be Gallipoli. Pursuing a military adventure of his own, Winston Churchill, then First Lord of the Admiralty, became convinced that the war could be brought to a speedy conclusion by striking at the soft underbelly of Germany. A combined Allied force would seize control of the Dardanelles, take Constantinople from the Turks, and by opening the Black Sea enable the western allies to link up with Russia and put Germany out of the war. The Allies would have to take the Gallipoli Peninsula because its guns guarded the Dardanelles, and attempts by a naval force to get through had failed.

It was a bold plan, atrociously executed. Command of the operation was given to General Sir Ian Hamilton, whose intelligence for the forthcoming campaign consisted of a few inaccurate maps, a tourist guide book and an out-of-date manual on the Turkish order of battle. At least the Australians were to have General Bridges in charge – they could have been unlucky like the British and French troops who landed at Cape Helles. Their commander was General Sir Aylmer Hunter-Weston. When his staff told him that his men were suffering heavy casualties, Hunter-Weston said, 'Casualties? What do I care about casualties?'

In the light of the horrors that were to come and the thousands of Australians who were to die at Gallipoli, we cannot ignore the disturbing possibility that it all might have been unnecessary, that Churchill's plan could have been achieved without battle. Three historians – Professor Richard Breitman, Patrick Beesley, and Robert Rhodes James, all give credence to a story that the British had made Turkey an offer of up to £4 million to open the Dardanelles and allow British and French warships free passage. The offer came to nothing because the Turks wanted a guarantee that Constantinople would remain in the hands of the Turkish government, and Britain could not give this guarantee because she had promised Constantinople to the Russians.

This is a sensational and important claim, and before deciding that Gallipoli was not only a sacrifice but an unnecessary one, readers will want to examine the evidence and make up their own minds as to its credibility. The source for all three accounts is Admiral Sir Reginald Hall, head of Royal Navy intelligence before the war, and then Britain's principal code-breaker. Hall's version is that in January 1915, without telling Churchill or the War Cabinet, he made contact with the Turkish government through intermediaries, offered £4 million for a peace treaty and that the Turks were considering it when an intercepted German telegram immediately changed matters.

Hall's code-breakers, thanks to the *Handelsverkehrsbuch*, captured on the outbreak of war on a German merchant ship in Australia, had been reading the telegraphic traffic between Berlin and the German and Turkish forces in Turkey. Hall says that on 13 March his staff decoded a message which revealed that the coastal forts along the Dardanelles had virtually run out of ammunition. When he took this decrypt to Churchill and Lord Fisher, the First Sea Lord, they decided that the Royal Navy would 'go through tomorrow'. Hall then felt obliged to tell them of his negotiations with the Turks and they ordered him to break them off immediately. His account concludes, 'Ironically enough, when the gallant attempt [to force the Dardanelles] on March 18 proved unsuccessful, the Cabinet were asking me to spare no expense to win over the Turks. Unfortunately it was then too late.'

Beesley says that his may seem an unbelievable story 'but then so are most of the stories about Hall, and most of them are true'. Over the years there has been corroborating evidence from Beesley and others. Beesley says that there *were* some sort of *sub rosa* negotiations going on with the Turks at that time, that a telegram as described by Hall *was* decrypted then – he says on 13 March, others on 15 March, the British Cabinet papers on 19 March – and Carden, the British admiral in charge of the Dardanelles operation, *did* receive a telegram a day later from Churchill, saying that the Turkish forts were short of ammunition and that 'the operation should now be pressed forward methodically and resolutely by night and day'.

From Beesley, Rhodes James and others we can fill in further details. The negotiators on the British side were George Eady, a British contractor who knew the Middle East well and who had been in Constantinople at the beginning of the war, working for Sir John Jackson Ltd; Edwin Whittall, a member of one of the leading British merchant houses in Turkey; and Gerald Fitzmaurice, a former member of the British embassy at Constantinople. In the remote Thracian town of Dédéagatch, they met with Talaat Bey, the Minister of the Interior in Enver Pasha's Young Turk government, and showed him a letter from Hall authorizing them to negotiate a treaty with the Turks to withdraw from the war, remain neutral and open the Dardanelles to Allied shipping. In return, Britain would pay the Turkish government £4 million.

After consulting his government, Talaat Bey asked for guarantees for Constantinople's integrity and became impatient and suspicious when these were not quickly forthcoming. None of the British negotiators knew of Britain's secret promise that Russia would have Constantinople, so they were unable to reassure the Turkish minister. It is not clear whether the British ended the negotiations on Hall's orders or whether the Turks decided there was nothing to be gained by continuing them. Eady wrote in 1919: 'The whole country [Turkey] desired peace and their leaders would have accepted almost any terms had we been able to assure them on Constantinople. They knew full well that signing away that city would mean signing their own death warrants.'

The Australians and New Zealanders (now called ANZACS, for Australian and New Zealand Army Corps) knew none of this, and on the morning of 25 April 1915 these troops, untried and untrained for amphibious assault, landed at Gallipoli – for reasons still largely unexplained – a mile north of the beach originally chosen and began to scale the heights. One soldier later wrote home, 'We had come from the New World for the conquest of the Old.' Or as a poet put it:

Cease your preaching! Load your guns!
Their roar our mission tells,
The day is come for Britain's sons
To seize the Dardanelles.

The Australians thought that the Turks would be like the 'Gyppoes' and that at the flash of the Anzacs' bayonets – their 'penknives', as they called them – they would turn and run. They were shocked at the reality. The Turks were unprepared because they did not imagine anyone would attempt to land where the Anzacs had, and until reserves could arrive only a company of 200 Turks had to hold the position. But so hard did they fight and so accurate was their rifle fire that by nightfall dead Anzacs lay all over the cliffs and on the beaches, or bobbed lifeless in the waves.

With the arrival of Turkish reserves all hope of a quick Anzac victory vanished. The battle slowly settled into a deadly stalemate, interrupted by furious assaults on the Turkish lines, all of which failed, and then a return to stalemate. Bridges, shocked by the confusion on that first night, called for a withdrawal but was overruled. This cost him his life: he was hit by a sniper on 15 May and died a few days later. By then the Anzacs realized that there was to be no resolution and a sullen resignation set in, as summer brought dysentery, flies, the smell of rotting bodies and boredom. Then, in September, Keith Murdoch arrived.

Murdoch, later to father the international media magnate, Rupert, had come second to Charles Bean in a poll conducted by the Australian Journalists' Association to decide who would be the one official Australian war correspondent to cover the fighting overseas. Murdoch, disappointed at losing, decided to head off for Europe anyway.

In Cairo, he wrote to General Sir Ian Hamilton a wheedling letter, begging to be allowed to visit what he called 'the sacred shores of Gallipoli' and promising 'faithfully to observe' any conditions Hamilton might impose. Hamilton was reluctant to give his permission – Bean was already there representing the Australian press – but finally decided no harm could come of it as long as Murdoch signed the standard war correspondent's declaration: that he would communicate only through military channels and for the duration of the war would not impart any military information unless he first submitted it to a censor. Murdoch readily signed.

He arrived at Gallipoli on 2 September 1915. He spoke with some soldiers, declined Hamilton's offer to provide him with transport to go anywhere and see anything, and then returned to headquarters on the island of Imbros. He had been ashore at Gallipoli for only four days but he was already under the spell of the Anzacs. In his first censored dispatch to Australia he described them as 'stoical but not contemptuous'. They had an elation that 'knew no fear'. They had died 'charging with the light of battle in their eyes'.

Back on Imbros, Murdoch met Ellis Ashmead-Bartlett of the London *Daily Telegraph*. Ashmead-Bartlett had covered the Russo-Japanese war and was an experienced correspondent. He hated the restraints General Headquarters imposed upon him – he was allowed no criticism of the conduct of the operation, no indication of setbacks or delays, and no mention of casualty figures. Over the months Ashmead-Bartlett had grown sour, hostile and pessimistic. Now he poured out to Murdoch's sympathetic ear a gloomy description of the way the campaign was being conducted.

Murdoch must have realized that, almost by accident, he was in possession of information that would rank as one of the great stories of the war, a real scoop. He quickly agreed with Ashmead-Bartlett that the only way to get the story out would be to break his word to Hamilton and get an uncensored account back to Britain. Ashmead-Bartlett wrote it and Murdoch set out to take it to London. At Marseilles Murdoch was met by a British officer with an armed escort who forced him to hand over the story. But there was nothing the War Office could do to seize the information which Murdoch carried in his head.

In London he sat down and wrote everything he had seen himself and everything he could remember of what Ashmead-Bartlett had told him. He wrote it in the form of a letter to the Australian Prime Minister. It was an amazing document – a mixture of error, fact, exaggeration, prejudice, and the most sentimental patriotism, which made highly damaging charges against the British general staff at Gallipoli and against the trusting Hamilton, many of them patently untrue.

This would not have mattered, however, as Murdoch did not intend that his letter should be published – if he had he would have sent it to his newspaper, the *Herald*, in Melbourne; the information was only for his Prime Minister. But matters were now taken out of his hands. Over lunch Murdoch told the editor of *The Times*, Geoffrey Dawson, the contents of the letter. Dawson persuaded him to repeat it for the chairman of the Dardanelles Committee who, in turn, passed it to Lloyd George, then Minister for Munitions.

Lloyd George, who had opposed the Gallipoli campaign, urged Murdoch to send a copy of the letter to the British Prime Minister, Asquith. Asquith then used the weapon Murdoch had handed him in an inexcusable manner. Without checking its more outrageous allegations, without even asking Hamilton for his comments, he had it printed as a state paper and circulated to the members of the Dardanelles Committee. When the Committee met on 14 October, Hamilton was sacked.

Murdoch felt that his actions had been vindicated – 'I have a perfectly clear conscience as to what I did . . . It compelled the wavering and hesitant Coalition Cabinet to make up its mind.' But Hamilton considered he had been betrayed by what Murdoch had written – 'No gentleman would have said it, and no gentleman will believe it.'

It is instructive to look at what Murdoch had to say in his letter to the Australian Prime Minister because in it we can see the origins of what became the legend of Anzac. Overall, it is hard to disagree with Hamilton's criticisms of Murdoch's approach. Hamilton wrote that Murdoch had divided his letter into two parts: 'First an appreciation in burning terms of the spirit, achievements and all the soldierly qualities of the Australian forces. Secondly, a condemnation, as sweeping and as unrelieved as his praise in the first instance is unstinted, of the whole of the rest of the force . . . He has allowed himself to belittle and criticise us all so that [the Australians'] virtues might be thrown into even bolder relief.'

Some of the Australians who had fought with the British Army in the Boer War had returned with little respect for their British colleagues. They had been astounded to see units march into the mouths of machine guns while the regimental drummer beat step and officers armed only with swords lead frontal attacks. They concluded that the British might be brave but they were also stupid.

Now Murdoch went further. The British troops were suffering from 'an atrophy of mind and body that is appalling . . . The physique of those at Suvla is not to be compared with that of the Australians. Nor, indeed, is their intelligence . . . They are merely a lot of childlike youths without strength to endure or brains to improve their condition . . . After the first day at Suvla an order had to be issued to officers to shoot without mercy any soldiers who lagged behind or loitered in an advance.'

As far as the British officers were concerned, epithets Murdoch used to describe them included 'selfish', 'rude', 'disgusting' and 'chocolate soldiers'. He accused them of 'wallowing' in ice on board the staff ship *Aragon*, 'a luxurious South American liner', while the 3rd Australian General Hospital on shore had 134 fever cases and no ice.

This is a classic case of 'Pom bashing', combined with such lavish praise for Australians that it is embarrassing to read. Unlike the British, according to Murdoch, Australians were 'game to the end'. He wrote, 'I could pour into your ears so much truth about the grandeur of our Australian army, and the wonderful affection of these fine young soldiers for each other and their homeland, that your Australianism would become a more powerful sentiment than before.

'It is stirring to see them, magnificent manhood, swinging their fine limbs as they walk about Anzac. They have the noble faces of men who have endured.

Oh, if you could picture Anzac as I have seen it, you would find that to be an Australian is the greatest privilege the world has to offer.'

This comparison of stupid, spindly British soldiers with noble, manly Australians ignores the fact that the core of non-commissioned officers (NCOs) in the AIF were British regulars and that one in three Anzacs had in fact been born in Britain, including perhaps the best-known 'Australian' at Gallipoli, John Simpson, 'the Man with the Donkey'. Simpson rescued many a wounded soldier before he was killed on 19 May 1915 on one of his many missions to Shrapnel Gully. He was born John Kirkpatrick of Scottish parents in South Shields, came to Australia as a merchant seaman only four years before the war, deserted his ship at Newcastle, changed his name to Simpson and enlisted in 1914.

Murdoch's sentimental depiction of the Australians as outstanding troops led by uncaring British idiots was picked up by Ashmead-Bartlett, first in his journalism and later in his books. Alan Moorehead continued it in his *Gallipoli* and it was brought to its refined best in Peter Weir's film of the same name in 1981. In fact, Weir's sentiments about Gallipoli echo Murdoch's and Weir talks about his own first visit there in cathartic terms.

But there is little truth in the legend as portrayed, and the story that Australians were great troops misused by British bungling, needs to be qualified. The Australian troops were no better and no worse than the many others taking part in the campaign. (Australians tend to ignore the fact that as well as themselves and the British there were French, Senegalese, Gurkha and Indian troops at Gallipoli.)

There were acts of heroism and cowardice among them all. Murdoch does not mention, for instance, that at the landing on 25 April the Australians were led ashore in many cases by teenage British midshipmen, at least two of whom were cadets in their first term at the Royal Naval College, Dartmouth, and one of whom was still only thirteen, the other fifteen; their holsters for their service revolvers were almost half the size they were. True, a British officer was awarded the Military Cross for summarily executing three British soldiers for alleged funk on the battlefield, probably the origin of Murdoch's 'shoot without mercy any soldier who . . . loitered in an advance'. But the Australians were not innocent of similar behaviour.

Charles Bean's diary for 26 September 1915 says specifically that in some instances Australians had to be driven ahead by the threat of being shot from behind by their officers. He records one case he witnessed himself: 'I have seen it happen once to an Australian NCO whom Col. McCay threatened to shoot and I know that others have done the same.'

Historian Robert Rhodes James says that the idea that the Australians were the most aggressive fighters in the whole peninsula is 'difficult to substantiate'.

He quotes one Australian as saying of his fellow troops, 'They were the most matter-of-fact, hard-shelled and cynical mob, who at no time exerted themselves more violently than they had to.'

The respect that the Turks supposedly had for the Australians' bravery is not born out by the account of one Turkish officer, Lieutenant Colonel Mustafa Kemal, later famous as Atatürk, the man who led his country out of the Middle Ages. He recorded that he saw Australian troops in the vicinity of Lone Pine on 27 April 'almost as a body come out in front of their trenches and waving white hats, white handkerchiefs and flags, seek to give themselves up. These scenes were watched by myself and my whole staff with the naked eye.'

A New Zealander, Lieutenant Colonel W. G. Malone, of the Wellington Battalion, thought more highly of the British troops than the Australian. 'It's a relief to get in where the war is being waged scientifically and where we are clear of the Australians. They seem to swarm about our line like flies . . . They are like masterless men going their own ways.'

In fact, the general opinion at Gallipoli was that the best troops were the New Zealanders. General Birdwood regarded them with admiration and confidence and one historian said, 'They combined the *élan* and dash of the Australians with the meticulous professionalism of the British troops.' Historian John Keegan says that New Zealanders were to win a reputation as the best soldiers in the world during the twentieth century.

Even the attacks involving the Australians at Lone Pine and the Neck – the climax to Peter Weir's film, which cemented the Australian view of the British as bone-headed muddlers – were no reflection on British planning. The plans were fundamentally sound, but the execution was appalling. And it was not the British troops who were at fault, but – as many senior Australian officers realized at the time – the leadership. Colonel Monash wrote to his wife on 26 September 1915: 'There are some things which don't get into dispatches . . . during the first forty-eight hours after the landing at Suvla, while there was still an *open road* to the Dardanelles, and no opposition worth talking about, a whole army corps sat down on the beach, while its leaders were quarrelling about questions of seniority and precedence . . .'

The Australian casualty figures, compared with the war in Europe, were not horrendous. A total of 7,818 Australians were killed during the Gallipoli campaign. This pales in comparision to 45,033 Australians killed on the Western Front or the New Zealanders, of whom 8,566 served at Gallipoli, but who recorded 14,720 casualties because some of their wounded returned to action two or three times. The Turks suffered too. They lost 300,000 men killed, wounded or missing at Gallipoli. In Europe, after only eight months of war, the total number of German soldiers killed, wounded or missing was two and three quarters million. No wonder Gallipoli is hardly recalled in European accounts of the First World War.

None of this is intended to demean the Australian soldier, who, as we shall see, could be among the best in the world, but to emphasize that the Anzac legend was not created as a celebration of his fighting qualities or because Gallipoli was a great victory, which it clearly was not. If this had been the intention then the taking of Damascus by the Australian Light Horse ahead of both Allenby's British troops and T. E. Lawrence's Arabs – a triumph that made Lawrence hate Australians for the rest of his troubled life – would have been much better material with which to work to create a legend.

Better still would have been Monash's brilliant tactics on the Western Front in the critical last months of the war with his hardened, victorious Anzac troops – by then 'the finest fighting machine ever seen'. Yet the Australians who fought on the Western Front appear doomed to live forever in the shadow of Gallipoli. No, we have to look elsewhere to discover why Gallipoli became such a legend and why it has endured for so long.

A good starting point is to ask what we were doing at Gallipoli in the first place, fighting the Turks, a people who could not possibly pose any direct threat to Australia itself and with whom we had no quarrel? Some Anzacs instinctively realized this. Les Leach admitted in 1997 that he felt more in common with the ordinary Turkish soldier than he did with English officers, so when he and his mates thought no one would see them they slipped through the trenches and shared a cigarette with the Turks. 'We had enough fags to go round and they were short so it was the only decent thing to do,' he recalled before his death, three months short of his 100th birthday.

Leach was there because 'I thought it would be an adventure, a lot of fun.' Australia was there because Britain, the Mother Country, had called us. We went because 14 years after Federation, Australia was still a British-Irish community in the Pacific, still part of Greater Britain, still with little sense of an independent identity and with only a subdued sense of nationalism. We were 'Britain's sons'. We had fought for her in China during the Boxer Rebellion, in New Zealand against the Maoris, in South Africa against the Boers. 'It seemed only natural to fight for the Empire to which we all belonged against the Germans and the Turks.'

It was madness, of course. Peter Bowers said his father only once told him a story about Gallipoli. 'It was about a bandsman named Crane who started playing his trumpet one night and one by one the rifles and guns on both sides shut down until the only sound was the clear sweet notes of the trumpet. As the last note faded into the strangely still night, the cacophony of killing resumed.'

Nothing stimulates nationalism like a blood icon and that is what Gallipoli became. The fact that it was a bungled defeat, probably unnecessary, a tragic waste of life, and someone else's fault did not matter; it made the icon even stronger. And the fact that it was Britain which had planned the fiasco was the beginning of the prolonged breaking of ties between Australia and the Mother Country that accelerated during the Second World War and finally snapped when Britain joined the European Community in 1973.

So over the years Anzac Day became not only a moment to remember the tragic sacrifice of our youth but to reinforce Australia's growing sense of a national identity independent of the British super-state. Before Gallipoli, Australians had been conditioned to believe themselves part of the worldwide Greater Britain. They had done their bit whenever they had been called upon to do so. But when the blood-letting failed to win Britain's recognition that Australia had grown up, it became a symbol of all that was wrong with the imperial relationship.

Yet this still does not explain why every year on 25 April thousands of young Australians in their late teens and early twenties come halfway across the world to gather for the dawn service at Gallipoli and stand in silent tribute to those of an earlier generation – then about the same age – who died there. Sydney columnist Richard Glover says it is because marking a defeat is in keeping with one of the most persistent themes in the Australian legend – the lauding of failure. 'Any country can make a hoopla about its victories. What makes Australia unique is the way it has always preferred to remember the brave-but-defeated, the underdog and the loser.' This is true. Something like Anzac Day would be unimaginable in the United States.

Photographer and author John Williams, whose father was an Anzac and fought in France, believes that Australia's Gallipoli legend is 'a load of nonsense . . . pure myth. But if you say that in Australia they almost want to kill you.' He adds that it is a harmful myth because it turned Australia into a country where military and sporting heroes were more important than cultural ones.

That may be so, but it is hard to ignore the fact that whether you like it or not war played an important part in forming Australia's view of itself. The Anzacs were young martyrs who died not for Britain and empire, but for Australia. Fourteen years after Federation they went to Gallipoli as troops from six separate states – the New South Wales 1st Infantry Brigade, Victorian Light Horse Regiment, the West Australian 10th Light Horse Regiment, and so on. But because of those who gave their lives in a hopeless cause, the survivors came back as Australians. It was the birth of a nation, and one can only hope that this thought provided some comfort to the parents of the

Anzac whose very Australian headstone stands where the first landing took place. It reads:

Died aged 18 near this spot
April 25th, 1915.
Did his best.

CHAPTER 5

THE WAR YEARS: FRANCE

The Great War left an astonishing every second Australian family bereaved, a country in mourning until the next time round. How does a young nation cope with such grief and recover from the loss of the flower of her young men? Perhaps it doesn't.

Back home in Australia it had not been until 28 April that the nation learned that the Anzacs were ashore at Gallipoli, and then only through a terse statement in the House of Representatives. True, there had been two joint announcements from the Admiralty and the Army in London. On 27 April they said that the Allies had disembarked forces at various points on the Gallipoli Peninsula 'despite serious oppositions from the Turks behind strong entanglements'. And on 28 April they revealed that Sir Ian Hamilton was commander of the operation.

The landing had been been made 'under excellent conditions', the French had taken 500 prisoners, and 'troops from Egypt' were taking part in the battle. But there was not a single mention of the Anzacs until the Assistant Minister of Defence made the downbeat announcement in Australia that the Allied Forces at the Dardanelles 'included a portion of the Australian troops which had been in Egypt'. Understandably, the public was not stirred. One of the most important events in the nation's history was occurring in a news vacuum.

So Australian civilians, thousands of miles away from the war, got on with their normal lives. The Sydney Stock Exchange was reassuringly stable. The department store David Jones was selling men's black calf lace boots for six shillings and ninepence and winter suits for 45 shillings. The theatres were full. Julius Knight was playing in *The Life of a Guardsman*; Frank Harvey and Violet Page in *The Man who Stayed at Home* – both plays with a military theme. The autumn racing carnival in Melbourne was attracting big crowds. Surfing as a sport was just beginning to gain popularity. War? What war?

The problem was censorship, the need for the British government to keep the harsh facts of the battlefront away from the public lest support for the war declined. Charles Bean, a conscientious reporter, kept a diary, regularly visited all sections of the front, and was respected by the Anzacs. But everything he wrote was censored by Hamilton's staff and was sent via London, where it was often censored a second time before being forwarded to Australia. And Bean himself felt that while the war was on it was his patriotic duty to write positively about the Anzacs.

Reports of the war as they appeared in the Australian newspapers were compiled from Bean's reports, statements from Birdwood's staff and the War Office in London, supplemented by censored dispatches from Ellis Ashmead-Bartlett, war correspondent for the *Daily Telegraph* of London. Understandably, they were all upbeat: 'Glorious Entry into War . . . Brilliant Feat at Gaba Tepe . . . Historic Charge . . .' Australians had no reason to believe that anything had gone wrong until early in May when the first casualty lists began to appear.

From then on, day after day, week after week, newspapers around Australia carried these lists on their main pages, one name to a line, an unending rollcall of the flower of Australian youth: the scions of some of the country's oldest families, the best brains of the professions, and sturdy boys from the bush, the mines, the shearing sheds, the cattle stations – all dead.

After a few months of this there was no longer any talk of the war being a great adventure, an obligation to help the Mother Country. It had become a grim personal war to which every family with a serving relative was committed. Soldier's letters were censored but occasionally an indication that all was not well would creep through. One Anzac wrote to his young brother: 'As you love me, keep out of this.' The brighter readers of newspapers noted that maps of the battlefield showed little movement and this, together with the casualty lists, caused significant disquiet beneath the outwardly patriotic surface.

The authorities were aware of this and when Ashmead-Bartlett went on a lecture tour of the United States and criticised the conduct of the war in Gallipoli, he was not allowed to go on to tour Australia until he gave an undertaking to stick to a censored script. Two plainclothes police officers accompanied him in Australia, ready to arrest him if he broke this agreement. So when the evacuation from Gallipoli occurred, it brought not the shame of defeat, but relief that for the time being at least, the slaughter had stopped. Little did Australians know what awaited their boys in France.

Compared with France, Gallipoli was a picnic, according to Fred Kelly, of Sydney, who enlisted when he was only sixteen and saw service on both fronts. He spent two months in the trenches at Gallipoli and was then a gunner at Fromelles, in France, where 5,533 of the Australian 5th Division were casualties in only 27 hours of charges across muddy fields into German machine-gun fire. 'We were slaughtered,' he said.

Amazed that he survived the war, he returned to Australia vowing never to leave it again. He did not. He also never joined any ex-servicemen's organization, preferred not to talk of his war experiences, and never attended an Anzac Day march. He died in 1998, aged 101.

Another Anzac who never marched on Anzac Day was Frank Williams, who carried the scars of the battlefields in France until his death in 1964. He had a permanent purplish hole in his chest from shrapnel, trench foot, a scar from a barbed-wire tear down his side, and recurring nightmares which caused him to wake up screaming. 'It was terrible. You'd be up to your waist in filth. Terrible. Stupid. Dead people. Bits of men,' he would tell his son, author–photographer John Williams. John says, 'Then he'd start to cry. My old man was deeply ashamed when he cried. And he couldn't talk about it without crying so he didn't talk about it.'

The war in France made Frank Redman so disillusioned that he was still looking back in anger when he turned 103 in 1997. 'A million died at Passchendaele, including thirty-eight thousand Anzacs,' he said. 'A million men. It was hell. There were dead and dying everywhere. The whole thing was mad, completely mad. It's all nonsense to say that the soldiers looked on it as glory. They didn't. It was a waste of time, a waste of life. You can never recover from that type of experience.'

Four Queensland brothers – William, James, Jack and Alexander Kemp – all enlisted in the AIF. Alexander was the first to be killed, by a gunshot wound to the head at Gallipoli. He was 19. Six months later, Jack was also shot at Gallipoli, but he survived only to die in France in 1916. He was 21. Nine months later, William was killed in France, too. He was 27. The Australian government decided that the Kemp family had done their bit and the surviving brother, James, aged 25, as in the Hollywood film *Saving Private Ryan*, was discharged 'for family reasons' and shipped home from France. The men's mother began wearing black when Alexander was killed and wore nothing but black for the rest of her life. Her daughters said her grief and anger were unbearable.

Little of what was happening in France got back to either Britain or Australia. The soldiers did not want to write about it because they knew it would distress their families and, even if they had the censors would have cut it from their letters. The war correspondents could not write about it because the censors would not let them. But one, Philip Gibbs of the *Daily Chronicle*, explained to the British Prime Minister, Lloyd George, what life in the trenches was really like. The next day Lloyd George told the editor of the *Manchester Guardian* how moved he had been by Gibbs's account. 'If the people really knew, the war would be stopped tomorrow. But of course, they don't know and can't know. The correspondents don't write and the censorship would not pass the truth.'

What could the correspondents have written? That human flesh, rotting and stinking, had become mixed in with the mud, and if the soldiers dug deeper to avoid shelling then their spades went into the soggy flesh of men who had once been their comrades. That soldiers knew their fate and sometimes marched towards the front line baa-ing like sheep. That the French often threw black troops from their colonies into exposed positions in order to save white troops from slaughter; that the British High Command had tried to use Australians and Canadians in a similar manner so as to save British troops. (Bean queried this use of Australians and was accused by British officers of excessive partiality, but the chief censor agreed that 'most of his criticisms of the Mother Country were justified' and that Bean was motivated by 'high-minded patriotism'.)

Or they could have coaxed a description of life in the trenches out of a literate soldier, someone like Frenchman René Naegelen: 'Three of us were crouching in a hole under a barrage of artillery fire. Then a flame, a blast . . . Was I killed or wounded? I cautiously moved my arms and legs. Nothing. My two friends, however, lying one on the other, were bleeding. The bowels of one were oozing out. The other had a broken leg; there was a red spot spreading on his breast, and he was rolling his panic-stricken eyes. He looked at me silently, imploringly; then unconsciously he unbuttoned his trousers and died urinating on the gaping wound of his comrade.'

Or they could have described life in one of the many châteaux behind the lines where the staff officers sat down to dinner served by little Waacs with the General Headquarters colours in bows in their hair, while a band played ragtime and light music. Not far away Australians were dying in enormous numbers – 23,000 casualties in seven weeks' fighting in the valley of the Somme in the late summer of 1916; 38,000 at Passchendaele in October–November 1917.

Or they could have exposed the lies of the High Command. During the night of 19–20 July 1916, the AIF undertook its first major mission in France. Bean knew the Australians were soon to see action but had not been told when. When an officer told him before breakfast that the 5th Division 'had their little show last night' he borrowed a car and reached the division's headquarters at lunch time. There he learned that the action had been a disaster – British estimates of German strength had been totally inaccurate, the British 61st Division had reported it had secured its objective as planned when it had not, and the Australians had walked into a wall of German machine-gun fire. Australian officers told Bean that the Division had lost between four and five thousand men for no apparent benefit. They said that this was as bad as the Gallipoli landing. But when Bean later collected the official communiqué, it read: 'Yesterday evening, south of Armentierès, we carried out some important raids on a front of two miles in which Australian troops took part. About 140 German prisoners were captured.'

Bean was disgusted by the communiqué's 'deliberate lying' and recorded in his diary what had really happened. But in his dispatch for the Australian press he included no criticism of the action. He said only that shell fire was intense and continuous, that Australian losses had been 'severe', but their efforts had been 'worthy of all the traditions of Anzac'. The censor objected to the word 'intense' and cut it out. A year later Bean noted in his diary that the condition of the Australians returning from the front had deteriorated – 'One tall, thin, white-faced youngster [was] especially looking like a dead man looks.' He added: 'The army has no right to squander men in this fashion.'

How did they endure it? Could we today? They endured because the survivors of Gallipoli had become battle-hardened warriors and they had a cheeky confidence in their own abilities. Given an Australian commander they believed they could march on Berlin. They were proud of their mateship and egalitarianism. Manning Clark writes in his *A History of Australia* of a British staff captain who ticked off an Australian private for failing to salute him. 'The Australian patted him on the shoulder and said, "Young man, when you go home, you tell your mother that today you've seen a real bloody soldier."' British historian John Keegan says that they endured because they belonged to the largest empire the world has ever known and like all members of that empire they saw themselves as a chosen people 'whom nothing in sight could cast down from their high place'.

We would not have to endure today, according to the *Sydney Morning Herald*, because the relationship with Britain has changed and Australian participation in a European war is inconceivable. 'Today, our soldiers stand to arms on stranger fields in states which did not exist when the first AIF was enlisted. Today we have our own policies to shape, our own strategic frontier to defend, our own responsibility for our own national security.'

In Australia, as casualty lists lengthened and returning wounded were able at last to tell of life in the trenches, recruiting began to dry up. By autumn 1916, the number of casualties in the AIF demanded 7,000 replacement recruits a month and only half that number of men were coming forward. In view of this, was it coincidence that the government declared that henceforth 25 April would commemorate the landing at Gallipoli, would be a public holiday known as Anzac Day, and would be marked by a march? Historian Joan Beaumont writes: 'Spontaneous though much of the celebration was, it was clear that promoting the exploits of the Anzacs was a useful recruiting tool for the British and Australian governments.'

The Prime Minister, William Hughes, said that there was only one answer to this shortage of volunteers – conscription. This issue was to split Australia like no other during the war. Anticipating this, Hughes announced that the proposal would be put to a referendum. The question would be: 'Are you in favour of the Government having, in this grave emergency, the same compulsory powers over citizens in regard to requiring their military service, for the term of this war, outside the Commonwealth as it now has in regard to military service within the Commonwealth?'

The first and most obvious division was between the Protestants and the Catholics. The Anglican Synod in Melbourne decided that it was so convinced that God was on the side of Britain that everyone should support conscription, which it euphemistically called 'universal service'. The Catholic Archbishop of Melbourne, Dr Daniel Mannix, the Irish-born son of a tenant farmer, hit back. He not only opposed conscription (and the White Australia policy and capital punishment) but launched an appeal for the victims of the Easter Rebellion in Dublin where 64 Irish rebels, 132 British soldiers or policemen and 300 civilians had been killed.

Other leading Irish-Australians supported Mannix. The assistant Minister of Justice, J. Fihelly, told a rally in Queensland that every Australian conscripted into the army released a British soldier to harass the people of Ireland. Another minister, W. Lennon, drew roars of applause at an Irish Association meeting when he said, 'The time has come when the Irish should speak out, and refuse to allow their country to be the doormat of England forever.' The Anglo-Australians said that all this amounted to treason – Dr Mannix was Australia's Rasputin, secretly preparing a social revolution for the country; he and his supporters should be tried and deported.

The Prime Minister opened the 'Yes' campaign with a huge rally at the Sydney Town Hall, reminding the audience that Australia remained free only while the British Empire was unconquered. Compared with the Mother Country, Australia had not pulled her weight. Britain had five million men under arms. 'If we had done as much we should have enlisted five hundred thousand instead of a little more than half that number.'

Hughes issued a manifesto appealing to the women of Australia. 'What is your answer to the boys at the front? Are you going to leave them to die? . . . Are you going to cover with shame and dishonour the country for which our soldiers are fighting and dying? . . . Prove that you are worthy to be the mothers and wives of free men.'

John Curtin, a future Labor Prime Minister, then secretary of the Anti-Conscription League, arranged for a poster saying that a vote for 'Yes' was

'The Blood Vote'. Underneath was the start of a poem by W. R. Winspear, said later to have been instrumental in swinging the vote to 'No':

> 'Why is your face so white, Mother?
> Why do you choke for breath?'
> 'Oh, I have dreamt in the night, my son,
> That I doomed a man to death.'

The other division in the country over the conscription issue was between country and city. Country people thought that city slickers were doing well out of the war while volunteers from the country who had spearheaded the recruitment drive were carrying more than their share of the war's burden. Opponents of conscription who turned up in country towns to argue their case were tarred and feathered. And at every meeting the 'Would-to-Godders' incurred the contempt of both sides – 'Would to God that I could go, but I am prevented by ties/age/infirmity.'

Ironically, one of the major problems the 'Yes' campaigners faced was that the heavily-censored accounts of the war published in the Australian press gave such a rosy picture of the war that many Australians wondered why more recruits were needed. A high-ranking Australian officer who had returned from France during the conscription campaign was so distressed by this rosy picture that he called on the Premier of New South Wales to urge him to use his influence with the newspapers to change their tone. But it was too late. As historian Kevin Fewster pointed out in a doctoral thesis in 1980, 'Field censorship . . . disguised reality so effectively that the authorities found it impossible to reverse the picture when the need later arose.'

When the votes were counted, those who voted against conscription had a small majority – 1,160,033 for 'No' and 1,087,557 for 'Yes'. A majority of soldiers voted for conscription, but no one is certain to this day how the troops in the trenches voted because all soldiers' votes, including those in Australia, in transit, in Britain and on the front line were counted together. This was on the advice of Keith Murdoch, who told Hughes it would be inadvisable to reveal the strength of the front-line troops' opposition to conscription. Patsy Adam-Smith, who read over eight thousand diaries and letters for her book *The Anzacs*, found that many soldiers had written to relatives criticising 'my friends who stayed at home', but just as many had written to say, 'Tell Bob not to join up, the game's up to mud.'

Hughes was not happy with the result and the following year he tried again. A lot had happened in the meantime to make him think that he could win the second time round. He was convinced that the Industrial Workers of the World (IWW) with their anti-Australian, anti-Empire activities would swing the Catholics towards a Yes vote, and he accused the IWW and the Labor Party of preferring to foment revolution in Australia to winning victory in France. There was some evidence for this. Dr Mannix had said that Australians were dying in 'a sordid trade war' and that Britain and Germany were simply fighting over markets. Australia had done more than its share in the war, and the best thing it could do now was to keep up food supplies to Britain but put Australia, not the Empire, first.

There had been a mutiny in the AIF when 15,000 troops in Sydney refused to accept an increase of one and a half hours in their daily training. They had marched from their camp to nearby Liverpool and then on to Sydney, wrecking hotels and shops on the way. Military and civil police, assisted by about 250 returned soldiers, eventually restored order but one soldier was shot and killed.

There had been a general strike in New South Wales which threatened food supplies and virtually brought community life to a halt. It started over the introduction of a time card system to monitor the amount of work done by each railway employee and had then spread to virtually every essential industry – coal miners, gas workers, seamen, waterside workers, painters and dockers and meat workers – until some 95,000 were on strike.

The state government's efforts to crush the strikers caused great bitterness. A crowd of 100,000 people assembled in the Domain to support the strikers, but at the Sydney Cricket Ground, only a few miles away, thousands of men from the country, answering the government's call for 'loyalists' to help restore essential services, camped out awaiting the call to work. Well fed and given free beer, they soon restored miminal services and when strikers attacked two of them driving a 'scab' lorry, one of them shot a striker dead. Eventually, the strikers were forced back to work.

Hughes could not believe that Australians could be so disloyal as to go on strike when Australian servicemen were dying overseas and Britain was fighting for her life on the battlefields of France. He wanted Australia to have even closer ties to Britain and surrender some of its independence to an 'Empire Parliament' which would have Australian representatives. He reasoned that there had to be more to the general strike than industrial unrest. Were the Industrial Workers of the World, the new-to-the-scene Bolsheviks, the Sinn Feiners and the left wing of the Labor Party trying to foment a revolution?

Two IWW members had been hanged for the murder of a policemen and twelve others jailed for arson and sedition. Now Hughes banned the IWW and invoked the War Precautions Act to make anyone suspected of being weak on loyalty and patriotism liable to prosecution. Hughes thought that loyal Anglo-Australians were fed up and given a second chance to express their devotion to Britain they would vote 'Yes' for conscription, 'Yes' for King and Empire.

He was wrong. Even though the famous singer Dame Nellie Melba joined the 'Yes' campaign and sent a message to the women of Australia, this time an even larger majority voted against conscription. (Dame Nellie secretly thought her message would not do much good; she told Hughes privately, 'Very few Australian women use their brains.') It was a great victory for the anti-conscription forces, especially since – a telling blow to the government – almost half the soldiers at the front voted 'No'. These two anti-conscription referendums had wider significance. They forced Australians to think about their attitudes to nationalism and the bond to Britain and they highlighted for liberal intellectuals the conservatism of the Australian middle class. But they also proved the power of the Australian sense of fairness and personal liberty – if a man wanted to volunteer to go overseas to fight, then that was his business, but no one had the right to force him to do so.

The last year of the war moved towards its bloody end in France – 'We just go into the line again and again until we get knocked,' wrote one Australian soldier. 'Australia has forgotten us, and so has God.' But occasionally there was a triumph. On 21 April 1918, Australians shot down the German air ace, Captain Baron Manfred von Richthofen.

Sergeant Cedric Popkin and Gunner Rupert Weston, hearing a dogfight over the Australians' line on the Somme, set up a Lewis gun and fired on a German triplane which was chasing a British Sopwith Camel only about 100 feet above the ground. They got in a burst of about 80 rounds and appeared to have hit the German plane. It abandoned the chase, banked and came back straight at the gunners.

In 1969 Weston, then living in Sydney, recalled, 'We were in hard open ground with no cover. It was him or us. Just as he dipped his nose to open fire, the Sergeant opened up with another eighty-round burst, and I reckon this was the one that got him. The plane gradually lost altitude, hit the ground fairly evenly and ran along into a mound and stopped. When we got there, the pilot was dead. We didn't know who it was until an officer pulled him out of the cockpit and identified him as Richthofen.'

And then there had been cheering news of Australian successes in the Middle East. There the war had all the glamour and romance that the trenches of Europe lacked. It was a war of movement, of cavalry charges in vast desert spaces, a war fought to drive the Turks out of biblical lands they had ruled for centuries.

Commander of the campaign was General Sir Edmund Allenby, a blunt, heavily-built man known to his troops as 'The Bull'. Serving under him was General Harry Chauvel, in charge of the Australian Mounted Division, the renowned Light Horse. Chauvel was a tough, country-born Australian who rose to be the best leader of horse in modern times and one of his country's greatest generals. His men found two paths to glory in the Middle East, the well-known taking of Damascus and, much less well-known, the charge at Beersheba, one of history's great cavalry charges.

The charge took place in fading daylight on 31 October 1917 to capture the wells of Beersheba before the Turks could blow them up and delay the Allied advance. The charge succeeded because of its sheer audacity. The Turks expected the Australians to follow their usual practice of dismounting and fighting as infantry, and so held their fire for crucial minutes as eight hundred men of the Light Horse swept down upon them. Then, suddenly, it was too late. As the Light Horse came thundering on, the Turkish gunners were unable to wind down their field guns or re-aim their machine guns to match the speed of the charge and the Australians rode through into the trenches. There it was savage and merciless slaughter. Earlier, Australian troopers had been shot by Turkish soldiers after surrendering, and now their comrades wanted revenge. This is probably why the charge was so little known for years – none of the Light Horsemen who took part wanted to talk about it. Film-maker Charles Chauvel, a relative of Sir Harry, made *Forty Thousand Horsemen* in 1939, but this was inaccurate and it was not until Ian Jones produced *The Lighthorsemen* in 1987 that the full story of the charge emerged.

The Light Horse's role in the taking of Damascus, on the other hand, was quickly known, largely because of Colonel T. E. Lawrence, 'the Prince of Mecca', later famous as Lawrence of Arabia. Lawrence was supposed to be General Chauvel's political adviser, but Chauvel felt – correctly as it turned out – that Lawrence had some hidden agenda of his own. Lawrence – also with some justification – did not like the cavalier attitude of the Australians towards what he regarded as a high moment in history. Their enmity came to a head in a dispute over who deserved the glory of taking Damascus, thus ending three hundred years of Turkish rule. In *Seven Pillars of Wisdom* Lawrence claimed that he and the Arab army entered the city in the early hours of 1 October 1918. 'We drove down the straight banked road through the watered fields, in which the peasants were just beginning their day's work. A galloping horseman checked at our head-cloths in the car, with a merry salutation, holding out a bunch of yellow grapes. "Good news; Damascus salutes you".'

But the Australian 3rd Light Horse had passed right through the city the night before. Chauvel, worried that Lawrence might rob the Australians of the credit, put this on record with a telegram to Allenby dated 1 October: 'The Australian mounted division entered the outskirts of Damascus from the north-west last night. At 6 a.m. today the town was occupied by the Desert Mounted Corps and the Arab Army.' Lawrence never forgave him and wrote in *Seven Pillars* that the Australians 'ran the campaign as a point-to-point with Damascus as the post.'

Len Hall, of the glamorous 10th Light Horse, did not care what Lawrence thought of him and his fellows. He was content to be still alive. He had enlisted at 16 because the Light Horse needed a bugler, but when he arrived at Gallipoli they made him a machine-gunner. After Gallipoli had been evacuated he rode with the light horsemen in the charge that captured Beersheba, but was lucky to escape with his life when a bomb dropped from a German plane landed among his gun crew and killed 9 of the 14 gunners. He recovered from his arm and shoulder wounds to ride with the Australians into Damascus. He saw Lawrence there. 'A clever man, Lawrence,' he said after the war. 'Spoke the local lingo as good as the Arabs. They loved him.'

When he was riding off to embark for the war in 1914, he saw a pretty Perth girl waving from the crowd. On an impulse he plucked the emu plume from his slouch hat and thrust it into her hands. Five years later, marching victorious through Perth, the same girl stepped out of the crowd, touched his arm and said, 'Excuse me, sir, would you like your plume back?' They got married and lived happily for the rest of their lives, which in Len's case lasted until he was 101.

In France the career of the engineer-soldier, General Monash, reached its high point. Monash had joined the Victorian militia in 1887 at 22 and held a commission as a part-time soldier. After Federation the various state militias were brought together as the Australian Citizen Force and those who volunteered were sent overseas as the AIF. It was at Gallipoli, where Monash commanded a brigade, that he first began to ponder on the military stalemate produced by infantry attacks on entrenched positions. He saw it as 'simply a problem of engineering'.

In its modern form, the problem came from three technical innovations which immensely strengthened the defensive side in the military seesaw. Continuous lines of trenches protected by belts of cheap barbed wire, originally made for farmers, covered by machine guns secured in concrete, could not be penetrated in daylight by unarmoured infantry – as those soldiers mowed down in futile attack after attack could have testified.

The earliest scheme for a breakthrough was British – a 'rolling barrage' of artillery which cut the wire and forced the defenders from their trenches into deeper dug-outs. The infantry were supposed to follow close behind the moving lines of shell bursts and occupy the wrecked enemy trenches. But to get any further, the attackers had then to bring food, fuel, ammunition and reinforcements through the breach they had made in the enemy front line, and the ground they had to cross to do this had been made impassable by the shelling that had made the breach in the first place. In the meantime, the defenders would have time to bring up reinforcements along undamaged roads and railways behind their own lines, out of artillery range. There was a further complication. By the end of 1914 the professional armies of pre-war times had been mostly wiped out, and both the Germans and the Allies had to fill and expand their ranks with hastily-trained and less than enthusiastic conscripts.

However, none of this could have been known for certain without trying it out. The ghastly Somme battles that began in June 1916 tested both the new artillery plan and the new conscript British armies. They killed a lot of Germans but there was no breakthrough. The Germans had already tried gas at Ypres in April 1915, soon copied by the Allies, but that did not work either. The battles of 1917 decided nothing and the war looked like going on forever.

Enter General Erich Ludendorff, a deplorable human being and a military genius. He saw that breaching a continuous enemy line was – like the capture of a medieval castle – a job for elite specialists, not rank-and-file troops, so he created what came to be called *sturmtruppen* or *stosstruppen*, volunteers selected for self-reliance, physical fitness and high motivation, who were spared most of the tedium of everyday soldiering. In the Middle Ages they had been called 'the forlorn hope' because of the risks involved. Ludendorff tasked these men with penetrating the Allied lines and paralysing the command structure and communications net behind them. Artillery preparation was kept to a minimum, or omitted altogether, and the well-fed and adequately-rested storm troopers took the line of least resistance.

The Germans tried the new system in the last-chance offensive they called the Kaiser's Battle, beginning on 21 March 1918 against the British Fifth Army in Picardy, the first big objective being Amiens on the River Somme. The result was a British near-collapse and the first clear breakthrough of the trench defence system in the war. The situation was best summed up in the words of an Australian brigadier to the officers of the 10th Brigade. 'German divisions are pushing forward with great rapidity and there is a long-range gun shelling Paris. You'll entrain tomorrow morning and you are to go straight into action. You're going to have the fight of your lives. The fate of the war is in the balance.'

Monash, who had been taking well-deserved leave in the South of France, rushed back after he read in the newspapers of the German breakthrough. He caught up with the British commanders in their makeshift headquarters in a château, examining their maps by the light of a candle. Monash recalled: 'As I stepped into the room General Congreve said, "Thank heaven, the Australians at last."' He told Monash that the Germans were making straight for Amiens, a key communications centre, and had to be stopped.

Monash's first task was to get some of his troops into blocking positions as fast as possible. This he did with – a convoy of veteran London buses. Twelve hours after receiving his orders, Monash had over five thousand fresh, well-rested Australians in place, sending out patrols to get into contact with the Germans, and more men were arriving every hour. According to Charles Bean, one Australian called on his basic French to reassure a French villager, '*Fini retreat, Madame. Fini retreat – beaucoup Australians ici.*'

Weeks of hard fighting followed, but by the end of April it was clear that the Germans' spring offensive had been held; the last German attack petered out in July. By this time the Americans were pouring into France at the rate of half a million men a month, so the Allies considered that the situation was ripe for a general counter-attack. Two Dominion generals had been planning for this moment. One was Monash, the other the Canadian Sir Arthur Currie. Currie had been a schoolteacher and, like Monash, 'a Sunday soldier'. Then he became a real estate agent in Victoria, British Columbia, before going bankrupt. He had been among the first Canadians in France in 1914 and three years later had risen to command all four Canadian divisions in France.

Monash and Currie led soldiers who were natural breakthrough specialists – Australians, New Zealanders and Canadians, all volunteers selected for physical and mental hardiness before they left home. By now they were experienced as well. But, more important, their divisions were commanded on lines quite different from the British class-ridden system. British officers were almost exclusively public school men who served no time in the ranks. If a soldier was commissioned *from* the ranks – which happened increasingly as the public school officers were killed off – he was not sent back to his original unit.

The Australian, New Zealand and Canadian armies worked on the reverse principle. Almost all junior officers were promoted from the ranks and stayed with the same unit, leading 'their old mates'. Personal links and loyalties were therefore very close and strong. Monash and Currie were products of this system and Monash realized that it made for tightly-knit, highly-motivated and resourceful soldiers – ideal military material for a successful 'forlorn hope'.

Sensibly, Monash devoted a lot of his time to seeing that his soldiers lacked for nothing when away from battle. He devised a system in which no man did more than 48 hours continuous front trench duty in every 12 days. Every 48 days the whole brigade was relieved by the reserve brigade and went out for a complete rest for a clear 24 days. All the Australians in the front lines regularly got hot food – coffee, Oxo, porridge, stews – brought to them in giant Thermos-type flasks (boxes lined with straw) which Monash got his engineers to build. He set up a special group to collect the men's wet socks and wash and dry them so that every soldier had fresh socks every day. He built two divisional baths where two thousand men a day had a hot bath in giant brewery tubs and then exchanged their dirty underclothes for a clean set.

As well, Monash told his wife, 'In the *Ecole Professionelle* in Armentières . . . I run a "Grand cinema show and pierrot entertainment", charging the lads half a franc admission. They appreciate it all the more if they are compelled to pay – wouldn't trouble to come if it were free.'

The Australians compared their lot with that of their British cousins who manned most of the drab, essential support services backing up the glamorous elite. They were quick to note the differences in the relationships between soldier and officer and decided that they were prepared to trust Monash to lead them anywhere. Monash was given the chance to use these tough, loyal soldiers when – in spite of intrigues against him by Keith Murdoch and Charles Bean – he was given command of the Australian Army Corps, its first Australian-born commander, with Brigadier Thomas Blamey, the future field marshal, as his chief of staff. The four Canadian divisions were moved alongside them and for the opening battle of the great Allied counter-attack, Monash was allocated 168 tanks with British crews and several squadrons of the Royal Australian Flying Corps.

The attack began at dawn on 8 August 1918. By the end of the day the Anzacs had taken 8,000 German prisoners and captured 100 guns for a loss of 1,200 Australian casualties. The Canadians had similar success, an advance of about ten miles through a gap twelve miles wide. Ominously for the Germans, their counter-attacks failed to recover any lost ground. Ludendorff called 8 August 'the black day of the German army in the history of this war' and advised the Kaiser that Germany should seek a way out. King George V showed up in person four days later to knight Monash on the battlefield.

Monash was next ordered to prepare a breakthrough of the main defences of the Hindenberg Line. He now commanded 200,000 men, only half of them Australians, the rest British and Americans. The battles were repetitions of the 8 August success and by 8 October 1918, the 6th brigade, 2nd Australian Division were in a village called Monbreihan, in open country on the far side of the German main line. After nine weeks of continuous action, the Australians were withdrawn to rest and took no further part in the war. It did not matter. As Monash says in his book, *The Australian Victories in France in 1918*, they had 'confirmed the irretrievable collapse of the whole Hindenberg defences'. In view of this high praise it remains a puzzle that Australia prefers to celebrate the defeat at Gallipoli rather than the triumph of her troops in France.

Virtually alone amid the military leaders of the war, Monash's reputation in Australia never declined. In 1961, a Melbourne university was named after him. Recognition in Britain took longer and was more discerning. War historian John Terraine wrote in *The Smoke and the Fire* that only three significant innovations could be traced to Monash – the use of tanks much closer to the barrage line than had been thought possible; mutual support by tanks and infantry; and the airlifting of supplies to troops locked in battle.

Terraine says that Monash's great strength was his planning – he was trying to evolve a science of war which would be as exact as the science of engineering. 'He wanted to be able to rely on a battle in the same way as he would rely on a bridge.' And no matter what verdict history may eventually pass on the British Commander-in-Chief, Sir Douglas Haig, it was Haig who recognized Monash's talent by selecting him to spearhead the Allied counter-offensive in the summer of 1918, and who is reported to have told Blamey, 'You don't know how much the Empire owes to you Australians and the Canadians.'

There was a price to pay for being an army of elite stormtroopers and the casualty figures reflect this. The Australians suffered 70 per cent dead, wounded and missing, the highest rate of any national contingent that took part in that immense war. It was glory, no doubt, but glory at a terrible cost.

Peace on 11 November brought out thousands of people in every Australian capital city, although because of the time difference, the real celebrations did not take place until the next day. At least eight million soldiers had been killed and perhaps as many as 12 million civilians. All casualty figures are approximate because the bodies of over half a million men who fought in France and Belgium were never found, or if found, never identified. Australia had sent about 330,000 volunteers to the war and of those nearly 60,000 were killed, 150,000 wounded and 4,000 taken prisoner. And, a most amazing thing – the two major combatants, Britain and Germany, had managed to fight the war on other people's territory and it ended without a single enemy soldier standing on their soil.

Had it been worth it? The Governor of New South Wales, Sir Walter Davidson, thought so. He told thousands of Australians who packed Martin Place, Sydney, 'I come here as the King's representative to announce that the King's enemies have been defeated. We have won. We have finished the job off; we have finished the enemy off, and they cannot come back again – dogs that they are.'

He was, of course, wrong, and those 'dogs' did come back again twenty years later. In fact the consequences of the war can be quickly summarized – Europe ruined as a centre of civilization; Nazi Germany; Soviet Russia; Fascist Italy and another war in 1939. But at the time it could seem as if the First World War *had* been the 'war to end wars'. Strange then that the Australians who had answered the call so readily in 1914, who had contributed so much, and suffered so greatly, had – like all the other members of the Empire – not even been consulted about the terms of the Armistice. If the war had made the Mother Country more appreciative of Australia's effort, it was not yet apparent.

The truth was that, apart from the appalling waste of young lives, the war was disastrous for Britain and the Empire. It raised the national debt by a multiple of eleven and greatly depleted the country's overseas assets. Before the war these were worth more than the foreign investments of Germany, France and the United States put together. By 1917 President Woodrow Wilson could boast that Britain would soon be financially in American hands.

As for Australia, the Great War left an astonishing every second Australian family bereaved, a country in mourning until the next time round, a country where every tiny town had its memorial hall or column. How does a young nation cope with such grief and recover from the loss of the flower of her young men? Perhaps it doesn't.

CHAPTER 6

THE ENEMY WITHIN

'One night of darkness in Melbourne, comrades, and there would be a few new suits in the city. The workers would have a few gold watches, too. One night of darkness would be more than enough . . .'

– Jock Garden, Communist Party leader

When the euphoria of victory had faded away, Australians found themselves facing a terrible reality. Pneumonic influenza, the '1918 flu' or the 'Spanish flu', the worst medical catastrophe in history, raced around the world and killed 40 million people. So four times as many died in less than a year than were killed in the whole of the war. Coming on top of the most appalling man-made conflict, it was seen as a punishment from God.

In Australia, troop ships returning from Europe were held in quarantine, the whole state of New South Wales was declared an infected area, and schools and all public places of entertainment were closed. The wearing of protective face masks was made compulsory and citizens were encouraged to get inoculated. But by the end of 1919 at least 12,000 people had died from the epidemic. This was nothing compared with 150,000 in England and Wales alone, or 500,000 in the United States, and 15 million in India, but it stunned the nation.

Thurza Bishop was nine at the time. She lived in Narromine in the western part of New South Wales, where her father, who ran a wheat and fruit farm, was mayor. 'My best schoolfriend lived down the road and I went over there one afternoon to play. My friend said her aunt was sick but she didn't say what was wrong with her. When Dad came home I told him where I had been. He said, "It might be flu" and he grabbed me, took me outside, stripped off my clothes, hosed me down with water and then poured disinfectant over me, hosed me down again, and then burnt all my clothes. Everyone was terrified of the flu.'

Ellie Simpson's father, Reeve Palmerston Rundle from Victoria, was a recently graduated doctor in 1918 and went straight off to the war in Europe. Ellie remembers, 'He spent his time there in hospitals in Belgium attending influenza victims. Then he sailed for Australia with demobilized Australian troops on their way home. Influenza broke out as the ship approached Sydney, so the whole ship's complement, Dad included, went straight into quarantine at the North Head Quarantine station where they spent the next six months. Many died, but Dad survived. He used to tell us stories of how terrible it was, so close to home but with the possibility that you'd never get out of the quarantine station alive.'

In those pre-antibiotics days, Australians, always fond of home remedies, tried everything to keep the flu at bay – wrapping the chest with brown paper soaked in vinegar; sprinkling eucalyptus oil on the pillow; and sniffing salt water up the nostrils. Official advice was little better, the authorities deciding against banning crowds on beaches in the belief that the risk of infection would be counter-balanced by the benefits of sunlight and salt air.

Even when the epidemic passed, life remained risky. Trains collided, ships sank, mines collapsed and planes crashed. And all the while those eternal enemies of Australia – drought, cyclones and floods – took their toll. The economy was slow, investment for new capital equipment scarce, and the streets were full of ex-servicemen with missing limbs, mutilated faces and the white stick of the blind man. Jobs were scarce anyway, so what chance did a broken, disillusioned ex-Anzac have?

They wandered from pub to pub getting drunk and violent, or they sold matches and trinkets at street corners or from door-to-door. They sat on stools in department store lifts, the empty leg of their trousers pinned neatly back so it did not dangle and upset anyone, their crutches tucked in the lift corner. 'First floor, children's clothing, ladies shoes, and ready-to-wear suiting. Mind the doors.' They were angry that no one seemed to understand them. The national mood was sombre and apprehensive, caught by the haunting hit song of the year:

> I'm forever blowing bubbles,
> Pretty bubbles in the air.
> They fly so high, they nearly reach the sky,
> Then like my dreams they fade and die.
> Fortune's always hiding.
> I've looked everywhere.
> I'm forever blowing bubbles,
> Pretty bubbles in the air.

One reason for this apprehension was that up in the northern hemisphere,

Lenin and his Bolsheviks had started to build the Soviet empire, fighting off the forces of 26 countries which had intervened in Russia to stop them. Australians were part of these anti-Communist forces. HMAS *Swan* and HMAS *Parramatta* were in the Black Sea supporting a White Russian army, and Australian troops were part of a British Army which landed at Archangel in northern Russia to engage the Reds.

But the intervention failed, Communism spread to Australia, strikes and violence were everywhere and the nation was once again torn into loyalist and nationalist factions where anyone not one hundred per cent behind Britain, the Empire, the Union Jack and the Returned Soldiers' and Sailors' League (RSSL) was a traitor and a Communist.

Looking at the long list of strikes, lockouts and lethal violence that disfigured this period of Australian history, one can understand the apprehension – even fear – that gripped middle-class, conservative Australia. On occasions it must have seemed as if the nation was about to be engulfed in civil war, that a workers' revolution was about to break out and that a Communist Australia was a real possibility.

The strike that really put the fear of revolution into the middle classes was the Victorian police mutiny of 1923. It was the first and (so far) only time a police force in Australia had gone on strike. The police had been dissatisfied for years over wages, pensions and conditions. They were not allowed to form a trade union or join one which would give them representation in their negotiations. They were paid less than police in other states and their request for an increase of one shilling a day had been refused.

Matters came to a head on 29 October 1923 over a relatively trivial matter – the introduction of plain clothes 'watching detectives'. When the night constables refused to go on duty under these supervising officers they were dismissed. The day constables then refused duty and they were dismissed also. A total of 636 police officers – one third of the entire force – were now on strike and sacked. It was the start of the spring racing season and the carnival atmosphere of Saturday night changed when the crowds on the streets in the centre of Melbourne, excited by drink, realized that there were no police around. Fighting broke out and soon crowds were smashing shop windows and looting furs, jewellery, sporting goods, tobacco, clothing and cameras. A clergyman who held a Bible aloft and called for calm and order was knocked to the ground and trampled underfoot. 'Police strike results in unprecedented scenes,' said the *Age*. 'The blackest page in the history of a fair city . . . Ex-soldier brutally murdered in thronged thoroughfare . . . the worst elements of the population ran amok on Saturday and for a while Melbourne and all it contained was theirs.' The trouble went on for the whole weekend and did not stop until the government passed a Public Safety Bill, and enrolled one thousand special constables, including ex-light horsemen, who under General Monash restored order. The authorities won but a week's anarchy frightened everyone. If a lawless mob could take over the streets of Melbourne was revolution around the corner?

These frequent conflicts between the state and its employees, between bosses and workers, followed an almost predictable pattern. In the aftermath of the great strikes and lockouts of the 1890s, the Australian states had developed a system of legal arbitration to resolve industrial conflict and regulate employment that later became the envy of the world. Supported by the unions, these arbitration courts had gradually expanded their role to include the fixing of a 'fair and reasonable' wage in all industries. Employers resisted this, arguing that the market should be allowed to decide wage levels.

In the 1920s, when a major issue was at stake, both sides were prepared to challenge the arbitration courts. Unions would call illegal strikes, the employers would lock them out, bring in non-union strikebreakers and try to crush the strike. The strikers would attack the strikebreakers, the police would intervene to maintain law and order, and there would be chaos until one side or the other gave way or the courts managed to arbitrate a compromise.

The waterside workers, miners, meat workers, seamen, and timber workers were the most militant and violent in the pursuit of more secure employment, shorter working hours, better conditions, and a guarantee that non-union labour would never be employed, the 'closed shop'. But the police frequently matched them blow for blow. The 'Dimboola' incident in Western Australia was typical.

On 22 April 1919, 2000 striking waterside workers gathered at Fremantle wharves and managed to prevent non-union labour from unloading the SS *Dimboola*. The Premier of Western Australia, Hal Colebatch, said it was illegal for the strikers to occupy the wharves and announced that they would be forcibly removed. He ordered the police to train a military raiding party equipped with rifles and bayonets. On Sunday, 4 May he went to the wharves himself with this raiding party and they tried to erect barricades to keep the strikers away. It did not work. The strikers broke through, overwhelmed the 'scab' workers, threw them into the Swan River and then turned on the police. The front two rows of the mob were women – wives, mothers and sisters of the strikers – who had demanded they be allowed to take the lead in the riot.

The police gave the order to charge and mounted policemen rode the strikers down and drove them away. In the melee 33 men and women were injured and one striker received a bayonet wound in the leg from which he later died. Over the next few days running battles broke out between the strikers, the strikebreakers and the police. But two events tipped the balance in the strikers' favour. Ex-servicemen, disturbed by the police charge against women, came to the strikers' aid, and the Federal government turned down a request from Premier Colebatch for military intervention. The strikebreakers decided to withdraw 'in the interests of the community' and the strikers went back to work.

Other industrial riots were equally bloody. In 1929, coalminers in New South Wales refused to take a pay cut to assist the industry. The mine owners sacked the entire workforce. There was legal argument over whether this was a strike or a lockout but at the end of nine months, with no conclusion in sight and essential coal stocks running low, the state government decided to re-open the mines using non-union labour. The strikebreakers were billeted in neat rows of tents alongside the Rothbury Colliery, near Newcastle. The miners replied by calling out pickets, sometimes as many as 20,000 strong, to deter the strikebreakers from entering the mine. The police would then escort them through. On 16 December, 1000 miners armed with batons and staves attacked the colliery at dawn. In the ensuing battle, police fired their revolvers, injuring several pickets and killing one, Norman Brown, who quickly became a working-class hero.

> Oh Norman Brown, oh Norman Brown,
> the murdering coppers they shot you down.
> They shot you down in Rothbury town,
> to live forever, Norman Brown.

But, in the end, the miners, close to starvation, were forced back to work. Could the Federal government succeed in its plan to smash the unions? Not if the Communists could prevent it.

The Communist Party of Australia was formed on 30 October 1920, when representatives of the most militant labour groups – 26 men, 3 women and a baby – met in the Socialist Hall in Liverpool Street, Sydney, to begin planning 'the complete overthrow of the Capitalist system'. Twenty years later it had headquarters in every capital city, hundreds of suburban branches, leadership of many unions and such a stranglehold on the New South Wales Labor Party that the Federal executive of the party had to dissolve the entire New South Wales executive.

One of its leaders, Jock Garden, spoke frankly to his followers on how the Communists would change things. 'One night of darkness in Melbourne, comrades, and there would be a few new suits in the city,' he said. 'The workers would have a few gold watches, too. One night of darkness would be more than enough, and that would be all we require of you.'

Others were less apocalyptic. Stan Moran, who was born on 1 May 1899 ('The date was an act of God,' he said later), was a trade union leader of Irish descent – 'My grandmother went to Mass in the morning and out shooting at landlords at night.' He was a brilliant mass orator. 'Come the revolution, comrades, you'll have strawberries and cream for breakfast every day,' he told a meeting in Sydney's Domain. A heckler shouted back: 'But I don't like strawberries and cream.' Moran did not hesitate: 'Comrade, come the revolution you'll have strawberries and cream whether you like it or not. And you'll like it.' He died in 1998, his views unchanged. He said true socialism was still to have its day but acknowleged that in the meantime 'Communism has suffered a few setbacks.'

The fact that there was now a tiny Communist Party established in Australia and that it was being financed from Moscow – suspected at the time but only confirmed much later – inflamed the loyalists. When a small May Day crowd marched to the Domain, sang 'The Red Flag' and then burned the Union Jack, a loyalist arranged a counter march which brought at least 100,000 people to the streets to display their patriotic fervour, even though, by then, there was strong suspicion that the burning of the Union Jack had been arranged by an *agent provocateur*.

This antipathy to Communists shaded over into religion and intolerance of Roman Catholics. In modern secular Australia it is difficult to imagine how religious prejudice permeated all aspects of Australian life in the 1920s. Catholics and Protestants went to different schools, pursued different careers, and did not mix socially. Both religions discouraged intermarriage. The secretary of my father's Masonic Lodge got married late in life to a Catholic woman. Despite his years of service to the lodge, he was asked to resign, and when he died the lodge declined to turn out for his funeral.

This sectarianism even found its way into Test cricket. Before the First World War, the Australian team consisted almost entirely of cricketers of English descent. After the war along came Roman Catholics of Irish descent – Bill O'Reilly, Stan McCabe, Leo O'Brienne and Chuck Fleetwood Smith. Although the two factions were playing for Australia, they did not mix socially, spoke no more than necessary to each other, and changed on different sides of the dressing room. According to O'Reilly, once when the great Don Bradman, a Protestant, was captain of the Australian Test team he accused all the Catholics in the team of 'insubordination'.

True, many Irish-Australians made it appear as if they did put the welfare of Ireland above that of Australia, that even though they were born in Australia they remained more Irish than Australian. (As late as 1997, a Protestant artist, angry about the way Irish-Australians had come to dominate certain sections of Australian life, headed his article 'When Irish Eyes Need Punching'.)

In November 1919 Irish-Australians held a huge rally in Melbourne to affirm

Ireland's right to self-government, and when the British government announced the following year that it would not allow Archbishop Mannix to visit Ireland, 10,000 Irish-Australians turned out again to protest. There was no immediate response from the loyalists. They were quietly working on their own plans to 'counteract disloyalty' and 'maintain a strong national pride of race and Empire'.

On 18 July 1920, a group of distinguished army officers invited 'all loyal people in Sydney' to a meeting at the Town Hall. General Sir Charles Rosenthal, a tall, powerfully-built man who used to take on entire tug-o'-war teams himself, was in the chair. With him on the platform were 150 distinguished citizens – ex-army officers like Brigadier General Sir Edmund Herring, prominent conservative politicians, high dignitaries of the Church of England.

They made patriotic speeches calling for absolute loyalty to King and Empire; all disloyalists should leave the country or be strung up. They wanted to 'rid Australia of the foreign element', keep the country white, and 'put an end once and for all to any attempt to establish in Australia a black, brown and brindle autocracy'. They voted to form the King and Empire Alliance to achieve these ends; then all stood and sang the National Anthem and 'Rule Britannia' so loudly that it echoed down George Street.

The *Sydney Morning Herald* reported the next day: 'Scenes of extraordinary enthusiasm marked the great crowded meeting in the Town Hall last night when there was launched, in affiliation with similar bodies in other Australian states, an organization to be known as the King and Empire Alliance, having as its objects broadly the welfare of the British Empire and the counteracting of attempts to encourage disloyal doctrines.'

The US Consul-General in Sydney, E. P. Norton, was blunter and more truthful in his assessment of the Alliance. 'It looks upon the present state of affairs as being unusually disquieting . . . and is contemplating measures calculated to overcome the influence of Roman Catholics, Sinn Fein and labor elements. [It is] preparing for a clash between the various influences in Australia.'

The first issue of the Alliance's magazine, *King and Empire*, was published on 21 January 1921. This revealed that John Stinson was the president, Brigadier General George Macarthur-Onslow a vice-president and General Rosenthal the secretary. A quick glance through its contents makes its aims crystal clear. The foreword says that Australia faced two enemies. The first was the Bolshevik movement, which had started in Russia, spread over Europe and had now reached Australia. This movement advocated the fomenting of industrial discontent, the creation of class warfare and constant recourse to strikes. The second enemy was 'a number of our own race who are absolutely disloyal, enemies of Britain and the Empire who would glory in its disintegration and downfall'.

This enemy was not named but its identity – Roman Catholicism – was established elsewhere in the magazine. Under the headline 'Recent Utterances as Reported in the Public Press' there were items like 'Father Forrest, recently returned from Ireland, lecturing recently in Sydney, told his hearers that the British government was behind every evil deed perpetrated in disordered Ireland at the present time.'

Another read: 'Father T. J. O'Donnell, who was a chaplain in the AIF, in the course of a lecture made the following statement: "The Germans were magnificent people. Their soldiers thought that they were fighting for right, but like us they were merely fighting for capitalists. It would be a good thing to bring two million Germans to this country. The Australian Republic is sure to come, and when it does I will hail it with delight.'

Readers of the magazine were not to know that many members of the Alliance were also supporters of a secret army pledged to maintain law and order by military means should the need arise. Since this was only one of several secret armies that sprang up, ready for civil war, in Australia during these troubled times, and since some of them were to remain on alert for at least the next twenty years – a little known facet of Australian life – it is instructive to trace their origins.

Early in 1917, when the United States was on the brink of entering the war against Germany, Washington suddenly realized that the country's large German-immigrant population could pose a problem. The fledgling Bureau of Investigation (later to become the FBI), part of the Justice Department, did not have the staff to keep a check on citizens' loyalty, so it readily accepted an offer from Albert Briggs, vice-president of Outdoor Advertising Incorporated, to organise a secret, America-wide organization that would work with the Department to combat espionage, disloyalty and civil unrest. It was to be called the American Protective League.

By the time the war ended in November 1918, the League, by then called 'The Web', was active in every state, city and town in America. It boasted a membership of more than 100,000 and had been responsible for tracking down and dealing with hundreds of thousands of 'slackers, wobblies and dissidents'. It had attracted well-known Americans to its ranks – the Hollywood branch was run by Cecil B. De Mille – and had won the thanks of the Department of Justice.

The Web did not disband at the end of the war but simply switched its activities to combating Communism and remained active in this role until the end of the Cold War in 1990. Word of this secret volunteer army spread to

Australia, probably in reports from Sir Henry Braddon, who was Australia's trade commissioner in the United States in the last years of the war. (He later wrote a book in which he described The Web as being made up of 'leading men in all professions and avocations, who did their secret work most capably'.) So, late in 1917, the Prime Minister, Billy Hughes, sent a Melbourne businessman, R. C. D. Elliott, to see if a similar organization would benefit Australia, which lacked any national agency to tackle 'disloyalty'.

On 29 May 1918, a meeting in the Melbourne office of the acting Prime Minister considered Elliott's report and its endorsement by Australia's military authorities. It gave its approval to the scheme and authorized a Melbourne businessman, Herbert Brookes, to organize the Australian Protective League (APL). Brookes went further than the American model. He decided that the APL would have two sides – an official one linked to the security services, and a separate 'voluntary arm'. This would be set up by approaching the executive heads of 'known loyal societies and associations' and, after swearing them to secrecy, inviting them to form their own state organization. Behind the screen of these loyalist bodies a volunteer army of 'vigilantes' would be assembled.

The end of the war caused Brookes to change the focus of the APL. Instead of anti-war activity, the APL would now concentrate on the threat of Bolshevism, radical socialism, union activity, strikes and civil unrest. The government's role faded into the background and the APL took on a more civilian and more secretive aspect.

In those early days of this secret army, Brookes concentrated on Queensland, the only state in Australia that did not have a conservative government. His first important recruit was the Queensland Police Commissioner F. C. Urquhart, who was soon reporting: 'Upwards of thirty societies have given their adhesion . . . Yesterday three of their leaders came along and told me they would have sixty societies signed up . . . They wish to go pretty far – not only uphold the Constitution by peaceful means but to have a formidable striking force ready if required.'

In March and April 1919, the 'Red Flag riots' rocked Brisbane. At a march organized by the Queensland Trades and Labor Council, workers paraded many red flags in defiance of a government ban on them. Within days the APL had mobilized a force of 2,000 ex-servicemen and they stormed the Russian quarter in South Brisbane, 'to punish the mob which let us down in the war' and to 'root out the monster of Bolshevism from this country'. They laid siege to the Russian Club, stormed the office of the Labor newspaper, the *Daily Standard*, and had running clashes with workers over four weeks of intermittent street fighting. Two deaths were reported (but not confirmed) and 19 people were wounded by bayonet, bullet and bludgeon.

Then in March 1920, to the dismay of the loyalists, the Labor Party won power in New South Wales. The State Labor Party had been moving steadily to the left and was soon to adopt as its objective the socialization of industry, distribution, production and exchange. The first chance the loyalists had to show how they felt about this came in May when a Labor-Catholic rally was held to protest against the deportation of a Roman Catholic priest of German extraction. The rally was violently disrupted by bands of ex-servicemen, with disciplined assaults on speakers and the crowd. The Catholics noticed that Major General Rosenthal and an aide, later identified as Major W. J. R. Scott, seemed to be directing operations.

Scott, who turned out to have been Rosenthal's deputy when he was in charge of repatriating Australian troops from Europe, and was now the treasurer of the King and Empire Alliance, gave an interview to the *Sydney Morning Herald* about the ex-servicemen's role in breaking up the rally. He said that after Germans had been cleared out of Australia, 'the loyalty of certain citizens should be taken up.' The presence of such prominent figures as Rosenthal and Scott at riots was not surprising. Secret armies, all under the general umbrella of the APL, were to pop up all over Australia under different names throughout the 1920s and early 1930s – the Old Guard in Sydney, the White Guard and the White Army in Melbourne, among them – and historian Manning Clark has written that some of the most prominent citizens of Australia were members. These included Philip Goldfinch, a descendant of Governor King, R. W. Gillespie, a director of the Bank of New South Wales, Sir Samuel Hordern, of the retail firm of Anthony Hordern, and William McIlwraith, the largest grocer in New South Wales. Other sources name Sir Cyril Brudenell White, former AIF chief of staff.

D. H. Lawrence, who was in Australia at the time the APL was most active there, ran across it in one of its many guises in New South Wales and used it as a background for his account of fascism-versus-democracy in his novel *Kangaroo*, a fact uncovered by Sydney author Robert Darroch. Darroch makes a convincing case that Lawrence's characters Cooley and Callcott are based on General Rosenthal and Major Scott, while Labor historian Paul Tracey says that Willie Struthers is based on Communist Jock Garden, who was interviewed by Lawrence in 1922.

While the loyalists and their secret armies worried about Roman Catholics, unionists, strikers, Sinn Feiners and the Communists, a real, full-scale rebellion against constitutional authority in Australia was taking place in the Northern Territory. Until December 1918, there had been only two rebellions in Australian history. The first was the Rum Rebellion of 1808, in which officers of the New South Wales Corps imprisoned Governor William Bligh because he had tried to control their corrupt trade in rum. The second was the Eureka Stockade, a rebellion of miners on the Ballarat goldfields in 1854 against arbitrary government in which 30 miners and five soldiers were killed – an event seen by many as the birth of democracy in Australia.

Now a third revolt, the Darwin Rebellion, was underway. It is worth looking at the events leading up to this rebellion because they cover so many of the themes that make up Australian history this century – the conquest of the outback, the romantic legend of the cattle drover, the loss of Australian assets to overseas interests, the battles between employers and labour, and the workers' ethic of always supporting your fellow worker, with violence if need be. What happened in Darwin before the Rebellion can be seen as typical of the bosses-versus-workers clashes of that period, and one of the reasons why conservative Australians felt so threatened by growing union power, and why so many felt justified in forming secret armies.

Before the First World War, Darwin, the main town in the Northern Territory, had been the most exotic, isolated community in Australia. It had no air service, no rail service and no road service. It took 14 days to get there by ship from Sydney. Its citizens were Australians, Greeks, Chinese, Malays, Scandinavians, Patagonians, Russians, Syrians, Turks, Afghans, Italians and Japanese.

In the wet season, when blinding rainstorms made work difficult, drinking became a way of life. Drunkards stumbled along rubbish-filled muddy streets. One visitor described Darwin as 'the most squalid and contemptible place I ever saw'. In the dry season, however, it could seem like paradise: cloudless skies, the intense blue of the Arafura Sea, the exotic flowers, the exciting smells of Chinatown. And in the background, the Northern Territory itself, forbidding, hostile, dangerous, in parts a lunar landscape, in others a lush, productive land with enormous potential.

In 1911, the South Australian government, unable to handle the Northern Territory's declining population, stagnant economy and calamitous race relations, handed over control to the Federal Government. Its first act was to appoint an administrator for the Territory, responsible only to the Federal government. His brief was simple – to bring the place out of its morass.

His Excellency Dr John Gilruth was a tall, authoritarian Scot, a Victorian Tory. In the tradition of great colonial administrators, he pledged to bring the Northern Territory to its rightful place in the British Empire. The people welcomed him as the man with the magic wand who would make the place bloom, but within a year he was the most hated man in the Territory.

One of Gilruth's first acts was to cut workers' pay, offering to change the working hours to the earlier, cooler part of the day as an inducement. The workers, complaining that the hotel keepers refused to serve breakfast so early in the morning, went on strike. Gilruth promply brought in Chinese labour and forced everyone back to work. Flush with this victory, he then went ahead with his master plan – to find a British investor to buy the whole of the Northern Territory from the Australian government and turn it into an independent country with himself as Viceroy.

This grandiose scheme was not as crazy as it might at first sight appear. Gilruth already had an entrepreneur in mind – the enormously rich Liverpool family, the Vesteys. The Vesteys, headed by two brothers, William and Edmund, owned cattle stations all over the world, and their company, Union Cold Storage, was one of the most successful British enterprises the country had ever seen. William, Lord Vestey, had married a Chicago typist, Evelene Brodstone, who worked as the company's international troubleshooter.

She had concluded a deal with Gilruth just before the war to lease 36,000 square miles of pastoral land in the Northern Territory and the East Kimberley district of Western Australia – an area larger than the state of Tasmania – at a peppercorn rental. The actual payment was kept secret, but in later years, when there was a dispute with Aboriginal workers at the Vesteys' Wave Hill Station in the Territory, some details leaked out. For example, in 1980 for the Wave Hill Station, which had an area of 6,158 square miles, the Vesteys were paying an annual rental of 55c per square mile. Or, put another way, for a cattle farm of nearly four million acres, they were paying an annual rent of about £1,650.

Gilruth told the Vesteys that they could buy out the leaseholds in the Territory and own the land outright. He said he could convince the Australian government that £5 million would be a fair price. Once he was appointed Viceroy of this independent country he could make untold riches for the Vesteys. He knew the idea would appeal to the family. They had made their first fortune in Hangkow, China. As outright owners of the Territory they could make a case for ignoring the White Australia Policy and by importing cheap Chinese labour make a second fortune.

In the end, probably because they scented political difficulties, the Vesteys opted instead to build an enormous slaughterhouse and meat processing plant in Darwin, to service all the cattle stations of the Northern Territory and Western Australia. They invested millions in building it, opened it in 1917, closed it in 1920, blaming 'labour indiscipline', and then allowed it to fall into disuse. As a result, by 1927 in what should have been a major Australian primary industry, cattle exports from the Northern Territory were lower than at any time since the turn of the century. What had gone wrong?

Some historians say that the government failed to keep its part of a deal with the Vesteys – that it would build a railway between Alice Springs, in the centre of the Territory, and Darwin, so as to bring cattle to the Vestey meatworks. This not only meant that the plant could not operate profitably, it also forced the Vesteys to send their own cattle to meatworks on the Queensland coast. The only way to do this was by droving them in mobs of 1,000 to 1,500 head.

On these cattle drives, ten miles a day was considered good progress, so they sometimes took three years and became part of the romantic outback legend of Australia, a subject for books and films in which the drover was the archetypal folk hero.

But Australian historian J. H. Kelly offers a different and intriguing explanation. He says that the Vesteys realized that to keep the British Empire together some form of empire trade preference was inevitable. When this happened, tariff-free beef from Australia would sell in Britain at lower prices than beef from a non-empire country which would have to pay duties. So the Vesteys' most profitable beef industry, that in South America, would not be able to compete with Australian beef.

But if the Vesteys had not built a meatworks in Darwin, some other entrepreneur might have done so. Or a future Labor government might build a State-owned plant. So by investing £1 million to build the meatworks *and then not run it*, the Vesteys stopped Australian beef from competing with South American and, at the same time, made sure that no one else could compete with them in Australia.

How much of all this did the unions know? They were already angry at the region 'being handed over as a vast fiefdom to foreign concerns', so the Australian Workers' Union, a militant one covering many miscellaneous trades, sent an official to Darwin to confront Gilruth. He was Harold Nelson, a nuggety engine driver from Queensland, father of five children, and a formidable organizer. Gilruth and the Vesteys got a taste of his tactics when he waited until the Vesteys had contracted 10,000 head of cattle for the first killing season at the meatworks, then called out on strike for higher pay all the Sydney tradesmen brought in to finish building the plant in time. The Vesteys had no choice but to pay up.

When Gilruth, anxious to raise revenue to finance his fight against the unions, nationalized the supply and sale of all forms of alcohol in the Territory, Nelson organized a union protest. First he boycotted all Gilruth's nationalized pubs. Then, realizing that thirsty workers might weaken, he persuaded the waterside workers not to unload liquor from ships, and the railwaymen not to carry it on the local railway system. When Gilruth still refused to budge, Nelson pulled out all the workers at the Vestey plant. In a 16 month period, Nelson called 19 separate strikes, a demonstration of union solidarity that worried not only Gilruth but employers throughout Australia.

Gilruth called for non-union labour to break the impasse and recruited men from the local Greek community to unload beer and whisky from a ship which Nelson's unionists had refused to touch. As the Greeks left the wharves, the Australians attacked them and a full-scale riot broke out, in which knives, bottles and staves were used. The outcome was inconclusive, so Nelson and the Greek leader met later that day and agreed to settle the matter once and for all with a pre-arranged riot the following Saturday afternoon.

The Greek leader recruited reinforcements from fellow countrymen working as fishermen and pearlers. The Australians brought in miners and buffalo shooters from outside Darwin. It promised to be the riot of the year. Word reached Aboriginal settlements and soon hundreds of them began arriving from miles away and Chinese shopkeepers set up chairs at vantage points to watch the action.

But on the day, the Cable Guard, a local militia created to protect the telegraph relay station to London, took up positions as the two sides approached each other down the main street and, back-to-back and with fixed bayonets, kept the Greeks and the Australians apart. The riot was postponed.

Nelson went to see the Greek leader the next day and lectured him on worker solidarity. He must have been convincing because on the following Monday the Greeks not only agreed not to unload the ship but they made a large cash donation to help care for sick and wounded Australian soldiers.

With the Vestey slaughterhouse due to open soon, Gilruth had to back down. He took defeat badly and a strange exchange of roles between him and Nelson began to take place. Gilruth appeared less and less in public, Nelson more and more. Nelson and the union now ran not only the industrial life of the town but the social life as well, organizing concerts, balls and conferences. And when the Vestey slaughterhouse finally opened in April 1917, an event that should have been a triumph for Gilruth, it was Nelson who organized the celebrations. A thousand people were invited, a special train chartered, and a brass band played as the guests drank rum punch and ate savouries. Nelson made the speech and was then presented with a cheque for £120 by his grateful union members. Gilruth did not even attend.

But the newly opened plant was dogged by strikes and stop-work meetings and at one stage was only working one hour a day. On Gilruth's advice, the Vesteys cut short the killing season and closed the meatworks. Still Nelson did not let go. When the first refrigerated ship ever to call at Darwin arrived to pick up what meat the Vestey plant had processed, the waterside workers refused to load it unless they were paid an extra threepence (6c) an hour to compensate them for this new type of work. Since the cargo was worth about £1 million in Britain and virtually nothing in Darwin, the Vesteys agreed to pay.

There was a truce until the following season. Early in 1918, the Vesteys opened negotiations with Nelson for a trouble-free killing season with a guarantee of no strikes. Nelson said the workers would agree to that in return for a twenty per cent increase in wages. On Gilruth's advice, the Vesteys' reply was to cancel the killing season in its entirety. There was now no work at all.

Nelson had lost face and if Gilruth had left it at that, the outcome might have been different. But the barmaids employed in the state-run pubs asked him for a few hours off on 11 November to celebrate the end of the war. He refused, and when they took time off anyway he locked them out, demanding that they publicly apologise to him. And he chose this moment to announce an increase in the price of beer.

Nelson called a mass meeting of protest on Darwin Oval and in a fiery speech attacking Gilruth, got carried away. 'Why should we lie down under the tyranny of a despot,' he shouted. 'Let us rise and end this reign of despotism.' Whether he meant this literally or not we will never know, but on 16 December, led by Nelson, about a thousand men, soon soaked by monsoon rain, marched on Government House. Gilruth had been expecting them and had sworn in 27 civil servants as special constables. They were armed with only two revolvers between them, but all had batons.

Gilruth agreed to admit a deputation to negotiate. Nelson said all the workers wanted was for Gilruth to address them and justify his five years in office. Gilruth agreed to do so and came out onto the verandah of Government House, impeccably attired in a black woollen suit, stiff collar and tie. He began to speak in a quiet, cultured voice but the drumming of the rain drowned him out. The crowd pressed forward to try to hear him, and the fence began to give way. Nelson had a rush of blood to his head and shouted, 'Over the fence, boys.' The mob swarmed across the lawn, overpowering the special constables, as Gilruth fled back into Government House and locked himself in.

Nelson now reversed course. He calmed the workers down, said that they had made their point and shown that they were serious. There was nothing to be gained by violence and they should go home. But as word of the attack on Government House spread, Darwin took a festive yet revolutionary air. A second crowd gathered, found a brass band and marched past Government House to the Darwin Oval where it passed a motion calling on the Federal Government to dismiss Gilruth.

When news of all this reached the south, it appeared alarming. Sydney papers wrote of 'Another Bolshevik Revolution' and, ignoring Nelson's role, said it had been fomented by Russian émigrés. The Federal Government thought it had better act but, uncertain what to do, settled on the tried and tested British tactic – it sent a gunboat. HMAS *Encounter*, a warship with eleven six-inch guns, sailed into Darwin Harbour, picked up Gilruth secretly at night and sailed out again.

Meanwhile, Nelson had set up a 'Reform Committee' and, as if to confirm the convicton of alarmists in the south that the Russian Revolution had already arrived in Australia, the committee began a reign of terror, calling men out on strike, sacking others, imposing levies and enforcing its will by beatings, abuse and intimidation.

All small unions were now amalgamated into one, 'The Big Union', representing all workers in the north. When the Federal government appointed Gilruth's deputy to replace him as Administrator, the 'Big Union' refused to accept him and passed a motion demanding that the three remaining members of Gilruth's administration should leave the Northern Territory by the first boat. Otherwise there would be violence. All three agreed and sailed on the SS *Bambra*. The rebellion was over, the unionists had won. Never before had Australian workers known such power.

But in the end everyone lost. A judge appointed to investigate the rebellion found that Gilruth was temperamentally unfit to administer the Territory. Gilruth refused to defend himself until 1936 when he started work on a book telling his side of the story. He dropped dead after writing 18 pages. Nelson opted for a political career and was elected to Federal Parliament in 1922 and kept his seat until 1934. But Northern Territory members in those years did not have a vote so he was able to wield little power. The Vesteys closed the meatworks in 1920. The buildings soon deteriorated, Aboriginals camped in its offices and the only cattle that passed through its slaughterhouses were strays looking for food.

CHAPTER 7

BLACK AUSTRALIA

'[The white officials] grabbed us and put us in the back of a truck. As the truck left Phillip Creek everyone was crying and screaming. I remember mothers beating their heads with sticks and rocks. They were bleeding. They threw dirt over themselves. We were all crying in the truck, too. I remember seeing the mothers chasing the truck screaming and crying. And then they disappeared in the dust of the truck.'

– Lorna Cubillo, 1947

The Australian desire for progress, for egalitarianism, for a better life and 'a fair go' for all, did not extend to one important group, the original inhabitants of Australia, the Aboriginals. Defeated in the Aboriginal Wars of the nineteenth century, they had become a forgotten race – reviled, murdered, harassed, discriminated against, and subjected to cruel and unusual punishments. It remains one of the mysteries of history that Australia was able to get away with a racist policy that included segregation and dispossession and bordered on slavery and genocide, practices unknown in the civilized world in the first half of the twentieth century until Nazi Germany turned on the Jews in the 1930s.

There are striking parallels. In *The Mirror Maker*, Primo Levi sums up German excuses for the Holocaust: '[We] did nothing but adapt ourselves to a horrendous but already established procedure – an "Asiatic" practice composed of slaughter, mass deportation, ruthless exile to hostile regions, torture and the separation of families. Our innovation was purely technological: we invented the gas chamber.'

The figure of six million Jews killed during the Holocaust has been reached by comparing the known Jewish population of Europe and the German-occupied territories of the East before the war with the known population of the same areas after the war. No such comparison can be made for the Australian Aboriginals for two reasons.

First, the size of the Aboriginal population when the first white settlers arrived is only an estimate. Next, Aboriginals were not included in census-taking until 1967, when there were disputes over the definition of an 'Aborigine'. So any figure for the numbers of Aboriginals killed by white settlers in the wars and massacres of the nineteenth and twentieth centuries can only be intelligent guesswork. But experts I have consulted say that 50,000 would not be an exaggeration. It could be as high as 100,000. Given the small size of the Aboriginal population, the loss of even 50,000 of its people was devastating.

The killing began early on. In December 1790, after Aboriginals had speared one of his servants, Governor Phillip decided on a punitive raid on the offending tribe 'in order to convince them of our superiority, and infuse an universal terror'. He ordered Captain Watkin Tench to take fifty men and capture two Aboriginals and to kill and cut off the heads of ten others. Tench convinced him to reduce the number to be captured to six, of which two would be hanged and four deported to Norfolk Island. But if none could be captured alive, then all six would be shot and beheaded.

This was not government policy. But as writer Padraic P. McGuinness asked 200 years later, 'What did the colonial authorities think would happen when they populated Australia initially with a mix of criminals and other desperates, especially those such as the Irish who themselves had a history of oppression and dispossession . . . In Australia, the poor and ignorant settlers were forced into close relations with the Aborigines, upon whose goodwill they depended while having no understanding of how they thought and no chance of comprehending them ever. So the most elemental accommodations of rape, abduction and violence were the only means of getting on.'

In many states in the early days these settlers cleared Aboriginals from their land as casually as kangaroos. They shot them, poisoned them and clubbed them. In Tasmania they succeeded in wiping them out entirely. In the rest of the country over a period of 150 years the Aboriginal population declined from an estimated 300,000 to about 75,000. Smallpox, tuberculosis and malnutrition took their toll, but many were murdered by white settlers.

If the Aboriginals hit back, then they were punished with a reprisal raid disproportionate – as in Governor Phillip's case – to the offence. A settler in Queensland described how a raiding party of Native Police in the 1880s would carry out such a raid. 'A white man, an "officer and a gentleman", at the head of half a dozen black murderers, watches a camp of blacks all night. The cool dawn of the morning comes, and the slender smoke circles up among the trees by the waterhole as the unsuspecting blacks wake to prepare their morning meal. Suddenly a shrill whistle, then the sharp rattle of Sniders, shriek on shriek, rushing to and fro: then ammunition gone, the struggle at close quarters, and well-fed lusty savages, drunk with carnage, hewing down men, women and children before them.'

It is amazing that Australians managed to keep from the rest of the world the fact that they were massacring the Aboriginals. In 1968 anthropologist William Stanner gave a series of lectures, 'The Great Australian Silence', in which he took to task Australian historians for showing no interest in what had happened to the Aboriginals. And it was not really until 1998, with books like *This Whispering in Our Hearts*, by historian Henry Reynolds, and the Australian Broadcasting Corporation's TV series *Frontier*, that the silence was broken.

Reynolds took his title from a legal argument defending the right of white settlers to dispossess the Aboriginals by any means they chose. The argument was prepared in 1842 by a young settler, a barrister called Richard Windeyer, who had the courage at the end of his brilliantly reasoned case to question his own logic. 'But how is it that our minds are not satisfied?' he asked. 'What means this whispering in the bottom of our hearts?'

The puzzling point is that Australia was a Christian nation, yet most of its Christian leaders either encouraged – if not the murder – certainly the dispossession of Aboriginals. At the very least they proved indifferent to their fate. The few who spoke out were reviled, attacked, treated as deranged, had their careers ruined, or were driven into exile. The general attitude was: how dare they criticise the way Australians treat the Aboriginals? The white settlers come from a superior race, the Aboriginals from a weak, useless one. 'Is there room for both of us here?' an outback farmer wrote to the weekly *Queenslander* in 1880. 'No. Then the sooner the weaker is wiped out the better, as we may save some valuable lives by the process.'

Those who protested at views like this understood only too well the process of rationalization that made murder and dispossession possible. George Robinson, the Chief Protector of Aborigines at Port Phillip, Victoria, had made it clear in his writings that Aboriginals had a strong and clear view on land possession. He told of meeting an Aboriginal elder along the banks of the Murray River in South Australia. The elder had stamped on the ground and had exclaimed: 'Belonging to me. Belonging to me. My country.' The lawyer Windeyer would have none of this. Aboriginals had no right to the land – ownership belonged to the person who first bestowed labour upon it.

But defamation and demonization had turned Aboriginals into non-humans, the property of white landowners to dispose of at will, like slaves in the southern states of America, accounted for along with the animals when a sale took place. 'Three questions were always put by the buyer, of which the third was the vital one,' wrote goldfields warden William Owen in 1933. 'How many acres? How many miles of fencing? How many niggers? The niggers always went as part of the stock.' If Aboriginals were viewed as less than human, as slaves, as a form of animal, then they could be shot down with impunity. Mary Bennett, the pro-Aborigine feminist, understood human nature only too well: 'The criminal cannot forgive the victim he has wronged.'

Western Australia was probably the worst state for the murder of Aborginals. In 1834 British soldiers under the command of Captain James Stirling, the state's first governor, butchered a sleeping camp of eighty Aboriginal men, women and children from the Nyungar tribe near Pinjarra. Their leader, Yagan, had learned English and tried in vain to explain Aboriginal law and culture to the settlers. Reverend R. M. Lyon, a missionary who befriended Yagan, described him as 'a noble, princely character'. Yagan was eventually shot by an 18-year-old white boy, William Keats, and his head was cut off – a form of mutilation of Aboriginal bodies often practised by whites at that time – smoked to preserve it, and then sent to London as a trophy.

In 1997, a group of Aboriginal elders located the head buried in a Liverpool cemetery and travelled to Britain to collect it. The following year a bronze statue of Yagan was erected in a park in Perth, but some of the early Australian attitude to Aboriginals lives on and the head of the statue was sawn off several times.

The worst recorded massacre was in New South Wales in 1838. Angered at the loss of their land and at the kidnapping of their women by white settlers, the Kamilaroi tribe in northern New South Wales mounted a series of attacks on local farmers and their stock. When the farmers demanded action from the state government, Colonel James Nunn led an expedition of mounted police on a reprisal raid. They encountered a large group of Aboriginals at Snodgrass Swamp and over a period of three days killed over 300 men, women and children – a number unmatched in other recorded massacres of Aboriginals in Australia. Even worse, Snodgrass Creek was then renamed Waterloo Creek, recalling Britain's victory over Napoleon in 1815.

Massacres went on well into the next century and became so common they hardly made news. There were at least two in the 1920s. In the East Kimberley region in July 1926, a boundary rider ordered an Aboriginal called Lumbia and two women to leave the Nulla Nulla cattle station in the Forrest River area. When Lumbia and the women were slow to go, the boundary rider dismounted and began whipping Lumbia with his stockwhip. As the rider remounted, Lumbia hurled a spear at him, puncturing his lung and killing him.

Fellow boundary riders found the body late the next day, partially eaten by predators, and raised the alarm at Wyndham, the nearest town. A party of four whites, two mounted constables, two policemen and seven Aboriginal trackers in police employ left the following day to find Lumbia. However locals who noted that the group took 42 horses and mules and more than 500 rounds of ammunition knew immediately that this was also to be a punitive expedition.

The men moved from camp to camp along the Forrest River for the next week, killing as they went. When they entered a camp they first shot all the dogs, then the men, then the women and children. At one camp the women were chained to trees and forced to watch their menfolk being shot and the bodies burned. Then they were marched for several miles before being shot and burned themselves. Estimates of Aboriginals killed ranged from 20 to 100.

In August two years later, at Coniston station, 140 miles from the small town of Stuart in central Australia, two Aboriginals killed dingo trapper Fred Brooks, claiming that he had taken one of their women and had refused to return her or supply the gifts expected as part of the exchange. Brooks's two Aboriginal assistants reported the killing and a punitive patrol under Mounted Constable George Murray, a Gallipoli veteran, set out to arrest the culprits and punish their tribe, the Waribiri. Over the next three months, Murray's patrols killed between 30 and 70 Aboriginals, including many women and children. To save cartridges, the children were killed by a blow to the back of the neck.

Both massacres turned out to be historically significant. A Royal Commission appointed to enquire into the Forrest River massacre found for the first time in Australian history in favour of the Aboriginal victims and not their murderers. And at the trial of the two men charged with Brooks's murder at Coniston (they were acquitted) the court listened with deep unease as Murray, still a Gallipoli hero to most Australians, testified with what can only be described as nonchalance to killing 17 Aboriginals himself. 'You mowed them down wholesale,' commented the judge.

The trial made news in Britain and the Anti-Slavery Society established a sub-committee to monitor Australia's treatment of Aboriginals. Five years later when the press agitated for a punitive expedition to Arnhem Land to avenge the death of a policeman, public opinion would not stand for it. Coniston thus became the last officially-sanctioned punitive expedition against Aboriginals. How widespread they were can be gathered by what Northern Terrirory anthropologist Dick Kimber told me in 1983: 'Every Aboriginal in this part of the country can state, quite correctly, that at least one of his relatives has been shot by a white man.'

The massacres were over but the war went on in other ways. Since the Aboriginals were dying out, the Australian authorities argued, then those children among them who had some white blood needed to be saved. (How many this was remains a controversial point. It is likely that a substantial proportion of Aboriginals have white blood, just as a lot of Australians have some Aboriginal blood, especially those who come from the outback.)

But how should they be saved? In the United States, Native American children, 'Red Indians', had been forcibly taken from their parents and placed in institutions to 'civilize' them. Australia tried a different approach. In 1937 the Chief Protector of Aboriginals in Western Australia, A. O. Neville, a man generally recognized as a decent, progressive bureaucrat but who nevertheless believed in 'breeding out the colour' (commonly called 'f--king them white'), spoke at the first national governmental conference on Aboriginals, an occasion Robert Manne, Associate Professor of Politics at La Trobe University, Victoria, has described as 'a terrible moment in the history of the twentieth-century Australian state'.

Neville asked: 'Are we going to have a population of one million blacks in the Commonwealth or are we going to merge them into our white community and eventually forget that there were any Aborigines in Australia?' The key resolution at the conference, 'The Destiny of the Race', passed unanimously, called for the total absorption into the white community of all non-full-blood Aborigines. Taking part-Aboriginal children from their mothers and families by force was part of this ambition. Over the years various regulations had been invoked to make this possible.

In 1918, while the war in Europe was still on, the Australian government found time to pass regulations designed to segregate Aboriginals from the white population and reduce the number of children with mixed blood. It was now illegal for a white man to live with an Aboriginal woman. (No mention was made of a white woman living with an Aboriginal man because such a situation was considered unthinkable.) This met the approval of the Perth *Sunday Times*: 'Central Australia's half-caste problem must be tackled boldly and immediately. The greatest danger, experts agree, is that three races will develop in Australia – white, black, and the pathetic, sinister third race which is neither.' Control of all Aboriginal children was removed from their parents and given to government-appointed white superintendents.

This was just another part of a process that lasted from the late nineteenth century until the middle 1960s. So-called 'half-caste' children were seized by the state and placed in institutions where they suffered physical mistreatment and sexual abuse. To this day no one is certain how many were involved but Aboriginal authorities say at least 30,000. Only in the late 1990s did Australians come to learn and understand the effects of this policy and appreciate some of the suffering of what are now known as 'the stolen children', just another aspect of Australia's treatment of Aboriginals, what the Governor-General, Sir William Deane, called 'our legacy of unutterable shame'.

The 1918 law caused no outcry. Government figures released in 1921 suggested that there were only 75,000 Aboriginals left, the lowest figure ever, and that since colonization their ranks had been reduced by nearly 80 per cent. There is doubt that these figures were accurate. In the 1970s, a period of strong Aboriginal activism, many Aboriginal leaders I spoke with said that they had done their own, admittedly limited, census-taking in their own areas and that their figures for the number of Aboriginals suggested that the official figures had been understated by anything from 25 to 50 per cent. But the low official figures enabled the authorities to argue that since Aboriginals were dying out anyway, the new legislation was aimed at easing their passing and finding decent homes for their children, especially those who had some white blood or light-coloured skin.

This cannot be over-emphasized – the Australian government literally kidnapped these children from their parents as a matter of policy. White welfare officers, often supported by police, would descend on Aboriginal camps, round up all the children, separate the ones with light-coloured skin, bundle them into trucks and take them away. If their parents protested they were held at bay by the police.

'My father was a Victorian policeman from 1922 until 1946,' Lang Dean remembered in 1997. 'He would sometimes come off duty and, as was his custom, sit on a stool outside our kitchen and take his helmet off. On occasions he would be crying and sobbing like a child. I would be upset to see such a strong man cry and ask him why. He said he would not tell me as I was too young to understand. What he did say then was, "Son, don't ever be a policeman, it's a dirty job."

'When I was about sixteen years old, he and I were camping on a fishing trip and we were sitting around the camp fire. I had often thought about how Dad cried years ago so I asked him would he tell me the reason. He told me that when he went on duty those mornings his sergeant would would order him to accompany two welfare officers to Cummerangunja, a mission station, to give them bodily protection when they entered the nice clean homes of half-caste people and bodily remove nine-, ten-, eleven- and twelve-year-old children from loving mothers and fathers into commandeered taxis.

'They were then taken to Echuca railway station and sent to the far reaches of New South Wales and Queensland. They were farmed out to service to wealthy businessmen and graziers. No doubt a few were treated well but the rest would be thrown on the human scrap heap when finished with. So that was the reason my father cried on those days.'

Sometimes, to avoid harrowing scenes of parents clinging to the sides of the trucks, and to frustrate attempts to hide the children when the trucks drove into the camp, the authorities resorted to subterfuge. They would fit out the back of a truck with a wire cage and a spring door – like an animal trap. Then they would park the truck a short distance from the camp and lure the children into the cage with sweets scattered on its floor. When enough children were in the cage, they would spring the trapdoor and drive rapidly away.

Aboriginals tried to save their children by blackening their skin so that they did not look half-caste. 'Every morning our people would crush charcoal and mix that with animal fat and smother it all over us, so that when the police came they could see only black children in the distance,' witness No. 681 told the National Inquiry into 'stolen children' (1995–7). 'We were told to be on the alert and, if white people came, to run into the bush, or stand behind the trees as stiff as a poker, or else run behind logs or run into culverts and hide.

'Often white people would come into our camps. And if the Aboriginal group was taken unawares, they would stuff us into flour bags and pretend we weren't there. We were told not to sneeze. We knew if we sneezed and they knew we were in there bundled up, we'd be taken off and away from the area.'

One day in 1956 two Northern Territory welfare officers arrived at the Aboriginal camp at Utopia cattle station, 250 kilometres north-east of Alice Springs. It was their second visit. On the first they had tried to capture a seven-year-old boy, Peter Gunner, from his Aboriginal mother and family. But Peter had been concealed by the elders of the tribe under a pile of blankets and eventually the officers gave up and went away.

This time there was no hiding place. It was ration day – the day when the proprietors of the cattle station handed out tea, sugar, flour and other supplies to the Aboriginals in part payment for their work. In 1999 Peter remembered what happened next. 'I was standing not far from the homestead with other people from my family. I didn't see the welfare blokes and they suddenly grabbed me by the arms and dragged me towards the truck.

'I went mad, screaming to my family to help me, but they didn't move. The women were all crying. They couldn't do nothing about it. They were yelling and talking. My mother was there. She was crying and so were her sisters. I thought the men were taking me away to be killed.'

This feeling of helplessness in the face of white officialdom recurs in all accounts of the stolen children. In 1999 Lorna Cubillo remembered what happened to her and sixteen other children in 1947 at a camp at Phillip Creek in the Northern Territory. 'They grabbed us and put us in the back of a truck. As the truck left Phillip Creek everyone was crying and screaming. I remember mothers beating their heads with sticks and rocks. They were bleeding. They threw dirt over themselves. We were all crying in the truck, too. I remember seeing the mothers chasing the truck from Phillip Creek screaming and crying. And then they disappeared in the dust of the truck.'

Margaret Tucker, in her autobiography, *If Everyone Cared*, published in 1977, explained what it was like for her mother when Margaret and her sister were seized while at school in 1916. 'All the children who had been dismissed must have run home and told their parents what was happening at the school. When I looked out that schoolroom door every Moonahculla Aboriginal mother – some with babies in their arms – and a sprinkling of elderly men were standing in groups.

'Mrs Hill [the white schoolteacher], the tears running down her cheeks, made a valiant attempt to prolong our stay . . . We started to cry again and most of our school mates and mothers too, when our mother, like an angel, came through the schoolroom door . . . I thought: "Everything will be right now. Mum won't let us go." . . . She still had her apron on, and must have run the whole one and a half miles . . . As we hung on to our mother she said fiercely, "They are my children and they are not going away with you." The policeman, who was no doubt doing his duty, patted his handcuffs, which were in a leather case on his belt, and which May and I thought was a revolver. "Mrs Clements," he said, "I'll have to use this if you don't let us take these children now."

'Thinking that the policeman would shoot Mother because she was trying to stop him, we screamed, "We'll go with him, Mum, we'll go." I cannot forget any detail of that moment, it stands out as though it were yesterday. However, the policeman must have had a heart because he allowed my mother to come in the car with us as far as Deniliquin. She had no money, and took nothing with her, only the clothes she had on.

'Then the policeman sprang another shock. He said he had to go to the hospital to pick up Geraldine [the third sister], who was to be taken as well. The horror on my mother's face and her heartbroken cry! . . . All my mother could say was, "Oh, no, not my Baby, please let me have her. I will look after her."

'As that policeman walked up the hospital path to get my little sister, [we] sobbed quietly. Mother got out of the car and stood waiting with a hopeless look. Her tears had run dry, I guess. I thought to myself, I will gladly go if they will only leave Geraldine with Mother.

'"Mrs Clements, you can have your little girl. She left hospital this morning," said the policeman. Mother simply took that policeman's hand and kissed it and said, "Thank you, thank you."

'Then we were taken to the [Deniliquin] police station where the policeman no doubt had to report. Mother followed him thinking she could beg once more for us, only to rush out when she heard the car start up. My last memory of her for many years was her waving pathetically, as we waved back and called out goodbye to her, but we were too far away for her to hear us.

'I heard years later how after watching us go out of her life, she wandered away from the police station three miles along the road leading out of town to the Moonahculla. She was worn out, with no food or money, her apron still on. She wandered off the road to rest in the long grass under a tree. That's where old uncle and aunt found her the next day . . . They found her still moaning and crying. They heard the sounds and thought it was an animal in pain . . . Mother was half demented and ill. They gave her water and tried to feed her, but she couldn't eat. She was not interested in anything for weeks and would not let Geraldine out of her sight . . . At the sight of a policeman's white helmet coming around the bend of the river she would grab her little girl and escape into the bush, as did all the Aboriginal people who had children.'

Other mothers were equally stricken. *Bringing Them Home*, the 1997 report of the Human Rights and Equal Opportunity Commission into stolen children, tells of an Aboriginal woman so ashamed of being unable to prevent her children being taken from her that she carried on her person, until the day she died references testifying to her good character. And of an Aboriginal family which for thirty-two years carried out a ritual mourning ceremony every sunrise and sunset to mark the loss of their daughter.

Where the children were taken depended on how old and how light-skinned they were. (Either way, siblings would not be allowed to stay together because the authorities believed that what they called the 'split the litter' system made the children easier to control.) Some started out in Roman Catholic orphanages where they were well-treated – 'All the kids thought it was one big family. We didn't know what it meant by "parents" because we didn't have parents and we thought those women [the nuns] were our mothers.'

But as they grew older they were moved on to 'homes' run by churches and missionary societies. There they were beaten and sometimes sexually abused. The harrowing evidence to the *Bringing Them Home* inquiry of witness No. 436, with its echoes of Nazi concentration camp practices, justifies the comment of Robert Manne that government policy for Aboriginals at that time was 'the most shameful act in twentieth-century Australia'.

'When you got to a certain age, like I got to ten years old, they just told us we were going on a train trip,' witness No. 436 recalled. 'We all lined up with our little ports [school cases] with a Bible inside. That's all that was in the ports, see. We really treasured that [Bible], we thought it was a good thing that we had something. The old man took us from Sydney . . . to the Kinchela Boys' Home. That's when our problems really started.

'This is where we learned that we weren't white. First of all they took you in through these iron gates and took our little ports off us. Stick it in the fire with your little Bible inside. They took us around to a room and shaved our hair off. They gave you your clothes and stamped a number on them. They never called you by your name; they called you by your number. That number was stamped on everything.

'If we answered an attendant back, we were sent "up the line". Now I don't know if you can imagine, seventy-nine boys punching the hell out of you, just knuckling you. Even your brother, your cousin. They had to, if they didn't do it they were sent up the line. When the boys who had broken ribs or broken noses – they'd have to pick you up and carry you though to the last bloke. Now that didn't happen once, that happened every day. Before I went to Kinchela, they used to use the cat-o'-nine-tails on the boys instead of being sent up the line. This was in the thirties and early forties.

'We had a manager who was sent to prison because he was doing it to a lot of the boys, sexual abuse. Nothing was done. There was a Pommie bloke that was doing it . . . Every six months you were dressed up. Oh mate! You were done up beautiful – white shirt. The welfare used to come up, the main bloke, the superintendent to check the home out – every six months. We were prisoners from when we were born . . . Even today they have our file number so we're still prisoners, you know. And we'll always be prisoners while our files are in the archives.'

Some of the stolen children did not have even a short spell of reasonable life but went straight from the Aboriginal camp from which they had been abducted into so-called 'half-caste homes' in Darwin or Alice Springs. The aim was to keep them segregated from local 'full bloods'. Conditions in these homes were deplorable. One report in 1927 reads in part: 'The building is not only too small but is very much out of repair . . . the floor is rotten . . . the shower is out of order . . . In the kitchen, the stove is unfit for use and the sink, together with the table, collapsed, causing all the slops to fall on the floor.' In one case 76 babies, children and young adults were living in a cottage suitable for a single family. Several children had to sleep on the floor. There were not enough knives and forks to go around and the food was often unfit for human consumption.

At Alice Springs, the half-caste home, 'The Bungalow', consisted of a very rough frame of wood with some dilapidated sheets of corrugated iron thrown over it. A clergyman who visited it said, 'The children . . . lie on the floor . . . The rations scale has been deplorable . . . The whole place makes me boil that such a thing can be tolerated in a Christian country.'

The Prime Minister, Stanley Bruce, thought something should be done about the half-caste homes and made an approach to the South Australian government to try to persuade it to help. He wrote that there were 'a number of quadroons and octoroons' at the Alice Springs home, who could 'hardly be distinguished from ordinary white children'. He suggested: 'If these babies were removed at their present early age . . . to homes in South Australia, they would not know in later life that they had Aboriginal blood and would probably be absorbed into the white population and become useful citizens.'

The South Australian government was having none of this. It replied: 'To give effect to this suggestion would be greatly to the disadvantage of South Australia . . . These persons of Aboriginal blood almost invariably mate with the lowest class of whites and, in many cases, the girls become prostitutes.'

Robert Manne found in the National Archives of Australia the views of Dr Cecil Cook, the Chief Protector of Aborigines in the Northern Territory between 1927 and 1939 and the architect of Aboriginal policy there. Manne wrote in the *Sydney Morning Herald* in 1999, 'No one endowed the sorry business of child removal with a grander social and geopolitical purpose than . . . the progressive intellectual, Dr Cecil Cook.'

During his protectorship only 3,000 Europeans settled in the Northern Territory. They were heavily outnumbered by 18,000 'full-blood' Aboriginals and 800 'half-castes'. Cook decided that the half-castes were a menace to White Australia. 'Anxious brooding on birth rates convinced him that in one or two generations the territory's half-castes might outnumber whites,' Manne wrote. 'He was also concerned that some of these half-castes were the result of sexual encounters between Aboriginals and those of "alien" coloured blood. His nightmare vision was the emergence of a sinister new hybrid race in the north [of Australia] brought about by the intermixing of Aboriginals, Asiatics, Pacific Islanders and low-grade Europeans – "multi-coloured humanity" he once called it – which would in 50 years overwhelm white civilization in the Northern Territory.'

What could be done about this? Cook's plans set out in chilling detail what should happen. The police would continue with the policy of removing half-caste children from their mothers and putting them into homes. There they would be kept well away from 'full-bloods', even relatives. The girls would be given a rudimentary education, trained in domestic work, and about the age of fourteen, found jobs as servants in respectable white homes. The boys would be trained to work as farm hands and then placed on cattle stations where they would be paid a full 'white's wage' so as to distinguish them from 'full bloods' doing the same work, who were paid only in accommodation and food.

(A story told in Australian anthropological circles has it that a researcher on a Vestey cattle station in the Northern Territory listened to an Aboriginal stockman describe an epic overland cattle drive that had lasted many months but had been completed with minimum loss of cattle. The researcher asked whether the Vestey manager had shown his appreciation by giving the Aboriginal some form of bonus. The Aboriginal was puzzled until the word 'bonus' was explained to him, and then his face lit up. 'Yes, boss,' he said. 'They gave me a bag of lollies.')

Manne says that Cook was concerned not only with the social but also the biological future of half-castes. The powers conferred on him by the government gave him absolute control over their lives. He refused permission for half-castes and full-bloods to marry and became notorious in the Northern Territory for arranging some sixty marriages between half-caste girls and white men.

'Cook's thinking was fashioned by the fashionable pseudo-science of eugenics, which taught the virtues of state-engineered human breeding programmes. Cook believed that if the state encouraged marriages between half-caste females and white males, eventually, over four or five generations, the stain of Aboriginal blood could be bred out altogether.' Cook felt that the chances of 'breeding out the colour' were good. He believed that Aboriginals were the remote ancestors of Caucasians or Aryans, not of any Negroid race, and that therefore a systematic breeding programme with whites would eliminate the darker blood with no danger of a 'biological throwback' – two apparently white parents producing a black baby.

The end of the Second World War, fought for freedom and the dignity of man, made no difference to Australia's policy of trying to breed out the Aboriginal bloodline. Compassion would not be allowed to stand in the way of progress. Even those officials who accepted that Aboriginals had normal human feelings, went about their distasteful work with the thought that although the Aboriginals might not realize it, 'it's for their own good'.

It is clear from what they said at the time that the government officials responsible for this inhumane policy saw nothing fundamentally wrong with it. Many were decent, family men themselves. But such was their deep, ingrained racism that it never occurred to them that the grief of an Aboriginal mother over the loss of her child could outweigh the prospective benefit that would accrue to the child from being brought up in a white environment.

Even Paul Hasluck, a senior Cabinet minister between 1951 and 1969, a much-admired writer and historian of note, author of *Black Australians*, a history of Aboriginal-white relations in nineteenth-century Western Australia, showed a harsh side to his character when confronted with the policy of forcible removal of Aboriginal children. In mid-1949, the United Nations Association began to take an interest in the policy. The Administrator of the Northern Territory, A. R. Driver, admitted the policy was 'at variance with the complete ideals of the Universal Declaration of Human Rights' but insisted that it was justified.

Later that year, a senior patrol officer, Ted Evans, flew from Darwin to Wave Hill Station, the giant cattle property in the Northern Territory owned by the Vestey family, and took away five half-caste children. On his return to Darwin, Evans set down what he really felt about his job. 'The removal of the children was accompanied by distressing scenes the like of which I wish never to experience again.' The contents of the report reached an old friend of the Aboriginals, Dr Charles Duguid, who attacked government policy in a speech in Adelaide in October 1951, saying that the removal of children was 'the most hated task of every patrol officer'.

After reading reports of Duguid's speech, Hasluck, then Minister for Territories, asked for more information about child removal. A report from Darwin began: 'Aboriginals are human beings with the same basic affections that we have and the Aboriginal mother has a real love for her children, especially those of a tender age. We cannot expect the normal Aboriginal mother to appreciate the reason why her part-Aboriginal child should be taken from her.'

The report pointed out that there were in place certain checks against abuse of the policy of child removal – written permission had to be obtained from the Director of Native Affairs, who would take into account whether the removal of the child was in its best interests. The administrator at Darwin had recommended that no child under the age of four should be removed, except where the question of danger arose. Hasluck read this and disagreed. 'No age limit should be stated,' he directed. 'The younger the child is at the time of the removal, the better for the child.'

The white settlers in Australia did their best to stamp out the country's original inhabitants. They fought a war against them as ferocious as the Indian wars in America, if not on the same scale. Aboriginals held their own against the flintlock because it could be fired and reloaded only three times a minute, leaving an interval long enough for an Aboriginal warrior to hurl several spears. But when the Terry breech-loading rifle and then the Snider came along, the Aboriginals had no chance and war changed to massacre in which neither women nor children were spared. These massacres continued well into the twentieth century. The one at Coniston took place in 1928 but there may well have been others later that have not yet come to light. Why was there not, then or now, the same fascination with Australia in Europe – even if only in fiction – as there was with the conquest of the American West? (The Norwegian Rudolph Muus wrote more than five hundred westerns, the Frenchman George Fronval, more than six hundred.) The answer is that novels need heroes and there was no way any author could turn the white man's murderous conquest of Australia into an heroic epic.

In parallel with the murder of Aboriginals – but, as we have seen, continuing for much longer – ran a government policy of 'breeding out' Aboriginal blood. Full-blood Aboriginals were dying out anyway, the theory ran, but we must do something about the 'half-castes', those born – in most cases – of a white father and an Aboriginal mother. The answer was to take such children, as young as possible, from their Aboriginal mothers, bring them up in a white environment and then marry them off to white people. In this way the Aboriginal strain of blood would be bred out in a few generations. It did not matter if the Aboriginal mother were a good one and objected to her child being taken. This was a legally approved policy in the child's interest, so force could be used if necessary.

How much Australia in general knew of this theft of children is a disputed point. It happened in remote places, parts of the country that most metropolitan Australians never visited and cared little about. I grew up in Sydney while it was happening and knew nothing of it. I worked as a journalist in country New South Wales and then in Melbourne and Sydney and knew nothing of it. As far as I can remember, neither did any of my colleagues in these places. But then we never discussed Aboriginals and most of us had never met one.

There has to be more to it than this. Where were the historians, the writers, the film-makers, the social workers, the lawyers, the clergy, the campaigning journalists? Yes, they were there for the Aboriginals in the eighties and nineties. But when the stolen generation was being dragged from its family, kicking, screaming and crying – and this went on right up into the 1970s – they were nowhere to be seen.

The feminist Mary Bennett said Australia's aim was 'the extermination of the unhappy native race'. She went on, 'This policy is euphemistically described by Australian officialdom as the absorption of the native race and the breeding out of colour. We shall be better able to evaluate this policy when another race applies it to ourselves as the absorption of the white race and the breeding out of white people.' Why did more Australians not see this?

The answer can be found in a letter, written on behalf of the Prime Minister, John Howard, explaining why the Australian government had refused to apologize to the stolen generation. 'Such an apology could imply that present generations are in some way responsible and accountable for the actions of earlier generations, actions that were sanctioned by the laws at the time, and that were believed to be in the best interests of the children concerned.'

Neither excuse stands up. Stealing a generation of Aboriginal children from their families may have been legal but it was also immoral. Some of the white Australians involved in the theft may have believed that they were acting in the best interests of the children – the 'good intentions' defence – but the policy itself was not concerned with the interests of the children. It was concerned with 'breeding out' a race, that is, eliminating it.

Lawyers can argue about the definition of genocide but for power and simplicity it is hard to beat Hannah Arendt. Writing in *Eichmann in Jerusalem*, she said that genocide was the desire to make a distinct people disappear from the earth. That was the shameful desire of the Australian government for at least sixty years. At the time of writing (1999) the furthest the Australian government has been prepared to go has been to express its 'deep and sincere regret to Aboriginals . . . for injustices under the practices of past generations'. But it remains the intention of the present government not to say sorry or to offer any compensation for those practices.

CHAPTER 8

WHITE AUSTRALIA

'I didn't have a clue what was happening. What does an eight-year-old know? They packed a little bag for me and put me on a coach to Southampton. Then we spent six weeks on a boat. When we arrived someone said we were in Australia, but it could have been anywhere. When we got to the [Perth] orphanage I thought I was back at the same one.'

– Eileen Ashby, one of the thousands of British children sent to Australia under a forced emigration scheme designed to boost the country's 'white stock'

The 1920s saw Australia moving ahead, slowly and perhaps with less zest than a young nation should exhibit. Its population was still less than six million but in an age of technological progress Australia was not a backwater.

The Council of Scientific and Industrial Research, the plans for which had been around since the First World War, was given a proper start in 1926 by the conservative government of Stanley Bruce. Its job was to encourage that capacity for invention which had made Australia a pioneer in farming machinery, goldmining, wheat growing, and sheep raising. This later expanded to include industrial research.

Electric power kept pace with the growth of suburbs in the capital cities. But, a curious reflection on Australians' sense of priorities, while many of these suburban bungalows boasted electric kettles, radiators, toasters, irons and refrigerators, there was no sewerage system. Instead an evil-smelling nightcart, manned by cheerful 'dunny men', called once or twice a week to take away from an outside toilet – often inhabited by poisonous spiders and occasionally maggots – a full pan of excreta and replace it with a newly-tarred empty one.

In some parts of urban Australia, the dunny man was still at work in 1998. Les Sternbeck, who was one for 25 years, said in 1999: 'You got used to it after a while. And you got to meet some interesting people. You see, although I called very early in the morning I still surprised some people on the throne.

I don't know who got the bigger fright, but they were all very nice about it and apologized for keeping me waiting.'

Work started in 1923 on the Sydney Harbour Bridge, which would link the south side of Sydney with the rapidly expanding northern areas. The 'father' of the bridge was John Bradfield, a civil engineer working for the NSW Public Works Department, a visionary who saw as early as 1912 that Sydney would need a bridge, an underground railway system in the city and an electrified one in the suburbs. Like many men of vision, he was frequently frustrated by politicians who were always claiming that there was no money for such grandiose schemes. They are gone and forgotten. Bradfield, who died in 1943, not only lived to see his bridge become an international landmark for Sydney but is remembered by the Bradfield Highway – the name of the approach road and main bridge section.

But it was aviators who were the Australian heroes of this period; Charles Kingsford Smith, Charles Ulm, Hudson Fysh, Keith and Ross Smith, Bert Hinkler, R. Parer and J. McIntosh, and the Englishwoman Amy Johnson, pushed forward the frontiers of flying over the vast areas of ocean that cut off Australia from the rest of the world.

Amy Johnson stopped off to refuel at Charleville in south-western Queensland during her celebrated flight from England to Australia in 1930. There was only one place to stay in Charleville, the Corones Hotel, run by 'Poppa' Corones, the first Greek hotel licensee in Australia. Amy, already known as 'the darling of the skies', wanted to celebrate that the worst and most dangerous part of her flight was over and asked for a champagne bath. It took sixteen bottles to get a reasonable level in the tub and local legend has it that 'Poppa', renowned for 'making a prisoner of every penny', not only managed to rebottle the champagne when Amy had gone, but somehow finished up with seventeen bottles.

His hotel had other aviation connections. It was in a bar at the Corones that Hudson Fysh, Fergus McMaster and Paul McGinnes discussed forming an airline to service outback Queensland. By road from Longreach to Charleville took a week but only five hours by plane. Wealthy Queensland pastoralists put up £100,000 and the airline, formed in November 1920, was called the Queensland and Northern Territory Aerial Service, or QANTAS for short, the only word in the English language in which the 'Q' is *not* followed by a 'U'.

And it is said that it was 'Poppa' Corones who proposed that the airline's first planes were named after Greek mythological figures – Hippomenes, Atlanta, Perseus, Hermes and Pegasus – which suggests that Australians in the 1920s found the classics more acceptable than today, when Qantas planes are named after Australian towns and cities. In return for his suggestions, Corones got the

Qantas catering contract. While the aircraft was refuelling at Charleville, his staff would rush on board with food from his hotel, white table linen and silver service.

Welfare benefits for Australians continued to improve. In 1922 Queensland became the first state to introduce unemployment payments, earning itself the nickname 'the loafers' state', because Australians in work have never really approved of governments giving money to those without a job. Also in that year Queensland abolished capital punishment. Three years later the Queensland Arbitration Court reduced the working week from 48 hours to 44 and two years after that New South Wales introduced widows' pensions – £1 a week and ten shillings for each child under 14. It followed this a year later with child endowment, under which every mother, irrespective of the family's income, would receive five shillings a week for each child under the age of 14.

Women were beginning to claim their place in society. In Western Australia's state election in 1921, Edith Cowan became the first woman member of parliament in Australia. She introduced and saw into law the Women's Legal Status Act which enabled women to become barristers and solicitors. (By 1999 women law students outnumbered men in many Australian universities.) In 1925 another West Australian, May Holman, won a seat at the state election and became the first woman labour parliamentarian in the world.

The postwar years brought a gnawing worry in some circles in both Australia and Britain that the Empire was not heading in the right direction. Yes, it was the mightiest the world had known and it had brought a much admired system of law and government to many countries around the world.

But most of the inhabitants of the British Empire were not white but black, brown and yellow. In Britain there was nothing as crude as the *Bulletin*'s complaint about 'the Queen's nigger Empire'. But there were men of power in London who felt that the white countries of the Empire should consolidate and extend their influence. No one was prepared to say outright that there should be two levels of Empire membership – white and coloured – so they concentrated instead on the factor common to the white countries of the Empire: they were English-speaking.

Their principal forum was the English-Speaking Union which boasted among its members Winston Churchill, the newspaper proprietor Lord Northcliffe, and the American broadcaster, journalist and propagandist Lowell Thomas. The Union wanted to emphasize the common heritage of English-speaking peoples, draw the countries closer together and forge a shared sense of destiny. It was the Union, for example, which encouraged British impresario Percy Burton, to produce and take on a worldwide tour Lowell Thomas's 'Lawrence of Arabia', a lecture and lantern show, seen by four million people. The Union wanted to promote Lawrence as an old-style British hero, but also as a representative of the new, benevolent British imperialism.

The Australian Prime Minister, Billy Hughes, decided that to fulfill its destiny Australia needed thousands and thousands of British migrants, even though a pre-war immigration scheme had been a failure. The Australian government had thought to attract British migrants by promising them agricultural land, but on arrival many discovered that the land allocated to them was small and unyielding. Of the 237,000 British migrants who went to the state of Victoria between 1901 and 1914, 182,000 later left. Hughes was prepared to try again. At the Paris Peace Conference he had achieved international endorsement of the White Australia Policy. The next step was to flood the country with hardworking families from 'home' and, in particular, young men to replace those who had been killed in the war.

But not everyone thought like Hughes and the English-Speaking Union. Lord Leverhulme, of Unilever renown, believed that the White Australia Policy was holding Australia back. 'If Australia's wonderful resources are to be developed, the introduction of native labour to perform the donkey work is essential,' he said. 'The dangers of foreign labour are very much exaggerated, and if careful selection is made there need be no fear that the native would ever be more than a beast of burden.' American prosperity had been built on black labour, a similar system would serve Australia well and 'Negroes' would not need to be paid the same wage as was presently given to white workers.

Australia moved quickly to pre-empt such disturbing ideas. Early in 1919, it launched a scheme designed to attract British ex-servicemen. Then in 1921 the first batch of boys arrived from the Dr Barnardo's Homes (refuges for destitute children founded by Dr Thomas Barnardo in the nineteenth century) and the first Barnardo girls came two years later. In 1924 prominent farming and business people in Australia and Britain launched the Big Brother Movement to encourage British youth to migrate to Australia as farm workers (known as 'little brothers') prepared to accept the moral guidance of an Australian 'big brother'. The scheme was promoted through the New Settlers League of Australia and the theory behind it was that younger migrants were more adaptable than older ones and would meet the need for farm labour while preserving White Australia. At the same time everything possible was done to persuade individual British families to come to Australia. 'Australia is the land of the better chance,' the High Commissioner in London, Sir Joseph Cook, said. 'There's sunshine and laughter in Australia.'

There was for some. Others found themselves in the middle of a nightmare. The West Australian government had agreed to take 75,000 migrants from Britain. It promised them enough land to run their own farms, tools and advice to get started, a three-day holiday after they got off the ship at Freemantle, and transport to their new home. The homes turned out to be tin sheds; they were expected to build a proper house themselves. The land was covered with trees and had to be cleared. The tools were an axe and a pick. 'I asked an official where the lavatory was,' said one British woman. 'He handed me a shovel.'

Many who had enough money packed up and went straight back to Britain. Others tried to make a success of their new life. But there were no roads, no neighbours, no shops, no pubs, no doctors, no hospitals. There were flies, mosquitoes and deadly spiders. At the end of two years, one in three of the British migrants had simply walked away from their land. (They were not 'Pommy whingers', because a similar scheme for Australians resulted in an even higher failure rate, with more than four out of ten Australians giving up.) However, those who stuck at it did transform their lives.

Manning Clark told of a woman who convinced her husband to sell everything they possessed, including her engagement and wedding rings, to start a new life in Albany, south-western Australia. The first year was a misery. But then, suddenly, 'Cows were in milk, pigs slaughtered for meat, vegetables grew, crops were sown, schools and churches were built. The government planned a railway line, and to build roads and bridges to allow motor cars to come to the district. The wireless reduced the isolation and the loneliness, while feeding the restlessness.'

The success or failure of this 'block scheme' was quickly apparent. But what happened to some of the Barnardo boys and girls and some of those who took part in the Big Brother movement and other similar schemes – many of which continued until well after the Second World War – did not come to light until seventy years later. In essence, Britain press-ganged and deported between 80,000 and 100,000 of its own children, often without their parents' consent. In the homes and institutions they went to in Australia, some were physically and sexually abused. When they grew older many were falsely told that they were orphans. Only in 1998 did the British government acknowledge that Britain had sanctioned this 'forced emigration scheme'. It set up a Commons Select Committee to investigate and the Health Secretary, Frank Dobson, promised the committee would consider compensation and an apology. The committee heard evidence that the main motive for the scheme, in the dark days after the two world wars, was to give the children the chance of a better life. But a further motive was the determination to populate white Commonwealth countries, like Australia, with more 'pure white stock'.

What this coldly-calculated official scheme meant in human terms emerged in 1998–9 as some of the deportees told their stories in Australia and in Britain. Matthew Dalton had been evacuated during the war and in 1945, aged six, he was in the Sisters of Nazareth orphanage in Swansea. When his mother came looking for him the nuns told her he had been lost in the evacuation. 'She accepted it,' he said in 1998. 'Why wouldn't she believe nuns?' Two years later he was sent to Australia.

John Hennessey was sent to a Christian Brothers orphanage in Perth when he was ten. 'When the boat got to Perth, brothers and sisters were separated. I'll never forget their screams. When we turned sixteen we were sent off to farms to work and told not to return to the orphanage. We had no identity, no birth certificate, nothing. It was as if we didn't exist. In all my years there I never received a single cuddle.'

The Commons committee, which travelled to Australia to take evidence from former child deportees, formally accused the Christian Brothers order of lying and trying to cover its tracks over allegations of sexual abuse. It reported, 'We cannot accept this. In some cases a criminal investigation may be called for.' It said it wanted to press the Australian government to suspend the state law in Western Australia which prevents prosecutions for offences that took place more than six years earlier. Witnesses described the Christian Brothers to the committee as 'the Christian Buggers' and the Sisters of Mercy as the 'Sisters Without Mercy'.

Eileen Ashby was eight when she was sent from an orphanage in Cheltenham, Gloucestershire to the Sisters of Nazareth House at Geraldton, Western Australia. She remembered in 1997: 'I didn't have a clue what was happening. What does an eight-year-old know? They packed a little bag for me and put me on a coach to Southampton. Then we spent six weeks on a boat. When we arrived someone said we were in Australia, but it could have been anywhere. When we got to the orphanage I thought I was back at the same one.'

The nuns told her that her parents were dead and it was not until she was eighteen and had left the Geraldton orphanage that she went back to try to trace her roots. 'They had no record of where I had come from except a birth certificate with the wrong name on it. It wasn't until six years ago that I found my mother and nine brothers and sisters in Ireland. Mum had been sent to England when she was pregnant and left me there as a baby. She felt terribly guilty because she thought I was still there. She died before I could get back to see her. I missed out the beginning and I missed out the end. As a child I was bitter about everyone who had mums and dads and I wanted to know why nobody wanted to have anything to do with me. I felt I had been robbed of a family life. No one has ever really said sorry to me.'

Few of the one-time child deportees blame Australia for what happened to them. In fact most of them are strong patriots – a group of forty who came to Britain to find their roots, on a trip organized as a gesture of goodwill by their Australian orphanage, marched into the arrivals hall at Heathrow singing the

Australian national anthem, 'Advance Australia Fair'. Mary Cooper, originally from Newcastle-upon-Tyne said, 'Australia is a beautiful land, I have had a great marriage and I have a wonderful family. Life was hard at times but there are hard times in everyone's life. If I did it all over again, I wouldn't change it.'

Many of the Barnardo's children made good in Australia and were undoubtedly materially better off than if they had remained in Britain. In response to criticism in the British press, Michael Jarman, director of child care at Barnardo's, wrote to the London *Sunday Times*: 'At the time, Barnardo's believed that it was providing these young people with an enormous opportunity, as well as expanding its capacity to help more children in Britain. This idea is greatly at odds with our emphasis today on working to keep families together. We cannot take away the hurt and resentment of those who feel badly about how they were treated. We can, however, seek to alleviate some of their distress. That is a responsibility Barnardo's had readily accepted.'

Others were less forgiving. Gerald Denley, a Coventry man who knew a young boy sent from a British orphanage to Perth, managed to trace him in 1998. 'All his life he had thought he was a nobody but I managed to find out where he was born and get his birth certificate and his parents' marriage certificate. To him I am now his father and grandfather to his three children. When I asked him if he was in favour of a monarchy or a republic, he said, "The bastards threw me out. Australia is now my country and I vote for a republic."'

On to this Australian scene on 23 January 1929 at 5.25 p.m. came the author himself. I was born at home, at 8 Newcombe Street, Sans Souci, a southern suburb of Sydney on the shores of Botany Bay. In attendance were the local doctor and a midwife. Home was a red brick bungalow with a red tile roof. It had a bedroom, a lounge room, a living room, a bathroom with a chip heater, two verandahs, a garage and an outside lavatory. It had a small front garden and a large back one in which my mother kept chickens and grew tomatoes.

At the bottom of Newcombe Street, the main highway, Rocky Point Road, ran in a more or less straight line all the way – with a few name changes – to Central Railway station in the heart of Sydney. A steam tram trundled along the middle of Rocky Point Road to the nearest railway station, Kogarah, where a train service, electrified thanks to Sydney engineer John Bradfield, took you into the city centre. Also at the bottom of Newcombe Street was a small shopping centre called 'The Loop', presumably because that was where the steam tram looped around from Rocky Point Road on the last leg of its run down to Sans Souci proper, where, if you were interested in travelling further south, a punt would take you across the Georges River.

The Loop had a grocer's shop, a delicatessen (roast rabbits 6d each), a fruit and vegetable shop run by Italians (in other suburbs, fruit and vegetable shops run by Australians sometimes carried the xenophobic slogan 'Shop here before the day goes' – a poor pun on the old English description of Mediterranean people), a draper's shop, a chemist, a butcher's shop, a hay and corn store, an ice-works, a bread shop (with the bakery at the rear) and a Greek milk bar which sold milk shakes, soft drinks, ice-cream, sweets and chocolates. Milk, fresh bread, and ice (for the household ice-chest, an essential item in pre-refrigerator days) were delivered by a horse-drawn cart. Similar suburbs existed in their thousands all over Australia in the late 1920s.

So did similar families. My ancestor James Phillip Knightley, a Protestant, left Clapham, south London, for Australia in 1863. He settled in Parkes, western New South Wales, where he married Martha McGuinness, the Roman Catholic daughter of an Irish convict. His son, another James, married Mary Dodd, the Australian-born daughter of a Lancashire Protestant who had married an Irish Catholic. By the time my father, Phillip James, came along in 1901, the eldest of a family of thirteen, the Knightleys did not know if the family was English or Irish, Catholic or Protestant, so the girls were brought up Catholics and the boys Protestants – except for Uncle Keith, who looked distinctly Aboriginal and refused to go to any church. My mother's side were Protestants from Kent. Her father, William Iggleden, arrived at the turn of the century, a music hall entertainer who had lost a lung from tuberculosis and hoped the warmer climate of Australia would help him keep the remaining one.

My father was a signwriter, my mother worked as a clerk in a gas company, but after she married my father she never went to work again. Since the dream of every Australian couple in those days was to own their own house, they lived frugally. Breakfast was the leftovers from the previous night's dinner, fried up with an egg: 'bubble and squeak'. Luncheon was sandwiches. Dinner was a serving of meat, usually mutton, which was cheap, and two vegetables. Chicken was a luxury, reserved for special occasions like Christmas. My mother baked her own cakes and scones, and made most of her own dresses. My father repaired his own shoes. If they wanted something badly, they put it on the 'lay-by' – the shop put the object aside for you until you had paid for it by weekly instalments. The shop then handed it over. Hire purchase, with its gratification before saving, was considered, if not sinful, then imprudent.

My father did have one financial weakness – he bought for a song water-colours and oil paintings of Australian bush scenes. The lounge rooms of friends and neighbours displayed prints of English landscapes under snow; ours had originals of gum trees, sheep, mountains and lazy Australian rivers. My father had never heard of Tom Roberts, the Australian painter who had limped back from Britain in 1923 having failed to make it in London and was at the time being savaged by the local critics for his scenes of the outback that today justify his title 'the father of Australian landscape painting'. But my father would have supported him in spirit. Australia needed Australian painters, who painted what they saw around them, not Down Under Englishmen who painted chocolate box images of 'Home'.

My parents had two seats permanently reserved at the local cinema for every Saturday night. They got good value for money – two films, usually one British and one American, a newsreel and two cartoons. At the interval they each had an ice-cream. They often left a few minutes early because the programme always ended with 'God Save the King' and since my father, an early republican, refused to stand up, my mother wanted to avoid any scenes, especially with ex-servicemen.

They were fond of dancing – the flapper era had just begun – and about once a fortnight went to a local dance hall, where they whirled around to tunes like 'Ain't She Sweet' and 'I Can't Give You Anything But Love, Baby'. The progressive barn dance gave them a chance to meet all the other dancers. The evening always ended with the lights dimmed for 'Goodnight Sweetheart'. The churches thought modern dancing was the gateway to hell, typical of the modern approach to sex and contraception. But by now my parents had stopped going to the local Church of England anyway, and a copy of Marie Stopes's *Married Love* had a prominent place on the bookshelf.

On Sunday afternoons, friends or relatives visited for afternoon tea, or my parents visited them. The menu was always the same – tomato sandwiches, scones, and nutloaf or fruit cake. Once a year they would attend a big family picnic in the Botanic Gardens at which fifty or sixty relatives would gather. Everyone would contribute food, gossip, songs and games. The afternoon would end about sunset with a community singing of 'Till We Meet Again'.

Living simply like this, my parents scraped together enough to get a mortgage to build number 8. It cost £45 for the land and £800 to build the house. When I was born they were eight years into paying off the mortgage, managing so comfortably that they were able to afford a Chevrolet Straight Six sedan. They read about trouble on the New York Stock Exchange, but like most other Australians in 1929, they were not too worried. Australia was God's own country and America was a long way away.

CHAPTER 9

THE DEPRESSION YEARS

From 1931 to 1932 Australia came closer to civil war than at any other time in its history. Armed groups on the right and the left prepared for a showdown. Weapons and supplies were stockpiled, military tactics devised, mobilization timetables worked out, training camps established.

The Depression came late to Australia. It took nearly a year for the 1929 stock market crash in the United States to shake the Australian economy and this made the shock even greater. How could a land that appeared to offer so much now not even be able to feed its people? Martin Flanagan, a journalist, says, 'Whole families were going under. A decade later my father was a PoW of the Japanese on the Thai-Burma railway, but he once said that the Depression was worse.'

That was certainly true in emotional terms. It was a bitter lesson, an end to optimism, to realize that your life was not in your own hands, that hard work, self-reliance, independence, mateship, family solidarity and compassion for your fellows was not enough, and that you and your country could be at the mercy of economic forces that neither you nor your government could control. By the time the Depression was over few people had much faith in anything at all, for how could they know it would not happen again? This loss of faith had a knock-on effect down the generations. The baby boomers, born in the wake of the Second World War to parents who had been young 'battlers' were shoved ahead as no earlier generation had ever been, because their parents were determined that their children would never suffer similar economic or social hardship.

It was bewildering. Could a country where a job was considered a constitutional right really have one million people, in a total workforce of a little over two million, without full-time employment? Men who had been brought up to believe that there was always work somewhere for those prepared to look for it found to their distress that this was no longer so. (In the 1990s Australians who were so ready to criticise workers like wharf labourers for their apparent greed should have considered the legacy of the Depression, when workers lurched from work that was arduous beyond belief to unemployment and near starvation.)

Even those who had a job became worried that they might lose it and took steps to prepare themselves. Dr John Wright remembered in 1999, 'My father had been a successful real estate agent and then lost everything. He took a job delivering bread for a big baker to households all over the wealthy suburbs of Sydney, using a large, brightly-painted cart drawn by a horse. He left for work early in the morning and did not get home until late at night. The big advantage of the job was that he could keep any bread left over from the delivery run. But he must have been worried about the family finances because he had another job in the evenings, a family concern. We all sat around the dining room table and packed something called "Chex" tablets.

'The tablets came in large jars, accompanied by boxes of pamphlets which showed a tree of life. On every branch hung a complaint which the tablets would cure – constipation, arthritis, rheumatism, indigestion, headaches and "women's complaints". We would wrap the pills in rice paper, add a carefully-folded pamphlet and put the lot into a packet. Then on fine afternoons we would deliver the packets to letter boxes, our family group doing each side of the streets until our supplies were exhausted.

'There was a third family business as well – making neckties out of cheap cloth, skimped in both length and width and mostly polka dot in pattern. Years after the Depression was over the ties would turn up in odd places around the house, a reminder of how bad things had been. Yet we were better off than many. I can remember men coming to the door begging for cash or food and offering to do any jobs about the house. Even as a small kid I felt very vulnerable to things I could not understand. We were very frightened by Dad's talk of banks closing their doors on people who were trying to draw out their money. How could this be? I suppose a lot of this stayed with me.'

Men from country districts scraped together the train fare to the state capitals, or 'jumped the rattler' on a goods train in the hope of getting a job in the city. They passed men from the city carrying their swags, going the other way to the country in the hope of finding work there. All over Australia educated men chopped trees for firewood in the outback while skilled farm hands went from door to door in the cities selling mothballs, pins, studs, matches, soap and holy pictures. One of them was Albert Jacka, the first Australian to win the Victoria Cross in the First World War.

Families who could not pay the rent or meet their mortgage repayments were evicted. They fell back on relatives. Two families could live almost as cheaply as one, so they often moved in together, the children sleeping four or five to a room or on outside verandahs. Or they tried to find lodgings. 'I came home from the dentist one day to find all our furniture on the street,' a Melbourne woman recalled forty years later. 'It was for non-payment of rent. It was a very traumatic experience in my childhood because after that we never had a home. We went from boarding house to boarding house and we never knew from one week to the next whether we would have enough money to pay the board.'

Many with no family to turn to gave up looking for work and stood on corners and begged, or – unbelievable for Australia – fossicked through rubbish bins for discarded food. At night they slept in hessian tents erected in parks, on racecourses, sand dunes and beaches, or huddled in caves, shop doorways and empty factories. Every city had its shanty town with primitive shelters made out of hessian or flattened oil tins, lacking sanitation and infested with vermin. But a form of collectivism prevailed. The shanty town in Sydney, known as 'Happy Valley', was on the shores of Botany Bay. 'We'd help the fishermen pull in their nets and they'd give us some of the catch that wouldn't bring much at the market,' an old man remembered in 1978. 'The Chinese market garden would give us their cheaper grades. We shared it all out according to the numbers in the families. There wasn't a lot of ill-health. We were getting plenty of sunshine and plenty of fish and vegetables. He who hath nothing and he who expecteth nothing shall not be disappointed.'

If the unemployed could show that they had been without work for at least fourteen days and had no resources, they could apply to the state government for sustenance or 'susso'. This was usually paid in the form of coupons, collected from the local police station, which could be used to buy meat, groceries and bread from special shops.

Being a 'susso' was a demeaning experience for an Australian man. First he had to admit to himself that he could no longer support his family. Then he had to queue at the Government Labour Exchange to register as unemployed. These queues often stretched for blocks and filled the streets. It could take days to get inside the exchange, there to be mocked by public servants. Then there was the humiliation of using the special shops and thus revealing to your neighbours that your family was on sustenance. Many would do anything to avoid this.

A miner and his family from Cessnock coalfields moved to a beach near Newcastle and pitched a tent there. Every day father, mother and two children would take turns to go scavenging for food. They found a local baker who would give them loaves of bread too stale to sell – 'as hard as bloody stone' – and a dairy farmer who usually had a pint or two of milk that he did not need. The other family members spent the day prising oysters from the rocks and catching fish. They turned out to be among the lucky ones – they saw out the Depression on a diet of fish and oyster chowder.

This might not have impressed a nutritionist but it was better than the food than many a family on 'susso' managed to find. Some desperate parents sought admission to hospital for their children and said that the sustenance allowance of from five shillings to seven shillings for each person each week – the amount varied from state to state – was insufficient to avoid malnutrition over a long period. Men who saw their children wasting away became angry at the authorities and every day there were scuffles in the streets of Sydney between the unemployed and the police. Sometimes they turned into riots.

The unemployed could not understand that newspapers and politicians seemed to blame them for not working. The *Daily Telegraph* said the unemployed needed to be taught 'a deeper sense of individual responsibility' so that they would learn to stand on their own two feet. Many with jobs resented the arrival in their districts of travelling groups of men seeking work, fearing that they would undercut their wages. In Cairns, northern Queensland, locals attacked a camp of unemployed at the showground, injuring several and forcing the others to leave. Politicians mouthed platitudes – they knew that the unemployed were having 'a rough spin', but they must realize that the government did not have 'a bottomless purse', and they would just have to learn to 'make ends meet'.

In city after city, the unemployed took to the streets to show what they thought of empty rhetoric. In Adelaide, one thousand men and women carrying placards and banners marched on the Treasury to protest against the withdrawal of beef from the government ration issue. When they attempted to storm the building with iron bars and sticks, the police waded into them with batons. The riot, the worst in Adelaide's history, lasted an hour and left the streets bloodstained and strewn with the injured. In Sydney, four hundred unemployed rushed the side gate to Parliament House, to try to get inside to make their protest to MPs. In Melbourne, the Chief Commissioner of Police admitted that when his mounted troopers confronted demonstrations by the unemployed he put his telescope to his blind eye. But it was the anti-eviction riots in working-class suburbs that were the most bloody.

On 19 June 1931, forty policemen arrived at a house in Newtown, an inner Sydney suburb not far from Sydney University. They were there to serve an eviction notice on an unemployed tenant for non-payment of rent. Serving an eviction notice was normally the work of one policeman, but after the Depression started and the number of evictions jumped, several groups sprang up to protect tenants. One of the most militant of these was the Communist-led Unemployed Workers' Movement. The police had been tipped off that the

Movement had fortified the house with barbed wire and sandbags, so they turned out in force.

A pitched battle followed. Watched by hundreds of bystanders, the police stormed the barricades to be met with a hail of rocks and bottles. When they finally got into the house, hand-to-hand fighting with iron bars, chairs and batons followed. Finally the police drew their revolvers and opened fire. A bystander was hit by a stray bullet but the firing had its effect and the Movement members surrendered and were dragged off to waiting patrol wagons. The clash was widely reported and more anti-eviction riots erupted in all the cities.

The Federal government, a Labor one, was at a loss. What should it do? Wages had been cut and then cut again. The Australian pound had been devalued. But banks were collapsing, even state-backed ones. The Government Savings Bank of New South Wales, with 1.3 million depositors and the second largest savings bank in the world, had to suspend trading in April 1931 when withdrawals for the year exceeded deposits by more than £6 million.

The Prime Minister, James Scullin, and his Treasurer, Ted Theodore, turned to London. They asked the British Treasury for help and on its recommendation they invited a director of the Bank of England, Sir Otto Niemeyer, to come to Australia and advise them on a course of action.

It would be difficult to imagine a worse choice. Niemeyer was an establishment economist, educated at St Paul's School and Balliol College, Oxford, who had risen rapidly in the Treasury. As well as being a director of the Bank of England he was the chairman of the financial committee of the League of Nations and a director of the Banque de Pays de l'Europe. He had all the beliefs and prejudices of a traditional City of London banker. It was not surprising, then, that the announcement of his visit caused apprehension in those sections of the press sympathetic to Labor.

The *Labor Daily* said the Federal government was 'being bluffed, well and truly, into handing over our present and our future into the clutch of the foreign Jews'. Australia should beware because the English capitalists were trying to reduce the wages of the Australian worker to the level of the lowest Hindu or coolie Chinese. The *Australian Worker* recalled Britain's debt to Australian troops in the war. 'Our workers spilt their blood to save England's war lords. Now they must go hungry to fatten England's money lords.'

Niemeyer lived up to everyone's expecations. He arrived in Australia in July 1930 after a pleasant sea voyage. He dined at the Melbourne Club, played golf at the Metropolitan Club, went to the races at Flemington, and drove up into the hills of Ferntree Gully. The cream of Melbourne society jostled to entertain him. He squeezed in some business appointments – he saw the Under-Treasurer of Victoria, a couple of businessmen, and the heads of four major banks. He was polite and affable but an exchange with the Speaker in the House of Representatives, Norman Makin, revealed what he thought of Australia's standing in the Imperial relationship. Makin said, 'I hope you're finding your visit satisfactory.' Niemeyer replied, 'That depends on whether you do as you're told.'

In a little over three weeks Niemeyer was ready to tell Australia what was wrong with it and what it must do to put things right. Oh yes, and he also had an idea or two of what was wrong with the Australian character and the Australian way of life. Australians had a natural optimism and this was very bad for them. The first thing that needed to be done was to strip the ordinary Australian of his belief that something would always turn up. He had to face the harsh fact that he was living beyond his means. Living standards were too high. The solution was to cut wages and cut them again. The government should set the example by stopping borrowing and cutting spending. All this and more Niemeyer put to a conference in Melbourne on 21 August 1930 attended by the Prime Minister and the state Premiers and their treasurers.

Yes, sir, they said. You're undoubtedly correct. We will raise no more loans overseas. We will balance our budgets. We will cut the wages of our public servants. We will undertake no new public works. With your help and that of the Bank of England we will pull through and continue to pay interest to the London bondholders. When news of what the conference had decided reached the public, radicals were outraged. The *Australian Worker* commented: 'It made one hang one's head in shame to witness the ignominious spectacle of Australia's leaders sitting like a class of schoolboys to be lectured by an emissary of British moneylenders and told how they should govern their own land.'

Niemeyer set out to tour the country to promote his terms to save Australia, the so-called 'Melbourne Agreement'. At a civic reception in Adelaide he said it was absolutely essential that Australians met their interest payments to the London bondholders. But the very same paper that reported his remarks, the *Evening News*, also carried an item about starving children. 'Mr W. Hughson, organizer of the Granville Relief Society, said there were six thousand unemployed in the Granville electorate and the amount of distress was appalling. A relief worker had heard a little girl of five sobbing bitterly in the street and he investigated and found that the only food the girl and her unemployed father had had that day was four small biscuits.'

Meanwhile the Australian press was doing Niemeyer's work for him. The *Sun* published an editorial saying that three English banks – the Bank of Australasia, the Union Bank, and the ES and A Bank – could make Australia bankrupt and would do so unless Australians balanced their budgets. This was cabled to London where it not only damaged Australia's credit but drove down Australian stocks on the London Exchange.

One politician was determined to reject 'Niemeyerism' – Jack Lang, the 'Big

Fella', the charismatic Labor leader in New South Wales. In the state election campaign that September (1930) Lang stood on a platform that had only one main point in its manifesto – the Australian standard of living. He told election meetings packed with cheering unemployed – 30,000 people turned out at a meeting in Waverley – that Labor would do anything necessary to defend that standard. He referred to Sir Otto Niemeyer as 'Sir Oracle Otto'. Sir Oracle had said costs in Australia had to come down. 'Not a Labor premier has protested. No one has offered the suggestion that costs in India are low, but that does not make the Indian coolies prosperous, well fed or healthy. No one has asked him if his purpose is to make all Australians live like the coolies of Asia.'

Jack Lang had his own solutions. Firstly, he felt that Britain had exploited Australia. Australia's war debt had been incurred in fighting for the British Empire in the First World War. Britain had had the cheek to bill Australia for every penny it spent on the upkeep of Australian troops from their day of enlistment until the day they were discharged. All the other Dominions had been relieved of these costs from the moment their soldiers had embarked for war service overseas. Further, the British government had been willing to write off many of the war debts owed to it by foreign countries but continued to hound Australia for payment in full.

'Italy's war debt of £400 million was compounded for a token payment of £4 million a year for the next sixty years. France has failed to pay up £150 million of Treasury Bills. Britain is paying the United States only three per cent. But Australia is being charged five per cent. The United States has reduced the interest rate from four and a half per cent to three per cent. Why shouldn't Britain do the same for Australia instead of sending a bailiff here?'

Support for this view came from an unexpected quarter in Britain. While the *Telegraph* called Lang 'the mad dog Premier of New South Wales', Lord Rothermere of the *Daily Mail*, then the newspaper with the largest circulation in the world, thought that Lang had a point. So he cabled all the leading newspaper editors in Australia: 'I am entirely in favour of the cancellation of all war debts between the constituent members of the British Commonwealth. I believe by this elimination a great step forward will have been taken towards the extinction of all war debts between the Allies and their associates who fought together in the Great War.'

Encouraged, Lang pushed forward with his proposal that Australia's loans should be renegotiated with a longer repayment period, at a reduced interest rate. The Depression meant that governments could no longer afford to pay both wages and interest rates, so they had to choose between them. He felt that it was his government's duty to see that no Australian starved to death because the banking system had failed. 'People are much more interested in getting enough for breakfast than worried about how their credit rating is in Threadneedle Street.'

Lang and Labor swept to power. At Labor celebration parties around the state, supporters roared out the hit tune 'Happy Days Are Here Again'. Lang told them that it was a victory for those Australians determined not to allow the British to treat Australia like a colony. He announced that his government was no longer going to pay a single penny in interest to London until the bondholders agreed to reschedule the loans.

If this had been purely a New South Wales matter then the outcome might have been different, but the bonds had been underwritten by the Commonwealth government, and if New South Wales or any other state defaulted on an interest payment, then the Commonwealth was obliged to collect it on the bondholders' behalf. Labor left-wingers set about stirring things up. Jock Garden, the Scots migrant who was one of the founders of the Australian Communist Party, and some of his colleagues at the Trades Hall in Sydney announced that they wanted a policy of 'repudiate everything', or 'pay nothing to the Kingdom of Shylock'. In London, newspapers failed to appreciate that this was not official Labor policy, and scare headlines further damaged Australia's standing. The London *Daily Telegraph* said: NSW AND FINANCIAL CRISIS: LABOR DEMAND FOR WAR DEBT REPUDIATION: NIEMEYER ECONOMY SCHEME REJECTED: EXPULSION OF ITS SUPPORTERS THREATENED. On 21 March 1931, the Westminster Bank in London asked the Australian High Commissioner whether the government of New South Wales proposed to make payment of £186,000 due in interest on 1 April. Lang told the High Commissioner to reply, 'No, it does not.'

While Lang was frightening the City of London, his Prime Minister had been creating trouble at Buckingham Palace. James Scullin had come to London to attend the Imperial Conference. But he had another mission to complete – pleading the case for an Australian Governor-General. The Governor-General, the Crown's representative in Australia, had always been an Englishman. Now Labor wanted to put forward Sir Isaac Isaacs, the Australian-born son of Jewish immigrants of non-British origin, who had fought his way from modest beginnings as a schoolteacher in outback Victoria to become a High Court judge – an Australian egalitarian success story.

Scullin, himself the son of a railwayman, soon became aware that the British authorities wanted nothing to do with Isaacs. Even Scullin himself was not trusted. He was anti-Imperialist and was believed in London to have a pro-Irish outlook. Some in Whitehall thought that there was already a Sinn Fein government in Australia and the nomination of Isaacs must be part of a plot to take Australia out of the Empire. Isaacs was not acceptable and that was that. General Birdwood, the commander of the AIF, would be a much better choice and was known to have hopes of the appointment. But the Establishment was outmanoeuvred by Scullin. He invoked the principle, accepted by the Imperial Conference, that the dominions could approach the monarch directly concerning vice-regal appointments.

So Scullin now demanded an audience with King George V. Lord Stamfordham, the King's private secretary, was given the task of fobbing him off. When they met on 11 November 1930, Stamfordham was arrogant and patronizing. By pushing Sir Isaac Isaacs's case so forcefully, he said, the Australian government had put a pistol at His Majesty's head, something no gentleman would do. Scullin now turned the pistol to point at Stamfordham's head. If he was denied an audience or the King did not listen to his advice, he said, he would call an election in Australia on this single issue.

George V was forced to agree to see Scullin and at a 45-minute audience the Australian Prime Minister put the case that Australia was now old enough to have a Governor-General who was one of its own. The King protested, 'We have sent many Governors, Commonwealth and State, and I hope that they have not all been failures.' Of course not, Scullin said, but he repeated his determination to allow the Australian people to decide if the King felt his request was unreasonable. The King said he did not want to be the centre of public controversy. He was a constitutional monarch and therefore, he declared, 'I must, Mr Scullin, accept your advice.'

Next day, Stamfordham wrote to the British Prime Minister, Ramsay MacDonald, explaining the King's decision: 'His Majesty is well aware how easy it is to light and fan the flame of agitation by an ill-disposed minority – especially when, as in this case – constituted of Trade Unions, Communists and Irish, not of the highest class.' The following year Sir Isaac Isaacs became Australia's first native-born Governor-General, thanks almost entirely to Scullin. The King had his revenge. In making the announcement it had been customary to say, 'The King has been pleased to appoint', but this time the King deliberately left out the word 'pleased'. The Victorian *Labor Call* did not care. It celebrated with the words: 'Australians are equal, if not superior, to any imported pooh-bahs.'

However, there were those in Australia unhappy at the whole Isaacs affair. They felt it marked a distinct drawing away from the ties of King and Empire. And it had been brought about by a man of Irish descent, one who had spent part of his time at the Imperial Conference on a visit to his Irish relatives in the village of Ballyscullin. Sectarian fears now added to worries about Labor's attitude to the London bondholders, the growth of the 'repudiationists', and the possibility of economic disgrace. Robert Menzies, a member of the Victorian Parliament, later to play a leading role on the Australian political stage, did not endear himself to the hungry unemployed by telling the nation what he preferred: 'If Australia were to surmount her troubles only by the abandonment of traditional British standards of honesty, justice, fair play and honest endeavour, it would be better for Australia that every citizen within her boundaries should die of starvation within the next twelve months.'

An atmosphere of fear and uncertainty gripped the nation. The prevalence of Communist agitators, the militancy of the unemployed and moral slippage manifested by flappers, the dancing craze, contraception and books on sex, made conservative Australians feel that their way of life was under threat. Labor was an easy target for blame. Labor apparently wanted to change society fundamentally – the New South Wales branch of the party had voted for a Soviet-style five-year plan of socialization. On the coalfields there was a Labour Defence Army organized and drilled by ex-AIF miners. Clashes with the police had become so violent that one MP spoke of 'impending civil war'. There were motions at the state Labor Party conference for the abolition of Empire Day and the custom in schools of saluting the Union Jack every morning, reciting: 'I honour my God, I serve my King, I salute my flag.' British-Christian-White Australia was being destroyed. Something would have to be done about it.

Between 1931 and 1932 Australia came closer to civil war than at any other time in its history. Armed groups on the right and the left prepared for a showdown. Weapons and supplies were stockpiled, military tactics devised, mobilization timetables worked out, training camps established. At one stage it looked as if the Federal government in Canberra was going to mount a *blitzkrieg* against the state government in Sydney. In the city's wealthier suburbs, many a man slept with a revolver under the pillow.

The right believed that Lang, Labor, Sinn Fein, the Communists, the unions, the unemployed and 'the cursed curs from overseas' were planning a revolution. Their fear obliterated their reason. One of the leaders of the right's secret armies wrote that the Lang government in New South Wales was 'dictatorship by a black, brown and brindle autocracy'. Labor leaders took their orders from 'the little yellow men of Shanghai . . . the Pan Pacific Secretariat – the outpost of Moscow . . . governed by the polyglot black, brown and brindle races of the Pacific. Lust, loot and rapine are their doctrines.'

True, there were signs that the workers were arming themselves for battle. Within the labour movement a number of paramilitary organizations sprang up – the Constitutional Guard, the Workers' Defence Corps (WDC), the Australian Labor Army (ALA) and the Ex-Servicemen's Defence Corps. The police believed that the WDC had a direct line to Moscow and was under the control of the Red Army. The truth was that if it had a direct line to any overseas organization it was to Sinn Fein. But one of its local leaders,

'Irish Paddy', who wanted to finance the WDC by bank robberies, was too drunk to move when the only serious operation he planned was due to get under way.

The ALA was potentially a much more serious threat. At a rally in the Sydney Domain on 29 March 1931 a crowd of nearly 40,000 pledged allegiance to it and promised to 'maintain the government in office, to uphold the constitution and to fight if necessary'. This formally identified the ALA with Lang Labor, the state government. The state government had responsibility for maintaining law and order through its police force and if the police were unable to do so, then it could swear in as many special constables as it wished. Historian Dr Andrew Moore says in *The Secret Army and the Premier*: 'While this was the case it was possible that a Labor government could form a special constabulary composed of militant unionists and Trades Hall radicals who would use an occasion of social emergency to carry out rather than subdue social revolution.'

But it was the Right which had mobilized its forces first, and the most potent of these was the Old Guard. Many Australians have heard of the New Guard. But the Old Guard remained a mystery for half a century. Its rules of absolute secrecy about its membership, its division into small cells so that few knew who its real leaders were, and its later destruction of its own records made certain that even though it existed from 1917 well into the 1950s, most Australians even today know little about it or its highly-placed leaders.

It is difficult to pinpoint the creation of the Old Guard. Dr Moore says it seems to have been part of a long-term secret army contingency plan initially conceived during the First World War and terminated around 1952. Or it could have been simply a military extension of the Australian Protection League (see Chapter 6) and the King and Empire Alliance, with many of the same wealthy, powerful and socially-exclusive men, often with military backgrounds, as office-bearers – Philip Goldfinch of the Colonial Sugar Refining Company; Alfred Davidson of the Bank of New South Wales; George Macarthur-Onslow of Camden Park, a member of the famous pioneering family; Sir Samuel Hordern, of the well-known retail firm; Major W. J. R. Scott, a member of the Street family, originally from Britain, whose descendants have filled many high positions in Australian public life.

Its leaders were certainly the pick of Australian society, an ethnic group of their own. They held interlocking directorships of many major Australian companies, lived in the best suburbs, married into each other's families, belonged to the same exclusive clubs and held the same political views. You could even claim that they looked alike – full in face and figure, well-groomed, tailored and barbered. They were fiercely anti-Communist, anti-Labor and anti-working class. They were vehemently Protestant, loyalist, King-and-Empire, Anglo-Australians.

The organizational abilities of the Old Guard's business members were matched by the knowledge of strategy contributed by its military wing – Brigadier General James Heane, Colonel G.C. Somerville, secretary of the Royal Agricultural Society, and Major W. R. Bertram, secretary of the Royal Sydney Golf Club. The 'old-soldier' link made it easy for these officers to make contact with similar organizations in the other states and forge working relationships. The one with Victoria was particularly strong. There, the League of National Security had among its leaders General Sir Cyril Brudenell White, of wartime fame, Colonel Edmund Herring, a future Chief Justice of the Victorian Supreme Court, and General Thomas Blamey, a future field marshall and, more important at the time, the police commissioner.

Clearly the Old Guard was no amateur organization and by early 1932 could boast of a membership of some 30,000. It was particularly strong in rural New South Wales, where it appears to have enrolled some 40 per cent of adult males. It was based on divisions, sections and companies, set on military survey maps. A company was usually five men, which matched the carrying capacity of the average car and made mobilization easier. How did all this go on without the newspapers of the time noticing and reporting it? The *Sydney Morning Herald* referred to the Old Guard on only one occasion between 1930 and 1932, and that in an oblique fashion.

The most obvious answer is that the conservative Australian press felt that the Old Guard's assessment of conditions in Australia made sense. Given the hostility of the banks, the Lang government might well run out of money and have to stop paying sustenance. That could well lead to a mass uprising of the starving unemployed. Communists might take advantage of this to bring about a revolution. So although the newspapers did not openly support the Old Guard, they respected its wish for secrecy and, says Moore, 'maintained an unofficial D-notice on material about [it]'.

Another reason might be that the Old Guard had close links with the police. New South Wales Police Commissioner Walter Childs and the head of the Criminal Investigation Branch, William 'Wee Wullie' MacKay, the giant Glaswegian, had a secret agreement with the Old Guard under which the police turned a blind eye to the Guard's activities and were ready to accept the Guard's support, under police orders, if the situation required it.

No one knew when this might be but rumours kept the Old Guard in a constant state of nervous tension. A Communist uprising was going to take place on International Unemployment Day, 6 March 1931. When the day dawned, the rumours were that the Communists had seized the two halves of the Sydney Harbour Bridge, still not joined. The Red Army was marching on Bendigo. Another army was approaching Mildura, seizing banks, wheat, horses and cattle on the march. Members of the Victorian League of National

Security were digging trenches on the main roads to thwart the Bolshevik advance on Melbourne. None of this was true – except the trenches part, and when the diggers realized the Communists were not coming, they had to fill them in again.

False alarms like this and the failure of the Communist threat to eventuate changed the nature of the Old Guard. By the end of 1931 its strength was greater than the manpower of the New South Wales police force and the Commonwealth armed forces combined, but it had nothing to do. Formed as a defensive organization, it now became more militant, more fascist. Although the Lang government had been democratically elected, its continued existence was not in the country's interest, so why not overthrow it? The Old Guard had the means. Did it have the will? Some of its wilder members thought not and in the middle of February, 1931, they formed a breakaway movement, the New Guard.

Unlike the Old Guard, the new version was not secret. In fact it courted publicity. Under its leader, a Sydney solicitor, Colonel Eric Campbell, it organized a series of publicity stunts to draw public attention and win support. But as an organization it was more Buster Keaton than Richard Hannay. Members from adjoining areas fought a wild brawl, each believing the other to be a communist mob. A detachment of New Guards set off from Sydney on 'Operation Bushfire' to fight a major fire at Cobar, about 500 miles away. By the time the New Guard arrived, the fire was out and the locals, resentful of what they saw as 'city-siders' on a political stunt, stoned their camp and drove them out of town.

The New Guard's best publicity move was to encourage one of its members, Captain Francis de Groot, a cavalry officer and probably an informer for Australian Military Intelligence, to sabotage the official opening of the Sydney Harbour Bridge on 20 March 1932. The bridge was considered one of Jack Lang's triumphs. Although a plan for a bridge across Sydney Harbour had existed since 1815, and Dr Bradfield, the eminent Australian engineer, had interested a British firm in tendering for it in 1924, no one had been able to raise the money – £5 million for the construction and £4.5 million in interest charges. In 1925 Lang had sent the New South Wales Assistant Treasurer, William McKell, and the Under Secretary to the Treasury, Clarrie Chapman, to London where, despite all the odds, they had managed to persuade the Westminster Bank to float a successful loan.

The completion of the bridge during Lang's new term as Premier of New South Wales gave him an opportunity to celebrate this success, and although many – including King George V – felt that such an important landmark should be opened by the King's representative, the New South Wales Governor, Sir

Philip Game, Lang was determined to do it himself. The build-up to the opening ceremony was as dramatic as the conclusion. First, steam engines with tenders loaded with pig iron were parked one after another along the bridge's electrified railway lines, thousands upon thousands of tons of deadweight, to put to the ultimate test the engineer's calculations that the slender steel arms from which the carriageway was suspended would take the strain. Crowds of spectators watched from a safe distance, agog in case the whole bang lot plunged into the harbour; but the steel arms held – the bridge was safe.

A broad silken ribbon was stretched at waist height across the bridge's roadway. Off to the side a wooden dais accommodated the official guests – politicians and their wives, leading Sydney citizens, military figures in splendid uniforms, representatives of the diplomatic corps. Jack Lang, tall and distinguished in a dark three-piece suit with a white shirt and cutaway collar, hat in one hand and scissors in the other, got ready to cut the ribbon.

Suddenly there was the sound of a horse's hooves clattering along the tarmac. Heads turned. Was this part of the ceremony, a theatrical interlude before the official cutting of the ribbon? Along the roadway from the city end of the bridge rode Francis de Groot, sword raised over his head, galloping directly for Lang. For a moment it looked as if one of the many assassination threats against him was about to become a reality. Then de Groot rode straight past Lang and with one professional slash of his sword, cut the ribbon and shouted, 'On behalf of decent and loyal citizens of New South Wales, I now declare this bridge open.' He then meekly surrendered to the police – who had been as stunned by events as everyone else – and was taken away to a mental hospital. The ribbon was replaced, Lang cut it again and the bridge was officially opened.

Ordinary Australians were divided over de Groot. Some felt that his act had damaged Lang's credibility, and when it emerged that de Groot was a member of the New Guard, Lang's supporters used this to argue the organization's danger to democracy. De Groot's statement when the police grabbed him – 'You can't take me. I'm a Commonwealth officer' – was interpreted as a claim that army officers were above state law. Others with no love for any politicians saw the whole thing as a great joke. One Sydney brewery rushed out advertising posters on the bridge's opening with the slogan, 'I'd sooner open a bottle of Toohey's Pilsner any day.'

The New Guard may have been a paper tiger but it set the pace in right-wing rhetoric. On 20 December 1931 – the day after a conservative government under the United Australia Party (UAP) had come to power in Canberra – the New Guard issued an ultimatum. All Communists were to leave New South Wales by 29 February or they would be forcibly removed. Unless Jack Lang vacated the office of Premier of the state by 26 January, Australia Day, the New Guard would remove him by force. Neither the Communists nor Lang took any notice and the New Guard did nothing.

It did not have to, because the UAP, led by J. A. Lyons, comprised a significant number of Old Guard and National Security leaders, and after seeking constitutional and legal advice, the new Federal government itself set about destroying the government of New South Wales. Both sides prepared for confrontation, violent if need be. Horace Nock, president of the Farmers' and Settlers' Association, was later quoted in a rural newspaper, the *Canowindra Star*, as believing, 'Never in Australia's history had any of her States approached so closely to bloodshed and revolution.' He was thought to be exaggerating at the time, but the reality was that the country was on the brink of a civil war between the Commonwealth government and the state of New South Wales.

The Federal government's first move was to rush through Parliament the Financial Agreements Enforcement Act. This empowered the Commonwealth to seize money from the revenue of any defaulting state government. It was as if Prime Minister Lyons had said to Premier Lang: 'OK. If you won't pay the London bondholders the interest you owe them, then we'll take the money from you and pay it on your behalf.'

Lang responded immediately by sending two state officials with an armed police escort to withdraw in cash one million pounds which New South Wales had on deposit – £750,000 with the Bank of New South Wales and £250,000 with the Commercial Banking Company of Sydney. They put the cash in the safes of the State Treasury. Then Lang engaged a small army of unemployed timber workers armed with axe handles to stand 24-hour guard over the Treasury building and authorized the recruitment and swearing in of 25,000 special constables from the State public service. 'If it's a fight they want,' said leading Labor figure and AIF veteran 'Digger' Dunn, 'then we'll give it to them.' The *Labor Daily* warned its readers, 'The war is on. Labor takes up the challenge.'

Deprived of a quick victory, Lyons's next step was to issue a proclamation ordering all New South Wales taxpayers to pay their tax not to the State but to the Commonwealth Bank. But to enforce this, Commonwealth officers would need access to the tax records and these were held in the locked, barred and guarded taxation offices. So the Commonwealth then gave the state income tax commissioner until 12 May to hand over all tax assessment notices or face action for contempt of court.

As Lyons and Lang stood eyeball to eyeball, Lang found support from a very unexpected quarter, one that could have swayed the balance in New South Wales's favour if the legal clash had turned into a military one. The New South Wales police, heavily influenced by MacKay, the notorious union-basher in the past, decided that their loyalty was with the state and Premier

Lang and not the Commonwealth. When news of this reached Canberra there was near panic. In July the previous year, worried about unemployment riots, the Commonwealth government had authorized the Defence Department to provide the New South Wales Police Force with steel helmets and 10,000 rounds of ammunition. This helped turn the New South Wales police into a small but formidable army.

Now, in preparation for the coming battle, large squads of police began drilling in the early mornings at Sydney suburban parks and ovals while the New South Wales Mounted Police practised cavalry manoeuvres in Centennial Park. To show the public where they stood – since the newspapers, until then, had said nothing about the police attitude – 1,500 policemen put on a grand parade, complete with drums and bands, through the streets of Sydney on 29 April, bringing an unheard-of response from the *Labor Daily*. It was, the paper said, 'a march of the Army of Democracy and Decency'.

It was more than that. It was a deliberate show of strength to intimidate the Old Guard. MacKay himself had planned the march route so as to take it past all the offices where the Old Guard leaders were to be found – the Stock Exchange, the CSR, pastoral companies, insurance firms, banks, the Union Club and the Civic Club and the Imperial Service Club. The conservative press now rushed into print to warn the police that their first loyalty was to the King and not to the elected government.

There was a spy from military intelligence mingling with the cheering crowds lining the route of the police march, and although he reported to Canberra that the police 'standard of drill was not very high and the staff work was faulty, especially at the saluting base', the Commonwealth government was not reassured. Troops of the 7th Light Horse were called up and billeted on the outskirts of Canberra to defend Parliament House in case Lang brought the battle to the capital. There was panic when a vehicle flying a red flag and carrying a man in uniform was seen driving through the streets. Had the Red Army arrived? It turned out to be an officer of the Salvation Army.

Tanks were seen rumbling at night through the back streets of Randwick, a Sydney suburb. The Royal Australian Air Force at Richmond, on the outskirts of the city, was ordered to be on standby, ready to adopt a police role in Sydney. All leave in the armed services was stopped. Plans were finalized for armed Navy personnel to guard Commonwealth buildings in Sydney, including the General Post Office, the Commonwealth Bank, radio station 2BL and telephone exchanges. The Newcastle Trades Hall Council passed a resolution that ships' companies should be asked not to take up arms against their class. A declaration of martial law for 9 May was being prepared.

The Old Guard was ready. On 26 February 1932 it finalized its plans for the mobilization of 10,000 of its members. There were to be rallying points at Windsor and Warwick Farm, both about an hour's drive from the centre of Sydney. The signal would be siren blasts from low-flying aircraft along with coded calls on radio stations. Wearing their makeshift uniform – blue patrol jackets with grey flannel trousers, arm bands and tin hats and carrying one day's rations – the men would assemble at designated points. Then private cars and trucks would carry them to Sydney, where they would operate under the Peace Officers' Act.

What was their function? Dr Moore concludes that the Commonwealth government had decided to put an end to the government of New South Wales and would do so by force if necessary. It was going to storm the offices of the State Taxation Office and remove its records. The elite of the Old Guard would be the shock troops and if the New South Wales police or Labor supporters resisted then the rest of the Old Guard and units of the Commonwealth armed forces would overcome them.

But Lang was also ready. 'If the Commonwealth peace officers attempt to usurp the functions of the New South Wales Police Force,' he warned, 'they will find themselves in the same position as any other person or body who attempts to ursurp the duties or functions of the Government of New South Wales.' The New South Wales Police Force was a formidable body and would have been more than a match for the Old Guard, especially if reinforced by Lang's special constables and the axe handles of the timber workers. But the Old Guard would, sooner or later, have had the support of the Commonwealth armed forces, and the two would have prevailed in the long run.

It did not come to that. In the end there was no civil war. That is what it would have been. The Governor, Sir Philip Game, simply sacked Lang and Lang went quietly. Game had followed the crisis with great concern. He was not the assassin Labor made him out to be and he worried constantly about the turmoil swirling around him. Educated at Charterhouse and the Royal Military Academy, Woolwich, he had served in Africa, India and Ireland. His appointment as Governor of New South Wales was meant to be a relaxing, post-retirement, socially-rewarding job. Instead it had turned into a nightmare.

He had been ready to believe all the anti-Lang propaganda he heard when he took up his appointment. But face-to-face with the Premier he had to admit that Lang impressed him. He decided that Lang was not as radical as made out, that he was not working towards 'socialism in our time', that his repudiation of interest payments was not the thin end of a socialist wedge, and that he might even at heart be a democratic constitutionalist. The two men – the servant of British imperialism and the radical Labor thinker – actually got on quite well on a personal level and were courteous and affable with each other.

This did not go unnnoticed among Lang's enemies and one wrote to Game, suggesting that if he wanted to know what conservative Australians thought of the governor's dealings with Lang, 'You should go to the Union club and hear yourself discussed.'

Game was also under pressure from the Sydney business and legal community and his social life began to suffer as some of them boycotted his banquets. There was even a demonstration of sorts against him on New Year's Eve, 1931, when a bus loaded with drunken revellers managed to drive into Government House without being stopped. He sought advice from Sir Philip Street, the Chief Justice of New South Wales. The Auditor-General had told him that Lang's circular instructing state government officials to stop paying government money into banks – in case the Commonwealth then seized it – was illegal. If this were so, should he not dismiss Lang and his government? Street replied, not very helpfully, that such a decision was ultimately for the governor himself to take.

Game appealed to the Dominions Office in London. He said he felt that the Commonwealth and state governments should settle their differences without his intervention. But he did not want to be seen to be condoning illegality. If he did act, would he be usurping the functions of the courts? The Dominions office prepared a reply which would have successfully passed the buck – Governor Game should obtain the advice of his law officers. This reply was never sent because in Sydney matters moved quickly to a head.

On Friday, 13 May, 'Black Friday' to Labor historians, Game put three options to Lang: demonstrate the legality of your order to public servants; withdraw the instruction; or resign. He let Lang know his own feeling, which was that if the only way he could maintain the essential services of the state was by breaking the law, it would be better if he resigned. Lang replied that he would not resign. Game immediately sent him a hand-delivered letter: 'I feel it is my bounden duty to inform you that I cannot retain my present Ministers in office, and that I am seeking other advisers. I must ask you to regard this as final.' Game then called on the Leader of the Opposition, Bertram Stevens, to form a government.

Lang faced a terrible dilemma. There is no doubt that he could have had a quarter of a million supporters on the streets of Sydney within 24 hours – he drew 200,000 to an election meeting a few weeks later. But he knew that the secret armies were poised to mobilize and that even if the New South Wales Police Force remained loyal to him despite his dismissal, the armed services would support the Commonwealth. The result would not be riots but civil war. He called in the reporters. 'Well, I'm sacked,' he said. 'I'm dismissed from office. I'm no longer a Premier but a free man.'

At the election the following month, conservatives made hysterical predictions of what would happen if Lang got back. Some newspapers banned Labor advertisements. A gang of New Guard toughs kidnapped a Lang supporter, a small market gardener, and branded the word RED on his forehead with acid. An anti-Lang propaganda campaign was launched, funded by businessmen. One of its slogans was: 'Can I take a chance that I may vote for a civil war?' The *Daily Telegraph* published a document purportedly found in the Labor rooms of Parliament House. It showed that the 'Lang Secret Service' planned to launch a revolution within forty-eight hours of re-election. It was a forgery.

Some employers threatened their staff with the sack if they voted for Labor and put anti-Lang pamphlets in their pay packets. Major-General Gordon Bennett, who was to make news in the Second World War, went on radio to warn: 'This is an election in which the people are asked to decide if they prefer honesty or dishonesty, confidence or chaos, a British democracy or a Moscow dictatorship.' Predictably, the Lang government was wiped out, retaining only 24 of the 55 seats it had won in 1930.

Most historians laud Lang's decision to leave office without a fight. Manning Clark says that time conferred a mantle of majesty on him. Dr Ross McCullin, the historian of the Labor Party, writes, 'The Big Fella went quietly. He did not want blood to stain the wattle either.' Moore says, 'Lang did not court his own dismissal. He merely stepped back from the precipice of armed struggle, leaving "responsibly" so that a civil war which was not primarily of his making was averted.'

Almost as soon as Lang had gone, the stock exchanges in Sydney and Melbourne rose rapidly and the London money market expressed its faith in Australia. Capital began to return to New South Wales and business prospects improved. There was upbeat talk of the worst of the Depression being over. And although no one wanted to talk about how close the country had come to the brink, there was a clear feeling that 'Aussie common sense' had prevailed, that no matter how violent the rhetoric, Australians were not true revolutionaries, that when it came to the crunch they were not prepared to spill blood in pursuit of a political ideal, no matter how seductive.

Jack Lang lived to be 99 and see most of his enemies depart before him. Although he had been expelled from the Labor Party, he was re-admitted in 1971 on the motion of one of his keenest admirers, Paul Keating, a future Prime Minister. At his funeral four years later, enormous crowds lined the streets of Sydney to watch the procession. He wanted to be remembered as the champion of the people and there is little doubt that he was.

The Old Guard, its job done, melted away, only to re-appear as 'The Association' during the Communist scares of the 1940s and early 1950s. Most of The Association's leaders were the same Old Guard members, greyer and stiffer of limb. The Association seems to have vanished in the mid-1950s although there were rumours of 'Civil Defence Leagues' and secret armies in various parts of Australia as late as 1991. In all probability the Old Guard, in some form or another, is out there still.

CHAPTER 10

BODYLINE

The bodyline tour's significance lies not so much in its sporting aspects but in its social and political ones. It defined a difference in Australian and British attitudes to life and the relationship between the two countries.

Even in the middle of the Depression and divided by political principles, Australians of all types and backgrounds still had one thing that united them – sport. That fine English sports writer, Simon Barnes of *The Times*, has said that a serious author is expected to laugh off sport as being beneath contempt. 'And while such a view is logical, it represents a terrible failure of the imagination. Sport, you see, has the power to seize the imagination: of individuals, of classes, of nations, of races, of the world. And nothing so powerful can be wholly trivial.'

This explains why, on 6 April 1932, Australia went into mourning. The nation's greatest racehorse, the mighty Phar Lap, winner of the Melbourne Cup, thirty-seven wins in forty-one starts, died in his stable in California, where he had gone to show the Yanks what a real horse could do. The post-mortem examination suggested that Phar Lap had been guilty of the same sin as many Australians – over-eating, in his case too much fresh lucerne. Of course, no one in Australia believed it. The envious Americans must have poisoned him. His body was shipped back to Melbourne, stuffed and placed in a glass case in the Museum of Victoria, then later moved to the National Gallery of Victoria, and finally to the new Melbourne Museum.

Never mind, everyone said when they had got over the shock, we still have Don Bradman and we still hold the Ashes and there is no way the English are going to get them back. Don Bradman? Ashes? For the enlightenment of readers from non-cricketing nations, the Ashes are a valueless trophy which England and Australia have competed for since 1882. They are reputed to be the ashes of a bail, burnt and kept in a small urn, like a religious relic, which in a way they are. The two countries had been rivals at cricket since the crew of HMS *Calcutta* played a match in Hyde Park, Sydney, in January 1804. By the 1830s the British Army and the New South Wales Corps were playing regular games. The first white Australian team to play in Britain toured there in 1879 (there had been an Aboriginal one in 1868, a great success) and the first England team to tour Australia came in the summer of 1861–2.

Over the years the Test matches, as they came to be called, grew to be more than just a game. Cricket was an English creation, 'a game which the English, not being a spiritual people, have invented in order to give themselves some conception of eternity', and the English regarded it as a symbol of the nation and its character, a reflection of Britain's power and confidence. One of the country's best-known poets, Sir Henry Newbolt, devoted a whole poem to cricket and its power to mould the man, one known to every English schoolboy, *Vitai Lampada*:

> There's a breathless hush in the close tonight
> Ten to make and a match to win,
> A bumping pitch and a blinding light,
> A hour to play and the last man's in.

The poem encourages an Englishman in sport, war and life to hold firm to the ideals of cricket, to 'play up, play up and play the game'. All English-speaking peoples know and understand the rebuke 'It's not cricket', to mean that something is not fair, decent and above board.

The game itself is complicated, difficult to explain to a non-player and, to anyone not versed in its subtleties, incredibly boring. It can take five, sometimes six, days to play a match which then might well end in a draw. Yet when generations of British Imperial servants took it around the Empire with them, it caught on everywhere. The locals adopted the game as their own and today play it better than the Mother Country. In the 1999 Cricket World Cup, England got knocked out early on and the last six countries were all old British dominions – Australia, India, Pakistan, South Africa, Zimbabwe and New Zealand. But the fact that they still come to Britain to play it suggests that British historian Peter Clarke is right, cricket has proved to be a more lasting legacy for Commonwealth unity than 'anything Joe Chamberlain achieved'.

Don Bradman? Simply the greatest batsman the world has ever seen, a true Australian hero who would undoubtedly have been the country's first president when it eventually, becomes a republic. Sadly, he died in 2001 aged 92. Australia mourned his passing and there were tributes from all over the world. He may have been not only the greatest batsman in the world, but the greatest sportsman. 'No person this century has remotely achieved such prodigious domination in any other sport,' says Frank Keating of the *Guardian*, 'Neither Jones nor Nicklaus at golf; not Laver, Hoad or Navratilova at tennis; not Owens, Lewis or Thompson at athletics, Pelé at football, nor Muhammad Ali nor Ray Robinson at prize fighting.' (The only possible rival to Bradman would be the Canadian ice hockey player, Wayne Gretzky, who retired in 1999 having won every possible statistical record in the game.)

When Bradman was on the critical list in a London hospital after an appendectomy, Buckingham Palace announced that King George V at Balmoral 'was being kept in constant touch with Mr Bradman's progress'. When Nelson Mandela walked into the sunshine after his years in prison in South Africa one of the first questions he asked was 'Is Don Bradman still alive?' Bradman's Test match average was a staggering 99.94 runs, a figure celebrated after Bradman's last Test match in 1948. In the 1970s the then head of the Australian Broadcasting Commission, Talbot Duckmanton, decreed that the address of the ABC would be PO Box 9994 in every capital city, and it remains so today. At the most recent count there are twenty-two streets named after Bradman in Australian towns, and probably the same number in England. He must be Australia's best-known citizen and has admirers all over the world. One English writer, Stephen Fay, summed up the sportsman's view: 'For cricketers what counts is that the man scored nineteen centuries against England and he still makes you think: God help the Poms.'

Bradman came from Cootamundra in south-western New South Wales but grew up in Bowral, a resort town in the southern highlands of the state. He sharpened his eye as a boy by hitting a golf ball against the brick base of the house water tank with a stump. From that he went on to hit the local bowlers all over the local grounds. Then it was Sydney, then to play for New South Wales, then Australia and then to show the world how good he was.

He became a hero to Australians for many reasons – his modesty, the way he shunned publicity. 'I'm sorry,' he was still telling sporting journalists in 1999, 'but if I give you an interview I'll insult a few thousand previous fellows I've refused.' Or was it his style? Commentator John Arlott wrote, 'He stood at the crease perfectly immobile until the ball was on its way, then his steps flowed like quicksilver out of trouble or into position to attack. He could still pull the ball outside the off stump accurately wide of mid-on's hand to avoid a packed off-side field.

'He still played the ball off his back foot past mid-off before that fieldsman could bend to it. He still hit through covers with the grace of a swooping bird. He could cut and glance, drive, hook and pull and he could play unbelievably late in defence. Those who had never seen Bradman until 1948 saw a great batsman; those who knew his batting saw an even new greatness.'

But the real reason he was a hero to Australians – as he was to Indians – was because of what he did to England. The record they most admired is that he scored more runs against England than any other batsman in history – 5,028. The Australian author Thomas Keneally, who was a schoolboy when Bradman was building up this total, wrote of those days, '[Our] only history was European history. Poetry cut out at Tennyson. If we spoke of literary figures, we spoke of Englishmen. Cricket was the great way out of Australian cultural ignominy for, while no Australian had written *Paradise Lost*, Don Bradman had made a hundred before lunch at Lord's.'

There were similar sentiments in India. Vishnan Prabhu, an Indian writer, said in 1996: 'On the very Monday at Lord's in 1930 when Bradman captured the imagination of cricket enthusiasts by scoring 254, Pandit Nehru was arrested and civil disobedience was at its height in India. When Bradman left the scene, Nehru was India's first Prime Minister. In those eighteen years of the Raj, I looked on Bradman as a kind of Avenging Angel, as he reeled off century after century against the common foe – England.'

The Test match series that consolidated Bradman's reputation was the so-called 'Bodyline series' of 1932–3. The series also confirmed Australians' view of the English cricket team, reinforced their existing prejudiced views of the English and added some new ones, revealed English views of Australians that had previously been carefully hidden, changed international cricket forever, and did more to hasten the rupture between Britain and Australia than any other single event.

Australia held the Ashes and England came out for the series coldly determined to win them back. Their chances did not look good. Bradman was in top form. In 1930 he had made the world's highest score in first class cricket – 452 not out in a match at the Sydney Cricket Ground. But the England captain, Douglas Jardine, had a plan. Jardine was the son of an English family which had served in India and, as was the custom at the time, had sent him back to England to boarding school while still a small boy. Lonely and missing his family, he found consolation in cricket. He was an aloof, arrogant man, absolutely convinced of Britain's superiority in everything in life that mattered, but particularly in cricket.

He had thought long and hard about Australian cricketers and believed that he knew their secret flaw. Like Winston Churchill, he thought that Australians came from bad stock and lacked guts and determination when the going was tough. All England had to do was frighten them physically and they would crack. He had two fine fast bowlers, Bill Voce, a big, powerful man, and Harold Larwood, a miner, both from Nottinghamshire. He would let them loose on the Australian batsmen and see what happened.

In a pre-Test warm-up match at the Melbourne Cricket Ground on 19

November 1932, Jardine told Larwood to put Bradman to sword and then went off trout fishing. Bradman came in with Australia 1 for 84. He hit Larwood's first ball for 4, ducked the second and then hit the third for another 4. The crowd loved it. Larwood was no threat to the boy from Bowral. But then Larwood trapped Bradman leg before wicket when he was 36 and bowled him in the second innings for 13. The series promised to be a battle between these two giants of cricket.

By the time the first Test started in Sydney early that December, Jardine had refined his plan, which he called leg theory, or bowling on the line of the body. A British cricket writer, sending a report back to his London paper and anxious to save on cable costs, abbreviated 'line of the body' to 'bodyline' and the name of the ensuing war was born.

The theory of bodyline can be briefly set out. The captain arranges his field with a cordon of five close fieldsmen extending from leg slip to silly mid-on. He places two more men near the long-leg boundary. The bowler then delivers short fast balls aimed straight at the batsman so that they bounce up to his chest and head. Defence against bodyline is difficult. If the batsman moves away from his wicket, he risks being bowled. If he tries to defend himself with his bat, he risks being caught by the leg-side cordon. If he goes for a hook over the top of the cordon, he risks being caught by the two fieldsmen in the deep.

But when Jardine gave his fast bowlers the order to open the bodyline attack in Sydney, Bradman was not there. He was, the team doctor said, in a 'run-down condition' and should rest. So it was left to his team-mate, Stan McCabe, another country-bred boy, to show that bodyline held no terrors for a confident batsman and that Australians were not yellow. McCabe came in with Australia 4 for 87. The first three batsmen had ducked, weaved and bobbed as Larwood and Voce rocketed the ball at the chest and head, frequently striking them. The Sydney crowd, the most vociferous in Australia, had shouted their disapproval but the Englishmen took no notice.

And McCabe, even though he was hit four times, took no notice of the English bowlers. He hooked, drove, cut and pulled the ball at will and was 187 when the innings closed. Old men in Australia still lay claim to having been at the Sydney Cricket Ground on that humid summer's day in 1932 when Stan McCabe showed Harold Larwood what he thought of bodyline bowling.

If Australia hoped that Jardine would now abandon his theory, they did not know their man. On the eve of the third Test in Adelaide he accused the Australian press of having invented the whole bodyline controversy. 'We know nothing about it,' he said. 'The practice is nothing new, and there is nothing dangerous about it. I hope it goes on being successful.'

The match opened in typical Adelaide summer weather. There was a ball-by-ball description of play being broadcast all over Australia. Manning

Clark writes: 'In hotel bars men drank their ice-cold beer as they listened to the scores, and the feats performed by their heroes under a hot sun beating down out of a brassy sky.' The drama began after lunch with the Australian captain Bill Woodfull facing Larwood. Larwood thundered down to the wicket and bowled one of his fastest balls. It landed short, kicked up and hit Woodfull right over the heart. Woodfull collapsed to the ground in agony.

As he did so, Jardine clapped and said to Larwood, 'Well bowled, Harold.' Woodfull got up and faced Larwood again. Jardine brought the leg-side fieldsmen even closer. The crowd was appalled and hostile. Shouts of abuse rang out from around the ground. England were pressing a wounded batsman. Before he was bowled, Woodfull was struck again and again. The Australian batting fell apart. Fingleton was caught behind and Bradman and McCabe went quickly, caught in the leg trap. Jardine's plan was working.

The England manager, P. F. ('Plum') Warner, went into the Australian dressing room to see how Woodfull was. Woodfull gave him short shrift. 'Of two teams out there, one is playing cricket, the other is making no effort to play the game of cricket,' Woodfull said. 'It's too great a game to spoil by the tactics you're adopting. I don't approve of them and I never will. If they are persevered in, it may be better if I do not play the game. The matter is in your hands. I have nothing further to say. Good afternoon.' Warner went back to the England dressing room, sat down, shook his head and, obviously distressed, said, 'He wouldn't speak to me.'

The Advertiser reported 'Sensational Cricket In Third Test Match Yesterday', and when the game resumed on the Monday morning, so did the sensational cricket. Bert Oldfield, the wicketkeeper and a tail-ender, faced Larwood, who bowled him a bouncer that hit him square on the forehead. He retired and was rushed to hospital with a fractured skull. Cricket is full of unwritten conventions and at that time one was that fast bowlers did not deliver bouncers to tail-enders. Was there no limit to what England would do to win back the Ashes? The Australian press thought not: 'The visiting English cricket team has sunk to depths previously unknown by any professional touring side.' Editorials called for 'something to be done'.

Their English counterparts laughed. The *News Chronicle* said Australia should stop complaining and find some batsmen skilled enough to handle the England onslaught. The *Daily Herald* announced that it was appalled by the 'undignified snivelling' of Australians just because England had won the Test. Jardine felt that his assessment of the Australian character had been justified. Labor newspapers complained about these 'sneers and jibes' in the British press and wondered why Australia put up with 'British imperialism' in all its forms.

On 18 January the Australian Cricket Board of Control cabled the Marylebone Cricket Club: 'Bodyline bowling assumed such proportions as to menace the best interests of the game, making the protection of body by batsmen the main consideration. Causing intensely bitter feeling between players as well as injury. In our opinion is unsportsmanlike. Unless stopped at once is likely to upset friendly relations existing between Australia and England.'

The MCC played it cool. It agreed that bodyline had caused controversy but it was up to the Australians to decide whether the tour should continue or not. If it decided to cancel it, then the MCC would bring its players home. This subtle rebuke had the desired effect. The Australian Board said of course it did not want to cancel the series. 'Plum' Warner and Bill Woodfull had a series of chats and made up their disagreement. The Australian Board and the MCC exchanged more cables and the series went on. So did bodyline. Australia had lost the argument.

The fifth and final Test was held in Sydney between 23 and 28 February 1933. Relations between the two teams had been further strained when Jardine complained to Victor Richardson that an Australian player had called him a bastard. This might be an innocuous term in Australia but Jardine assured him it was not in Britain. Richardson summoned the Australian team and while Jardine waited, asked: 'All right, which one of you bastards called this bastard a bastard?'

The Sydney crowd, however, was determined to show that Australians were not snivelling losers and bad sports and when Harold Larwood, a bowler and therefore not expected to make runs, scored 98, the Australians rose to their feet to cheer him as he walked back to the pavillion. England won and regained the Ashes. The MCC and Jardine had accomplished what they had set out to do, but at what cost.

Cricket was never the same again. In the middle of the bodyline controversy a fan wrote to the *Sydney Morning Herald* saying that if bowlers were allowed to target the batsman's body, the game would come to resemble baseball with players wearing padding and helmets. That is exactly what happened. Further, it is now acceptable to bowl bouncers at tail-enders and, ironically, one of the bowlers most prone to do this is an Australian, Glenn McGrath.

England did not abandon bodyline but never used it again in the way Jardine did in Australia that summer of 1932–3. Various reasons are on offer. In September the Australian Board of Control asked the MCC if it would now agree that a direct attack on the batsman by the bowler was against the spirit of the game. Yes, the MCC replied, but making no reference to the recent series, such an attack would indeed be against the spirit of the game. But not to worry. There would be a warm welcome for the goodwill tour of England in 1934 when the game would be played in the same spirit as in previous years.

But there is a more intriguing explanation for the demise of bodyline. It concerns an Australian Rules football player who also played cricket – Laurie Nash. Nash, from Launceston, Tasmania, was the son of a policeman who was sacked after the Melbourne police strike in 1923. From his father he learnt to say exactly what he thought and one of the things he was saying at the time of the bodyline tour was, 'I'm the fastest bowler in the world with the sole exception of Harold Larwood.'

He had results to back his case. Playing for Tasmania against South Africa in the summer of 1931–2 he took three for eleven off five overs, including two wickets in two balls. He missed his hat trick because his third delivery bounced short and broke the batsman's jaw. Selected to play for Australia in the final Test of the series against South Africa that same season, he took four wickets for eighteen runs in the first innings and one for four in the second.

When Jardine's bodyline tactics became apparent, the word in every pub in Australia was that Nash would be called up for Australia to 'give it back to the Poms'. He was not. His fans said it was because Australia's captain, Bill Woodfull, had taken the high moral ground and was determined not to retaliate. There is no doubt that Woodfull knew of Nash's potential. After the bodyline series, Nash's team, South Melbourne, was playing Woodfull's team, Carlton. Nash opened the bowling with a series of deliveries that mirrored Larwood's performance, even down to eventually dropping Woodfull with a ball to the heart, just as Larwood had done.

But he was not chosen for the subsequent 'goodwill tour' of England. Nash himself complained, 'they wouldn't give me a go', implying that the Australian cricket establishment, which was predominantly Protestant and Freemason, did not want a brash, working-class Catholic in its ranks. His relevance here is that he did play one more Test. In 1936–7 he was included in Bradman's team to play the fifth and deciding match against England in Melbourne. According to Nash's biographer, E. A. Wallish, Bradman used the fact that he had Nash in his team to negotiate a secret agreement with England's captain, Gubby Allen, that neither side would bowl bodyline. Australia won.

The bodyline tour's significance lies not so much in its sporting aspects but in its social and political ones. It defined a difference in Australian and British attitudes to life and the relationship between the two countries. Jardine remained an Englishman until his death in 1958 at the early age of 57. Larwood upped and emigrated to Australia, where he lived in the Sydney suburb of Kensington in happy obscurity. He died in 1995.

In 1983 an Australian television drama series, *Bodyline*, reminded an older generation of Australians of how strongly they had felt at the time. Sir Anthony Mason, a former Chief Justice of the High Court of Australia, revealed to a slightly shocked legal profession a secret he had kept during his entire career

– the bodyline tour and England's attitude had turned him as a youngster into an ardent, lifelong republican.

It took an Englishman, however, to assess the real importance of it all. Charles Willliams, a serious biographer (he chose General de Gaulle as his first subject) published a biography of Bradman in 1996. In it he advanced the thesis that Bradman was a pivotal figure in the making of modern Australia and put Bradman's achievements into the context of 'an Australia feeling her way gradually towards something that the world would recognise as nationhood'.

Things began to look up, but jobs were still scarce. Miners at the Kalgoorlie goldfields in Western Australia, worried that foreigners were taking their jobs, held a meeting to decide how to get rid of them. An Italian barman punched a local football hero, who fell, hit his head and died later in hospital.

Over a thousand miners then went on a rampage of violence and destruction, looting, burning and wrecking shops and houses owned by Italians, Yugoslavs and Greeks. A full-scale battle was fought at Dingbat Flat. Both sides used rifles, shotguns, knives, dynamite and jam-tin bombs. Two men, a Montenegrin and an Australian, were killed and six Australians wounded.

This race riot weighed heavily with the government when the first Jewish refugees from Hitler began to seek admittance to Australia. They were not exactly welcomed with open arms. In a statement to a conference on political refugees held at Evian, France, the Australian government said, 'It has been pointed out that the US and Australia owe their development to migrants from the Old World. Such migrations to Australia have been predominantly British and any large departure from this is not desired while British settlers are forthcoming. But realizing the unhappy plight of Austrian and German Jews, Australia has included them on a pro rata basis.'

What exactly this was became evident when the Australian Jewish Welfare Society tried to sponsor about 6,000 European Jews for entry to the country. Permission was granted only to those Jews who had a substantial amount of capital or a trade or profession in which there was a shortage of Australians, so that there would be no risk of the Jewish immigrant taking an Australian's job. There was an additional requirement – the applicant had to be of the 'right type'. This was not spelt out but was felt to be a reference to the White Australia policy. It was all right for the migrant to be Jewish but he or she should not be too swarthy.

In 1938, Sydney celebrated the 150th anniversary of its founding with a re-enacting of Captain Phillip's landing. But, significantly, a meeting of several hundred Aboriginals issued a protest, saying that the date did not mark 150 years of progress but 'a hundred and fifty years of misery and degradation imposed on the original native inhabitants by the white invaders of this country'.

Their protest coincided with the publication of a novel, *Capricornia*, by Xavier Herbert, hailed as 'one of the most important contributions to Australian fiction for a long time'. The novel's major theme was described as 'the injustice meted out by the conquering whites to the Aboriginal and half-castes of the Northern Territory'. White consciences were stirring.

I knew little or nothing of this. I read *Capricornia* a few years later but my parents did not take me to the 150th anniversary celebrations because they were not very interested in Australian history. Neither were my schoolteachers. Australian explorers – or rather British explorers 'discovering' new parts of Australia – were on the curriculum, but only after we had learnt about the Magna Carta. (Years later an Aboriginal told me that explorers on expeditions to 'discover' new parts of Australia would often hire an Aboriginal guide. In the course of taking the white men to where he understood they wanted to go, he had to be careful not to infringe tribal territory. This involved following fault lines, gaps between where one tribe's authority ended and another tribe's began. So early explorer's maps, instead of showing a straight line from A to B, often wandered all over the place as their guide tried desperately to avoid fatally offending his fellow Aboriginals.)

At home we had a comparatively easy Depression – although for the rest of her life my mother kept a tin of silver coins hidden in the loft in case of emergencies, and the family car sat jacked up on blocks in the back garden for a year because we could not afford to run it. Even in Depression times there was work for a skilled signwriter and once my father had rescheduled the mortgage on our house with the new Rural Bank of New South Wales, we managed not only to get by but, as economic conditions improved, to resume our annual holidays, even though we did not go far.

About the middle of each December my parents would pack our 1929 Chevrolet with everything we would need for two weeks and we would head north, cross the Hawkesbury River at Wiseman's Ferry and tackle the winding hills on the other side. This was the moment of terror for Margaret, my little sister. Had Dad remembered the tin of water? If he had forgotten and the Chevvy's radiator boiled, as it frequently did, then (Dad said) we would all have to pee into the radiator, and my sister took this to mean that she would have to perch on the bonnet like a mascot, in full view of any other passing car, a childhood nightmare that would have intrigued Freud.

If all went well, about three hours after leaving Sans Souci – the Chevrolet could hit 45 mph on the straight – we would be pulling into Aunty Polly's front yard at Booker Bay, on the Central Coast. She was not really our aunt.

In fact she was not a relative at all, but in those days of ultra-politeness, my parents insisted that I call all older people aunty or uncle.

Aunty Polly was a large woman with snow-white hair and, like a lot of Australian housewives at that time, she wore shapeless floral frocks and an apron, or pinafore, no matter what the occasion. It was like a uniform and it said, 'You men might have time to go drinking or fishing. We women are always busy with the housework.' In winter she shared the main house – weatherboard with a rusty, corrugated iron roof – with her thin, shy sister. Neither had ever married and their needs were so few that I do not think either had ever worked. What little income they had came in summer from holidaymakers like ourselves who rented the main house while Polly and her sister moved into the garage.

We were an undemanding family. The house had running water stored in corrugated iron tanks that collected rain water from the roof. Some summers when there had not been sufficient rain, the water in the tanks dropped to levels that threatened water rationing. Then we would watch anxiously every day for a 'Southerly Buster', a storm that came suddenly out of the south, bringing high winds, thunder, lightning and heavy rain.

When she bursts o'er baking streets,
Hear the shouts of glee,
From the burning, aching streets
'Here's the southerly'.

To have a hot bath you had to gather an armful of small pieces of wood, stuff them into the chip heater, light them and then gauge the right moment to turn on the cold tap – too soon and the water was lukewarm; too late and the heater exploded. The toilet was in the yard near the garage – most of the time. In those pre-sewerage days, the toilet was a hole in the ground over which sat a small portable shed. When the hole was full, Aunty Polly's handyman moved the shed, filled in the old hole and dug a new one. Part of his job was to ensure that there was a good supply of toilet paper, usually an out-of-date telephone directory or a mail order catalogue, nailed to the door.

Wherever it was in the yard, the toilet was home to maggots, poisonous redback spiders and, in my imagination, snakes. For the first few days I would creep down the yard at night armed with a torch and a large stick, but then I soon did the same as everyone else and borrowed one of Aunty Polly's Victorian chamber pots.

The kitchen had a sink with one tap, a big cast-iron fuel stove, and a beautifully burnished Primus made in Sweden. 'Where's Sweden, Dad?' I can still remember the smell of burning methylated spirits, the thump of the pressure plunger and the pop as the paraffin vapour caught fire. It was as good as an alarm clock and meant that soon last night's left-over vegetables would be sizzling in the fat from last night's mutton chops, the toast would be burning on the fuel stove, and breakfast was around the corner. Breakfast at Aunty Polly's was no snack. Two or three plates of cereal topped with chopped banana and dried fruit were followed by the fry-up of last night's leftovers topped with an egg or two, followed by five or six slices of toast with Vegemite and several mugs of tea.

We ate at a big pine table in the living room, which was always curtained and dark, with gloomy photographs of Aunty Polly's English ancestors staring down from the wall, the women with high-necked dresses and high-piled hair, the men straight and severe in Masonic lodge regalia. The room's rough hardwood floor was covered with linoleum that was more for protection than decoration because if you got a hardwood splinter in your big toe, it took a pair of pliers to get it out.

There were two double bedrooms with big brass beds and sagging mattresses, and four or five single beds on the open verandah, all hung with mosquito nets because Booker Bay mosquitoes, like a lot of other things in Australia in the 1930s, were the biggest in the Southern Hemisphere. You could step straight from the verandah on to the wharf which ran alongside the swimming pool. This was not a fancy suburban pool dug out of the ground and filled with chlorinated water, but a proper sea pool with its own little beach covered with soldier crabs and protected from sharks by green-slimed piles. The wharf was one of the things that made Booker Bay heaven for a small Australian boy. Your own wharf had a sign that said 'Private – No Fishing' and if you wanted to go into the nearest town, Gosford, you could stand on the wharf and wave your handkerchief at the Ettalong-Gosford launch and it would swing over, pick you up, and later drop you back.

Not that we went into Gosford a lot. Booker Bay had everything anyone could want. The milkman delivered every morning, the rabbito man came with his fresh rabbits once a week. The local shop was stacked to the ceiling with everything from Vegemite to Minties to Scorched Peanut Bars. Prawns and oysters for tea were not unusual. The prawns came from a big copper cauldron full of boiling water that the prawn fishermen set up on the beach at Ettalong about 7 a.m. and were sold by the pint. Oysters you chipped off the rocks yourself at low tide.

Real luxury was roast chicken. At home we had it only once or twice a year but at Booker Bay we had it every Sunday. And it was certainly fresh. Aunty Polly chased the luckless rooster around the yard, chopped off its head with a meat cleaver, plucked and gutted it, then roasted it slowly in the fuel stove. Served with roast potatoes and green peas and lots of bread and butter, followed by a slice of fruit cake and washed down with a cup of tea, it remains one of memory's great meals.

Buying some liquorice all-sorts in the shop one afternoon I met Renate and we walked back along the road – even though the tar was almost too hot for our bare feet – because we did not want to get any pricklies from the grass pavements. She was dark, petite, friendly and had a funny accent. We took to meeting every day until my mother wondered why I was so keen to volunteer to pop down to the shop. So I told her and made the mistake of mentioning the girl's accent, adding, 'It's because she's German.'

'German,' my mother said. 'We fought them in one war and it looks as if we'll have to fight them again. Don't you dare speak to her or I'll tell your father and we'll go straight back home.' Go home early from Booker Bay! For the next three days I avoided the shop. On the fourth day I met the German girl out walking on the beach with her mother and father. I moved closer to the water's edge to avoid them and as I did so I saw a puzzled and hurt look in Renate's eyes. I wanted to tell her: 'It's not my fault. My mother made me choose Booker Bay over you.' But I kept quiet and I never saw her again. Only later did I realize the irony – she was Jewish, and a refugee from Germany.

There were not enough hours in the day. My cousin Brian, who was eight, would sometimes come on holiday with us and the three of us had all sorts of projects under way. Brian wanted to join the Air Force when he grew up, so on this particular holiday we abandoned building billycarts and opted for a small fighter plane built out of scrap wood and corrugated iron. We thought that we were the only inventors to appreciate the value of corrugated iron as a building material but in 1998, John Brannan, of Harbord, Sydney, corrected me. In a letter to the *Sydney Morning Herald* he pointed out that corrugated iron was an Australian icon, used for more than 125 years for roofing, shearing sheds, water tanks and finally for the 2000 Olympic Stadium. 'As a boy,' he wrote, 'I made a canoe from discarded sheets of corrugated iron.'

It may have worked for a canoe but it turned out to be too heavy for an aircraft. We manhandled it on to Aunty Polly's garage roof and my sister waited down below to catch it if it headed for the street. Then Brian sat in the cockpit and I pushed him off. My sister ran away, the plane and Brian hurtled on to the grass, the plane disintegrated and a piece of the fuselage cut off the top of Brian's little toe. His mother was furious and he had to go to the ambulance station to get it dressed. But I told him later that no one had ever said that being a fighter pilot was easy.

We never lacked for entertainment. Aunty Polly did not even have a radio, but at the end of a day's fishing, swimming, building aircraft and playing cricket, we would settle down to play cards, fiddlesticks or tiddlywinks. Or Brian's mother, Aunty Min, would play the piano and we would all sing or recite poetry. One night cousin Brian said he had learnt a new one at school. He started to recite it and then blurted it all out before his mother could stop him. '*Captain Cook did a poop/ Behind the kitchen door/ The cat came up and licked it up/ And then it asked for more.*' Aunty Min was furious and banished Brian to bed without his Milo and next day she walked all the way into Ettalong to telephone his father. It must have done some good because Brian grew up to become a minister of the church and did lots of good work.

I always thought that my holidays at Booker Bay were unique but I know now that there were lots of Booker Bays all over country in the late 1930s, with their Aunty Pollys and Aunty Mins. Most have been developed, built on, or concreted over, and the lazy, timeless way of life they embodied has gone for ever. But they were an Australian heaven while they lasted.

The outlook in Europe grew darker. Was another war on the horizon, so soon after the war to end war? If there was, Australia, like the other dominions, was none too keen to fight it. In 1938 the Australian government warned the British Prime Minister, Neville Chamberlain, that 'almost any alternative' would be preferable to involvement in a war with Germany, if Czechoslovakia were to be invaded.

Several factors brought the Australian Prime Minister, Joe Lyons, to this conclusion. He listened to his wife, Enid, mother of twelve children and an anti-war pacificist, and although he himself had been too old to serve in the First World War, his tours of Australian war graves in France and Belgium in 1935 and 1937 convinced him that if the cost of peace was the dismemberment of Czechoslovakia, then this was a price Australia was prepared to pay. British historian Lawrence James comments, 'The other Dominions were of a like mind. They had provided over a quarter of Britain's fighting men in the previous war and were unwilling to make similar sacrifices on behalf of the Czechs.'

So Australia went along with appeasement for practical and emotional reasons. It saw only folly to plunge into a war in central Europe when communications between Britain and the Empire were under threat from Italy in the Mediterranean and Japan in the Far East, and when large numbers of British and Indian troops were tied down in operations in Palestine and on the North-West Frontier. As for the Opposition, Labor was committed to isolationism and pacificism and its supporters were prominent in the huge anti-war rallies held around the country in 1935, still remembered because of the strange affair of Egon Kisch.

Kisch was one of those Central European intellectuals, of German-Jewish origin, born in Czechoslovakia, who spoke many languages and followed a career that mixed journalism with political agitation. Like the Australian journalist, Wilfred Burchett, who came to prominence some years later, Kisch did not just report events, he tried to influence their outcome.

Kisch arrived in Fremantle on the P & O liner *Strathaird* in October 1934 on his way to attend an Anti-War Congress in Melbourne, to be held on Armistice Day, 11 November. The government banned him from landing, citing his 'subversive activities', and later reinforced the ban by giving him the notorious 'dictation test', usually reserved for cases the authorities wanted to exclude under the White Australia Policy. The immigration authorities were allowed to choose any European language for the test, and aware of Kisch's formidable skills as a linguist, decided on Scottish Gaelic. Kisch duly failed.

In Melbourne, still confined to the *Strathaird*, which was to return him to Europe, Kisch decided to put Australian law to the test. While he was on the ship, prohibited from landing, it was not easy to challenge his exclusion. But if he landed, the government would then have to deport him and such an action could be fought in the Australian courts. So while the *Strathaird* was docked at Port Melbourne, Kisch landed by jumping from the deck on to the wharf, breaking his leg in the process.

He was arrested but bail was quickly arranged. He then travelled the country on crutches to address huge anti-war rallies in Melbourne, Sydney and Brisbane, signing autographs, 'The Famous Jumper', and being lauded or reviled in the newspapers. His message to meetings was simple – don't fight another war; don't let Britain make a deal with Australian militarists for your bodies; join the movement against war and Fascism.

Meanwhile lawyers were having fine fun in court. The Attorney-General, Robert Menzies, persisted in his efforts to have Kisch deported. But Herbert Evatt, at that time a Justice of the High Court and a future leader of the Labor party, ruled that the government had failed to comply with certain formalities needed for excluding Kisch on the grounds of his alleged subversive activities. And the full bench of the High Court decided that Scottish Gaelic was not a European language as required by section 3a of the Immigation Act and that therefore Kisch had not failed the language test. This ruling infuriated a lot of Scottish immigrants and some of them wrote to the *Sydney Morning Herald* mocking the High Court's ruling. The newspaper unwisely published the letters and was promptly prosecuted for contempt of court.

Australians split into those who felt that the government would not have tried to ban Kisch without good reason and that there must be something in the 'subversive activities' accusation, especially as Kisch himself refused to say whether or not he was a Communist, and those who felt Australia's democratic traditions required that Kisch be allowed into the country to speak no matter what he had to say.

Kisch, who had loved every minute of the stir he was creating, eventually accepted a compromise put forward by Menzies – he would get his passport back, his legal costs would be met and at the end of his tour he would return to Paris forthwith. Since Kisch did not plan to stay anyway, this was a good deal for him and on 11 March he boarded the liner *Oxford* at Fremantle, gave the crowd that had gathered to say farewell the Red Front salute and sang, 'Die Strasse frei: Rot Front!'

Australia missed him. He had brought a whiff of European intellectual flair to a country starved of good debate. He was charming and amusing, a good raconteur and an amateur but impressive conjurer. But after his Australian adventure he faded rapidly away. He saw out most of the war he had tried to prevent in Mexico, returned to Prague when the Communists came to power, and died of natural causes soon afterwards.

The Australian government never revealed what it had on Kisch and some of the files on him are still secret or have been removed. But there seems little doubt that, although the Kisch affair has gone down as an example of Australian xenophobia, Kisch was actually a Comintern agent, probably working in its propaganda department, and since the links between the Comintern and the Soviet Intelligence Service were close, the Australian government was advised by the British Security Service (MI5) to consider Kisch a spy and treat him as such. Menzies could not say this without revealing the source of the information, which MI5 did not want exposed.

Menzies himself set sail for Europe not long after Kisch, accompanying his Prime Minister, Joe Lyons, to the celebrations for the silver jubilee of King George V. Menzies fell in love with England and everything English – 'the green and flowering things, a beauty no new country town in Victoria could ever possess . . . Trafalgar Square and one of the Wren churches by starlight . . . young Englishmen with the usual attributes of cleanness, good manners, interest in Test matches and the championship at Wimbledon . . . the grey bulk of Buckingham Palace.' He met the King and Queen and the two princesses and was delighted – 'This is a real family. We leave walking on air.'

Summing it all up he wrote, 'I am made to feel that an Australian Robert Menzies who is Attorney-General is a person of consequence.' To use a phrase favoured by the Australian newspaper proprietor John Norton to describe an Australian who had gone over to the English, Menzies had been 'well and truly duchessed'. He remained an ardent royalist for the rest of his life, one whose publicly expressed devotion to the Queen was so extravagantly expressed that even she was embarrassed.

Back in Australia he remained undecided about the threat of war. He could not make up his mind whether Hitler was a swashbuckler preparing for aggression and expansion or a German patriot determined to restore his country's status and self-respect. Menzies soon found out. In April 1939 he succeeded Joe Lyons as Prime Minister. Five months later the Second World War began.

CHAPTER 11

THE WAR YEARS: A FALLING OUT

The fall of Singapore changed the very basis of the relationship between Australia and the Mother Country – the Australians accused the British of incompetence; the British accused the Australians of cowardice, and the two have been arguing ever since over what really happened.

News of the German invasion of Poland reached Australia on the night of Friday, 1 September 1939. Historian Ken Inglis, who was ten at the time, had gone to the cinema to see Jeanette MacDonald and Nelson Eddy in *Sweethearts*. He remembers, 'The screen was filled with the lovers' technicolour singing faces when a handwritten message slid under them saying that German forces had entered Poland.'

Two nights later I listened with my parents to the British Prime Minister, Neville Chamberlain, announcing on the radio that Britain was at war with Germany. There was never the slightest doubt – even with my father, a long-time republican – that this declaration meant that Australia would be at war as well. That was the way the Empire worked. As the Australian Prime Minister, Robert Menzies, put it: 'Great Britain has declared war [on Germany] and as a result, Australia is also at war. There can be no doubt that where Britain stands, there stand the people of the entire British world.'

Well, yes and no. In 1998 Christopher Somerville persuaded some eighty Commonwealth veterans to explain why they had volunteered to go to war for Britain. Somerville was attempting to define what the British Empire was and why five million soldiers – of every colour and creed and from its every corner – were prepared to die fighting for it.

It is not difficult to understand the motivation of the white Empire, countries like Australia, New Zealand and Canada. Phil Roden, an Australian, said simply, 'I joined because I thought it my duty to do so.' Rod Wells, another Australian, said, 'The Old Country needs help. Let's go and show them what we can do.' Bill Broes, a British migrant who had come to Australia as a boy in 1926, captured the mood in his autobiography, *That Was Happiness*, 'My generation had been brought up to believe in the tradition of Anzacs and the high ideals of Britain, the champion of justice. I thought that it was the duty of every able-bodied man to enlist, although the spirit of adventure was a big influence too.'

Charles Bell, a New Zealander, said, 'It was pretty well put by our Prime Minister at the time: "Where Britain goes, we go".' Glen Niven, a Canadian, said, 'Britain was home. You saluted the flag and sang God Save the King at the drop of a hat. We were more patriotic, I think, than half of Britain.'

But the reasons given by volunteers from the rest of the Empire are surprising. Although many of these countries had active movements fighting for independence from Britain – India, for example – once war broke out many of their citizens wanted to fight for the Mother Country. Kofi Genfi II, from the Ghanian Gold Coast, said, 'We felt proud of being in with the British – we felt British.' Connie MacDonald, from Jamaica, said, 'We didn't want to be anything but British. England was our Mother Country.' Aziz Brimah, from the Ghanian Gold Coast, said, 'We felt that we were British, that we were safe under the British administration. That is why when they requested help for the British in Abyssinia, we surrendered ourselves and went.'

So these children of the Empire went to war for Britain out of gratitude, a sense of familial obligation, a call of duty. There was no overriding purpose, no common stance against the rise of Fascism in Europe or Japan's military adventurism in China. Yet it was the greatest mobilization of people in the history of the world – eleven million people fighting under the same flag.

The irony is that this massive coming together because of the tug of heritage began the process that led to the disintegration of the Empire, the very thing they were fighting to save. The war highlighted the Empire's racial prejudice, the British sense of superiority, their reluctance to feel the winds of change. Yet it was not all bad. The sun set on the British Empire, but as Christopher Somerville concludes: 'It blew the fragments of the old Empire into a modern Commonwealth shape . . . that newly-forming countries would become eager to join, rather than to leave.'

But not everyone in the Empire was hungry for battle. There was confusion over what the war was about. In Britain many Conservatives in high places still admired Hitler, and the determination of a substantial section of the British wealthier classes to escape any unpleasantness was obvious.

Many people in Australia, too, were appalled by the prospect of another war. It was all too soon since the 1914–18 war. Were Australians once again to be Britain's shock troops on the Western Front? Would they be expected yet again to die in their thousands at Pozières and Passchendaele? How much blood-letting could a small country – still only seven million people – stand? Did we get through the Depression only to die on some foreign battlefield?

This was not an isolated attitude, but was shared by some leading politicians. The Australian government had already warned the British Prime Minister, Neville Chamberlain, that 'almost any alternative' would be preferable to involvement in a war with Germany if Czechoslovakia were to be invaded, and other Dominions were of a like mind. Canada, in particular, felt that Britain had squandered the lives of Canadian troops in the First World War and was determined that this would not happen again. Of the two main Australian parties, Labor was committed to isolationism and pacificism, and the Prime Minister, Joe Lyons, as we have seen, after visits to the Australian war graves in France and Belgium, was against becoming involved in Europe.

But once Britain had declared war, Australia decided that she had to be there. Twenty thousand volunteers were quickly recruited for the AIF 6th Division to serve overseas, and conscription for the militia was introduced in October – without any reference to the people as in the First World War – despite strong trade union protests. Later a Volunteer Defence Corps, a 'Home Guard' consisting mostly of older men who had fought in the First AIF, was formed and integrated into the Australian Military Forces.

The National Security Act gave the government the broadest powers in Australian history – total control over all Australians and their property in defence of the Commonwealth and the prosecution of the war. All enemy aliens were interned – including some who were refugees from Hitler – prices were controlled, the press and radio made subject to censorship, a national register of manpower created (to 'ensure that each man is allocated to the task which best fits his training and occupation') and ten organizations, including the Communist Party, were declared illegal. (Later the Jehovah's Witnesses were added to the list.) The Menzies government was determined that this was to be an 'all-in' war, with no backsliders, pacifists, Communists, Fascists, anti-British foreigners or trade union shirkers to stand in the way of victory.

In all this, Australia was following Britain's lead. There the government was preparing for total war and was willing to go further in doing so than ever before. The first peace-time conscription in British history had been introduced in April 1939 and was later extended to include women, for the first time in the history of any civilized nation. The Emergency Powers (Defence) Act authorized the government to do virtually what it liked to prosecute the war, without reference to Parliament.

With everything ready, it all became like the first time around. The AIF sailed for the Middle East, everyone saying that they would be back in six months.

Nothing much seemed to be happening in Europe, where the Allies faced the Germans hunkered down in the Siegfried Line in what became known as the Bore War. At school in Sydney we sang 'We're gonna hang out the washing on the Siegfried Line/ Have you any dirty washing, mother dear?' and tried to better Dick Bentley's German accent and raspberries in 'Ven der Führer says "Ve is der master race"/Ve Heil! (Raspberry) Heil! (Raspberry) right in the Führer's face.'

Then Germans struck through Belgium, Holland and Luxembourg, the Allies fled for the Channel in disarray, and our whole world collapsed. By June 1940 what was left of the British army had been evacuated via Dunkirk, and France had fallen. 'This is the end of the British Empire,' the Chief of the Imperial General Staff, General William Ironside, told Anthony Eden.

It did not seem like that in Australia. Britain would come good, as it had in the past, because we knew that in adversity the British spirit was indomitable. Just how indomitable, we did not realize until the BBC TV series *Finest Hour* was broadcast in 1999. A Royal Navy rating told what it was like on a British destroyer during convoy duty in the North Atlantic. 'When a freighter that had been torpedoed went down there would often be survivors, merchant seamen, swimming around in the water. We'd be going all out after the submarine and we'd have to race past them, leaving them to drown. They knew we couldn't stop to pick them up. They understood that. I'll never forget one of them waving his hand as we went by and shouting at the top of his voice, "TAXI, TAXI".'

Or Bess Walder, a 15-year-old British schoolgirl, clinging to an upturned lifeboat with her friend after the *City of Benares* was torpedoed in the freezing mid-Atlantic in September 1940, on its way to Canada with ninety child evacuees. The two girls encouraged each other until twenty-four hours later HMS *Hurricane* picked them up, half frozen but still alive. 'I knew we'd make it,' Bess said, nearly sixty years later, unconsciously speaking for wartime Britain. 'We weren't in the business of giving in.'

Our boys were making their contribution in the Middle East. We knew that because not only did the Australian media have its own correspondents there but most of the correspondents for British press and radio seemed to be Australians as well. In fact, it is one of those odd quirks of history that Australians dominated the media coverage of the Second World War in nearly every theatre – Alan Moorehead (*Daily Express*, London), Chester Wilmot (BBC), George Johnston (*Time Life*), Wilfred Burchett (*Daily Express*, London), Ronald Monson (*Daily Telegraph*, London), Damien Parer (Department of Information and later Paramount), Richard Hughes (Consolidated Press), Kenneth Slessor (Australian Official War Correspondent), Osmar White (the *Herald* and *Weekly Times*, Melbourne), David McNicoll (*Daily Telegraph*,

Sydney), Lorraine Stumm (*Daily Mirror*, London) Stanley Johnston (*Chicago Tribune*), Denis Warner (*Sun Pictorial*, Melbourne).

There have been various theories to explain what makes Australians such good war correspondents. Murray Sayle, a former war correspondent himself, says, 'It's because Australians are good at camping.' This has a serious undertone. It suggests that Australian war correspondents eschew the comfortable and relatively safe life behind the lines to camp out with the troops, and that therefore they know more about what is happening at the front.

Parer had another explanation. He said that an Australian correspondent had greater sympathy with the ordinary soldier. 'He goes in with the troops and for a short time becomes one of them. He must look at these men with the open-eyed surprise of a child first seeing the world and *comprehend their greatness as they themselves cannot*.' (Author's emphasis)

This quality can be seen both in Parer's film work and in the stills of his colleague, George Silk, the New Zealand-born photographer who worked for the Australian Department of Information and later for *Life*. The only ones that look stilted and posed are those of generals and staff officers. The others – the ones of ordinary soldiers – stand out for their warmth and empathy. When they were in the field, Parer and Silk became Aussie diggers.

We may have known a lot about the war but we did not know everything. The Australian troops training in Egypt were under the command of General Thomas Blamey, former police commissioner in Victoria. As we have seen (Chapter 9) he had had a checkered career there – a member of one of the secret armies and notorious in trade union circles for his harsh treatment of demonstrators. There was a further shadow on his reputation – the 'copper caught in brothel' affair – a scandal so juicy in its own very Australian way that more than 70 years later the media still regularly resurrect it as one of the mysteries in Blamey's life.

In 1925 three plain-clothes constables from the licensing police raided a Melbourne brothel they believed was illegally selling alcohol. In one bedroom they came across a half-dressed man and woman and reported later that the man said to them, 'That is all right boys. I am a plain-clothes constable. Here is my badge.' The badge, number 80, turned out to be on issue to Blamey. Soon after the raid, the badge turned up in Blamey's letter box at the Naval and Military Club.

Blamey's explanation for all this was that he had handed his keyring, with his badge attached, to an old army friend so that the friend could collect some alcohol from Blamey's locker at the Naval and Military Club. He had arranged for the friend to leave the keyring and badge in Blamey's letter box. This admittedly somewhat thin story could have been confirmed by Blamey's naming of the friend, but he resolutely refused to do so because 'he is married and the father of three children'. And despite public outrage and calls in the State parliament for a commission of inquiry, that was that. Blamey simply did not care what the Victorian public or the press thought of him and he actually won the admiration of many Australian men for declining to 'dob in' his mate.

In the First World War he was at Gallipoli and at the Somme where he experienced the attitude of the British general staff to Commonwealth troops – they were respected, even admired, but they were there primarily to serve Britain's interests. Blamey was enough of a politician and had sufficient sense of history to know that by 1939 this relationship had changed. Australia had become a politically independent country which, to ensure its future, needed military independence as well.

When British and Australian forces came together in the Middle East it needed a man of Blamey's bloody-mindedness to stand up to the British and make them realize that the Australians were not available to be moved around and used to suit some short-term tactical plan. 'We have had some trifling difficulties in educating the British staffs to an understanding that the AIF is a national force, and is not available to break up into detachments all over the country,' Blamey reported with deliberate understatement. 'With a little quiet persistence we were able to get our point of view understood.'

In fact it needed more than quiet persistence. The British commanders did not appreciate the realities of Australia's independent status. It was Blamey, backed by the Australian war cabinet, who had to make it clear that he was not there to do the bidding of the British commander, that he owed a duty to the Australian government and that the British could not simply order Australian units about as they liked.

This involved him in confrontations with the British that in earlier times might well have seen him shot for refusing to obey orders. Churchill, meddling as was his wont with his generals' plans, at one stage complained bitterly to General Wavell, commander in the Middle East, that he was not making proper use of his forces: 'I do not understand why the Australians and New Zealanders who have been training in Palestine for at least six months, should be able to provide only a brigade for service in Egypt.'

So without even consulting the Australian government, Churchill directed Wavell to remove the remaining Australian battalions to the Nile Delta. Blamey countermanded this order – the Australians were not to move without his authority. As soon as word of this insubordination reached Wavell, Blamey received a message from British headquarters in Cairo. The exchange of cables captured the new nature of the military relationship between Britain and Australia.

'Our Archie [Wavell] in London has seen our Winston [Churchill] who has directed that Tubby [Allen, commander of the 16th Brigade] should move as directed.' Blamey's headquarters replied: 'This cuts no ice with us. The decision will rest with our Thomas Albert [Blamey] in Gaza and our Robert Gordon [Menzies] in Canberra.' So there. After a few more acrimonious exchanges, the British backed down.

We knew none of this in Australia. Nor did we know about Blamey's reputation among war correspondents as a playboy. It emerged only later that he could often be seen 'jazzing' in the fleshpots of Cairo and Tel Aviv and acting as host at lavish parties on the AIF houseboat. He brought his wife to join him in Cairo and when there were public protests in Australia, and the War Cabinet asked him to send her back, he refused.

One Australian war correspondent, Chester Wilmot, then with the Australian Broadcasting Commission, claimed to have strong evidence that Blamey was receiving 'kickbacks' from the businessman who had the exclusive contract to show films in the AIF cinemas. Wilmot was quietly working on confirming this allegation – and that was all it was – but it was to have major repercussions when the war moved to the Pacific.

In the meantime, the desert campaign gave Wilmot and his colleagues plenty to write about. The action took place in an empty arena, devoid of those houses, schools, hospitals and civilians that elsewhere get in the way of battle. It was war in its purest form, what Alan Moorehead called 'a Knights' tournament in empty space', because although it was a mechanized war, all the participants pretended it was being fought with cavalry – armoured division officers gave orders from their 'chargers' (tanks) and supply trucks moved off on the order 'Lead horses!'

Moorehead had a sense of history as well as an Anglo-Australian's arrogance, still strong in the approaching twilight of Empire. 'We were rich and powerful. We believed that we were a superior race of men . . . we were constantly moving among coloured men whose territory for many years past had been governed by the British . . . wherever we went in the Middle East we remained on what we liked to think of as "British soil". Like the children of very wealthy parents, it seemed quite natural to us that we should occupy the best houses and hotels, that we should have at our command cars, motor launches, servants and the best of food.'

Australia's later struggle for survival in the Pacific war has crowded out the memory of most of the events in the desert campaign. Some still remember 'Lili Marlene', the haunting love song the German Army under Erwin ('The Desert Fox') Rommel, brought to North Africa and which the British and Australian soldiers then stole and made their own.

In the Western Desert it was almost as if the Australians were determined to begin where they had left off in France in 1918. Within weeks of leaving their camps around Cairo, in the bitter cold of a desert winter, the 6th Division AIF was marching along the Nile Delta to the west. Soon the coast towns of Bardia, Tobruk, Derna and Benghazi fell to the Australians. In the dark winter of the London Blitz, cinema audiences cheered as newsreels showed long lines of Italian prisoners guarded by a few Diggers. Ivan Chapman of the 2/1 Machine-Gun Battalion, stationed in England, remembered how the matron of the hospital he was visiting burst into tears when she saw his slouch hat. 'Now I know everything's going to be all right,' she said. 'The Australians are here.'

But then General Erwin Rommel arrived and soon the Australians were the 'Rats of Tobruk', hanging on to the port of Tobruk on the Libyan coast during a seven-month siege in 1941. (The 'rats' came from a German radio propaganda broadcast – 'The Rats of Tobruk, those self-supporting prisoners of war.' The soldiers of the 9th Division proudly took on the epithet as if they had invented it themselves.) The siege was probably not as strategically important as made out at the time, but the way the Australians handled themselves was a boost to national morale. Lawson Glassop's novel, *We Were the Rats*, gets it about right, and the laconic words of one of the characters captures the romantic way the 'Rats' were regarded back home: 'Within the next few days Jerry's going to have his smack at us. They're not asking you to do much. They're only asking you to do what's never been done before – to stop Hitler's army.'

Greece and Crete were a different matter. Menzies had been told that an operation to reinforce the Greek expedition was being planned when he visited the Australian troops earlier in the year, but he appears not to have discussed this with General Blamey. Blamey, when finally told about the expedition, had strong doubts. He thought it a 'very hazardous' operation but did not tell an appalled Australian War Cabinet until it was too late.

The German push through Greece and finally Crete was a disaster for the British, Australians and New Zealanders. 'All Corinth was permeated by the smell of burning human flesh,' reported Robert St John, the Associated Press correspondent. 'Human flesh burns with a sickening sweet smell. It's a smell you never forget.' He wrote of trying to give a lethal dose of morphine to a man whose hands had been blown off and whose intestines were hanging out, and he listened all night to the whimpering of a five-year-old girl whose right arm hung in black, tattered shreds.

We heard nothing about this in Australia because the British and Australian censors cut it all out. They warned St John that Britain and the Empire must not be told how bad things were. In fact, they changed one of his reports to reverse the meaning. He had written that the evacuation from Greece had not been another Dunkirk; it had been much worse. The censor simply cut out the second clause. When he wrote that the Allied casualties had been 20,000 killed, wounded, or captured, the censor changed the figure to 3,000.

Kenneth Slessor, the poet and official Australian war correspondent, was angry about Australian losses and asked Blamey if the expedition had been worth it. Blamey replied, 'That's a very difficult question and I'm not sure I know the answer.' Slessor decided to take the matter further but there was no need. Commentary in Australian papers and on newsreels was so scathing that the Department of Information banned a summary of it which a journalist was preparing for the British press on the grounds that it might damage British-Australian relations. And the home front did not learn at the time that when Blamey had been ordered to evacuate Greece lest he and his senior officers be captured by the Germans, Blamey chose against all advice to include his own son on the evacuation plane.

The almost total absence of air cover was considered a scandal. Chester Wilmot said in a later broadcast, 'At no stage during the campaign was there anything like adequate support in the air to the force that had been sent to face the world's most powerful army on the ground.' The Australians vowed never again to fight without proper air support.

But probably the greatest cover-up of the campaign did not come to public attention until fifty years later. This scandal concerns Churchill, his attitude to Commonwealth troops, one of the greatest secrets of the war, 'Ultra', and the way it was protected at the expense of a fine soldier's reputation.

Churchill had always been fascinated by spying and intelligence gathering. He had been a spy himself once – he visited Cuba in 1895 during that country's revolt against Spain and reported on what he had observed to British military intelligence. When he became Britain's wartime Prime Minister he was absolutely dazzled by Ultra, the intelligence produced by the decoding of intercepted German radio messages.

This turned out to be a major weapon in the winning of the war, but it had its downside. It took a while for Churchill and his military leaders to realize that they could not always rely on the accuracy of what German generals told Berlin because they often exaggerated their difficulties in the hope of extra support.

Further, the need to keep Ultra absolutely secret had serious side effects. Churchill felt that the fewer leaders, military and political, who knew about Ultra, the better, even if keeping them in the dark meant short-term sacrifices. So he did not bring all the other Empire Prime Ministers into the exclusive Ultra inner circle. In fact, it appears he told only one – Jan Smuts of South Africa. Historian David Stafford, who chronicles Churchill's fascination with intelligence in *Churchill and Secret Service*, says Churchill 'revered and trusted the old Boer fighter and personally revealed it to him'. The implication has to be that Churchill never told the Australian wartime Prime Minister, John Curtin, because he did not trust him.

This obsession with secrecy and Ultra blighted the career of General Bernard Freyberg, the British-born commander of the New Zealand division defending Crete. The British Secret Intelligence Service knew from Ultra that the Germans were planning an airborne landing on Crete's Maleme airfield. This crucial piece of information was passed to Freyberg. But Freyberg was under a standing order forbidding him to act on the basis of Ultra alone. Unless he could learn from some other source of the planned German landing, he had to pretend that he did not know it was about to happen. As he himself succinctly put it years later, 'The authorites in England [preferred] to lose Crete rather than risk jeopardizing Ultra.'

So Freyberg had to cancel a planned move of troops to protect the airfield, the Germans duly landed there and Crete fell at the beginning of June. Some 15,000 Allied soldiers were killed, wounded or captured there and of the 7,000 Australians nearly half were taken prisoner. Freyberg's reputation never recovered and since the secrecy surrounding Ultra continued after the war, he died in 1963 without ever being able to explain himself publicly.

Bad news now came in waves. On 19 November, the Australian cruiser, HMAS *Sydney*, was sunk off the West Australian coast with the loss of all 645 officers and men. Nearly sixty years later the loss of the *Sydney* remains a mystery. All that is known is that she challenged the German raider *Kormoran*, which had been masquerading as a Dutch freighter, there was an exchange of gunfire, and the German ship went down, leaving survivors in lifeboats. The *Sydney* was last seen, badly damaged, turning towards the Australian coast.

Questions began immediately. How could *Kormoran*, a merchant ship converted into a raider, sink a heavily-armed and war-experienced cruiser like the *Sydney*? (She had taken part in the sinking of the Italian ship *Bartolomeo Colleoni* in the Mediterranean the previous year.) Why were there no radio messages from the *Sydney*? Was it really possible that the *Sydney* had gone down with *all* hands, that there was not a *single* survivor?

In the absence of satisfactory answers, all sorts of conspiracy theories thrived. The most persistent one was that the *Sydney* surprised the *Kormoran* in the middle of re-fuelling a Japanese submarine and that it was the Japanese submarine which torpedoed the Australian ship. Such co-operation between Germany and Japan, then still a neutral nation, would have been dynamite if revealed, so after sinking the *Sydney*, the Japanese and Germans machine-gunned all the Australian survivors, leaving no one alive to reveal what had happened. However, a committee of the House of Representatives said in 1999 that it was not convinced that a case had been made to show that the Japanese were responsible, and that there was no evidence to support the theory that Australian survivors were machine-gunned in the water. It is unlikely now that we will ever know what really happened.

On 7 December 1941 Australia's worst nightmare came true. Since the country's earliest days many had feared 'the Yellow Peril'. The Chinese and/or Japanese were thought to be so envious of Australia's riches and way of life that one day they would march south in their millions and invade.

British leaders scoffed at this idea. Japan might covet the oil riches of south-east Asia but would hardly risk a war with Britain and probably the United States to gain access to them. In 1939 Churchill said, 'A war with Japan! I do not believe that there is the slightest chance of it in our lifetime.' The Americans agreed. Although alarmed by the outbreak of war in Europe, they underrated the risk of a war with Japan.

The Chief of Staff of the American army, General George C. Marshall, told the seven press correspondents at a secret briefing in Washington, on 15 November 1941 that although the army was run down (it ranked only nineteenth among the world's armed forces, after Portugal, but barely ahead of Bulgaria) it could defeat Japan by bombing raids on its 'paper cities'. He never once mentioned the possibility of a Japanese attack on Pearl Harbor. It was, he later testified, 'the one place we felt reasonably secure'.

The shock of that Japanese attack on 7 December reverberated around the world. When the last Japanese plane roared off, five American battleships had been sunk and three damaged, three cruisers and three destroyers badly hit, 200 planes destroyed, and 2,344 men killed. For the loss of only 29 planes, Japan had virtually crippled the US Pacific Fleet at a single blow.

The significance of this victory was not lost on Australia, which now had a new Prime Minister, John Curtin. Intrigues in the Australia Party had forced Menzies to resign, and after a brief period with the Country Party leader, Arthur Fadden, as Prime Minister, three independent MPs who held the balance of power transferred their allegiance to the Labor Party. Curtin, the Labor leader, an austere man who had battled all his life against alcoholism, took over as the country's leader in October 1941. Three weeks after Pearl Harbor he announced a major shift in Australia's foreign policy, perhaps the most significant one this century – Australia would henceforth look to America for her protection rather than to Britain.

'Without any inhibitions of any kind, I make it quite clear that Australia looks to America, free of any pangs as to our traditional links or kinship with the United Kingdom,' Curtin said. 'We shall exert all our energies to shaping a defence plan, with the US as its keystone, which will enable us to hold out until the tide swings against the enemy.'

Curtin did not take such an important decision out of the blue. It followed the submission of an American plan to build up Australia's defences so that she could play a bigger part in the war. But had he heard whispers of a top secret Anglo-American document code-named 'WW1'? This recorded the terms of an agreement reached in a series of meetings in Washington between Roosevelt and Churchill in the aftermath of Pearl Harbor.

The two leaders agreed that Britain and the United States would concentrate on defeating Germany first and only when they achieved superiority in Europe would they turn their full force on Japan – *even if this meant temporarily abandoning Australia to the Japanese*. Australia was not consulted about this and the agreement was not conveyed to the Australian government, but there were sufficient hints for a leader of Curtin's acumen. For example, Churchill had cabled Curtin from Washington on Christmas Day, and although he did not reveal the formal agreement to defeat Germany first, he did say that the defeat of the German army was 'the dominant military factor in the world war at this moment'. The Sydney *Daily Telegraph* reported from London that 'the British War Office still believes that the conflict in the Pacific is a minor segment of the main conflict in Europe', and *The Times* referred to Japanese actions in the Pacific as 'distractions'.

It did not seem like that in Australia. Suddenly Japan appeared to be unstoppable: Singapore, Hong Kong, Manila, Borneo, Java. By 6 May 1942, the last Americans in the Philippines had surrendered at Corregidor; by 15 May, the British had been run out of Burma. In five and a half months the Japanese had seized the richest colonial area in the world, snuffing out in the process the British, French and Dutch empires.

Of these disasters the most significant as far as Australia was concerned was the fall of Singapore. It changed the very basis of the relationship between Australia and the Mother Country – the Australians accused the British of incompetence; the British accused the Australians of cowardice, and the two have been arguing over what really happened ever since. It marked the decline of Britain as a world power and the beginning of the rise of Japan. With the fall of Singapore, white supremacy in the East ended and nothing was ever the same again.

The nightmare began with the loss of the *Repulse* and the *Prince of Wales*. Churchill had sent these two British warships, among the most effective in the British Navy, out to the East in pursuit of one of his wilder strategic fantasies: to bluff the Japanese out of the idea of war altogether or, failing that, to persuade them to attack the Soviet Union. When the two warships, along with four elderly destroyers comprising Force Z (under the command of Vice Admiral Sir Tom Phillips) sailed into Singapore, Lady Diana Cooper, wife of the Resident Minister, remarked perceptively, 'A lovely sight – but on the petty side.'

Vice Admiral Phillips had no time for modern theories about air cover for battleships, which is just as well because he did not have any – the aircraft carrier HMS *Indomitable* had run aground in the Caribbean before she could join Force Z, and land-based British aircraft in Singapore could offer no cover because Phillips had neglected to say where he was going.

So two days after Pearl Harbor, there was Force Z blundering about in the Gulf of Siam within easy reach of Japanese bombers. Twelve hours after a Japanese submarine had sighted the British ships, seventy Japanese aircraft attacked with torpedoes and bombs and sank both ships in a little over an hour. Phillips signalled for more destroyers during the battle and then for tugs but never for fighter aircraft. (Eleven came up after it was all over.) There were 840 men lost, including Phillips, and the battle entered history as the worst individual disaster the Royal Navy suffered in the war.

Two war correspondents, O. D. Gallagher of the *Daily Express* and Cecil Brown of the Columbia Broadcasting System, survived the sinkings. Back in Singapore after first-hand experience of Japanese fighting qualities and thus better placed than most to assess the base's chances of resisting a Japanese attack, they were horrified at what they found.

The official attitude was that the Japanese could not be taken seriously. They were racially inferior and therefore incapable of efficiently waging war with Western inventions like battleships, submarines, tanks and aircraft, especially against an impregnable fortress like Singapore. Back in Australia we read reams of soothing nonsense about how strong Singapore was. 'Giant guns guard the jungle-fringed shores . . . Australian, British and Indian troops in the fly-infested forests are ready for anything, while clouds of planes daily patrol over neighbouring islands.'

The truth was that Churchill had decided that Singapore was expendable and, if the worst came to the worst, could be abandoned, along with Australia. There is the possibility that Churchill's decision on Singapore was made much earlier than has been previously thought. James Rusbridger, a former British intelligence agent, weapons systems designer and amateur historian, together with Eric Nave, an Australian code-breaker, published an account in 1991 in *Betrayal at Pearl Harbor* which claims that Churchill knew in August 1940 – almost 18 months before the fall of Singapore – that the island was undefendable.

Churchill was handed a report by Britain's Chiefs of Staff that said bluntly that Malaya and Singapore could not be held against a determined Japanese attack. Rusbridger and Nave said that Churchill considered this news so serious that he decided not to tell the Australian government. But while Australia was kept in the dark, Japan got to hear and rearranged its battle plans accordingly.

Churchill sent by special courier a single copy of the report to the

British Commander-in-Chief Far East, Air Chief Marshall Sir Robert Brooke-Popham, at his headquarters in Singapore. But the Blue Funnel ship, the *Automedon*, carrying the courier was captured by the German raider *Atlantis* off the Nicobar Islands on 11 November and the report seized. The Germans took it to Tokyo and gave it to the German naval attaché on 5 December 1940 who, with Hitler's approval, handed it to the Japanese High Command on 12 December.

Rusbridger and Nave added that it was not until Australia's representative in London, Sir Earle Page, was allowed to attend British War Cabinet meetings that Page learnt of the *Automedon* affair and told his government, to Curtin's subsequent fury. 'By failing to tell the Australians the truth about Singapore and Malaya and the *Automedon* disaster, Churchill allowed them to continue to pour more reinforcements into the island, believing it was an essential part of his Far East strategy.'

There is circumstantial evidence to support Rusbridger's and Nave's account, but convincing proof is lacking because three separate Foreign Office files that once existed concerning the loss of diplomatic material on the *Automedon* have been destroyed. On the other hand, no one has been able to prove their account wrong.

In the last days of 1941 and the early ones of 1942, the reality of 'Germany First' became clear to Australia. It was clear that Britain was not reinforcing Singapore and that she was concentrating the remainder of her available resources on protecting her imperial jewel, India. This was a bitter blow to Australians, who thought that *they* were the jewel in the Crown, and Curtin began to pepper Churchill with telegrams of protest at this 'inexcusable betrayal'.

Without knowing it, Curtin was treading on very delicate ground. In his seminal book on the subject, *The Great Betrayal*, Australian historian David Day explains why Churchill was so sensitive: 'Roosevelt had taken a considerable political risk in supporting the "Germany first" strategy at a time when the natural inclination of the American public was to concentrate on hitting back at the Japanese.' Churchill was worried that Curtin's actions were endangering the British-American alliance at a vital stage of the war.

So there was no sympathy for Australia's plight amongst the British establishment. In fact some high British officials seemed to delight in Australia's predicament. Oliver Harvey, private secretary to Anthony Eden, the Dominions Secretary, wrote in his diary in January 1942 that Australia was in 'the greatest possible flap . . . almost a panic' and was blaming Britain. Curtin was 'a wretched second-rate man' and, echoing Sir Otto Niemeyer's views about Australians, their natural optimism, and their penchant for living in a fool's paradise for a century, he wrote, '[She has] now suddenly woken up to the cold and hard fact that her very existence as a white country depends not on herself but on protection from Great Britain.'

Other British leaders agreed. Sir Ronald Cross, a former Tory Minister who came to Australia as British High Commissioner in 1941, before the Labor government, said Australians were 'inferior people' with poor nerves and urged that Britain take no notice of their 'rude squeals'. Churchill himself even went so far as to tell his medical adviser, Lord Moran, that 'the Australians came of bad stock', meaning convict, Irish and working-class English.

So publicly Britain continued to reassure Australia about Singapore and made token gestures towards reinforcing it, while diverting the armour and fighter planes it needed to the Soviet Union. By 19 January Churchill was more interested in Burma than Singapore and gave the impression to his War Cabinet that he was prepared for the worst. But all thoughts of saving troops or sparing the civilian population were to be suppressed. 'The battle must be fought to the bitter end at all costs,' he ordered. 'Commanders and senior officers should die with their troops. The honour of the British Empire and the British Army is at stake.'

As we know, it did not happen like this. General Arthur Percival, GOC Malaya Command, failed to implement a plan to seize Thai bases before they could be taken by the Japanese. The Japanese captured a marked British map. Three raw brigades of the 11th Indian Division, some without native-speaking officers, were roundly defeated at Jitra.

There followed a series of disasters exacerbated by incompetence, failure to communicate, poor equipment, and plunging morale. British troops retreated from positions the Japanese were not attacking, disobeyed orders to counter-attack and failed to follow up advantages obtained by the rare ambush. The last retreats led to a successful withdrawal by 31 January to Singapore Island. But once there, things grew even worse.

The city was bombed but there was no one to sound the warning sirens and all the street lighting remained on because no one could find the master switch. The Post Office cut the telephone circuit to the front when the regulation three minutes were up. A last-minute attempt to build defences on the island was delayed ten days by a dispute over the correct wages to pay the coolies.

The secretary of the golf club refused to allow guns to be mounted on the links until he consulted the committee. The white civilian population refused to believe it was all happening and carried on as usual – the daily tea dance at Raffles Hotel, the visit to the cinema to see Joel McCrea and Ellen Drew in *Reaching for the Sun*.

On 8 February, the Japanese crossed the narrow channel from Johore to Singapore. Once the Japanese were ashore, the outcome was never in doubt. Seven days later a British army with ample munitions surrendered to a Japanese force barely one-third its strength and down to its last one hundred rounds of ammunition per man.

The next day there were only Japanese correspondents and cameramen around to record the British officers, in their baggy shorts and faintly absurd toothbrush moustaches, signing the surrender and agreeing that British troops would help the Japanese maintain law and order until the occupation could be properly established.

This orderly hand-over of power from one imperial regime to another was marked by a display of friendly relations between many British and Japanese soldiers. Until prison camps could be prepared, many Australians were even allowed to wander around Singapore town after giving their word of honour that they would not try to escape, and some were recruited to help build the barbed-wire fencing around the Changi prisoner-of-war compound. But the dominant emotion among the Australian prisoners of war was shame and anger. In 1998, Sir John Carrick remembered, 'Surrender! A cascade of violent and confused emotions. They surrendered us. We would have fought on. Our leaders have failed us. We've failed our people.'

The surrender of Singapore shocked the world. Everyone, even Churchill himself, had come to believe the official propaganda – Singapore was a fortress capable of holding out for many months until it could be relieved by a British fleet. Churchill privately described its ignominious collapse – despite his orders for a fight to the death – as 'the most shameful moment of my life'. He was furious and blamed the Australians, thus sparking off a battle between Britain and Australia – particularly between their military commanders and historians – that is as bitter today as it was in 1942. Epithets flew. The British were 'inexperienced . . . lacked fight'; they had 'a retreat complex . . . a system based on effete conservatism and arrogance . . . poor leadership selected by old school tie'. The Australians, in British eyes, were 'very half-hearted' and 'threw their hands in as soon as things looked black.' They were guilty of 'desertion, looting and rape . . . cock-a-hoop . . . but yellow'.

Churchill was particularly angry about the capture of the British 18th Division which he had wanted to send to Burma but which he had ordered to Singapore as a sop to Curtin's demands. The men, unacclimatized and desert-trained, arrived just in time to march off the troopships and straight into Changi Prison.

The Australians stood up for themselves. Their commander, Major General Gordon Bennett, had made a hazardous and controversial escape from Singapore. Australia was divided on whether he had done the right thing by escaping and leaving his men to go into prisoner-of-war camps. In Britain, however, there was no doubt – he had not lived up to the proper code of military honour. The British were prejudiced against Bennett anyway because he insisted on sending to London a damaging report on the Malayan campaign. In it he admitted desertions by Australian troops but, at the same time, was highly critical of British military leadership. Churchill suppressed the report, saying it was unfit for publication.

The Americans held a number of inquiries into the disaster at Pearl Harbor but there was never a similar British inquiry into the fall of Singapore. There was, however, a War Office report on the behaviour of the Australians. This was considered so sensitive that it was never sent to Canberra, but over the years most of its allegations have leaked out. It was compiled from reports by British officers and from letters written by escapees and intercepted by the censors. The following extracts, taken from *Odd Man Out: Story of a Singapore Traitor*, by Peter Elphick and Michael Smith, are typical:

> 'The one and only reason for the sudden collapse of the Singapore defences was the Australian troops. That Singapore might have fallen in any event after a long struggle may be true but if Monday night–Tuesday morning, the soldiers who were allocated the task of holding the Buki Tinnah [sic] had not just thrown down their arms and run into Singapore it may have been a different story.'
>
> 'On the tenth morning an effort was made to drive the Japs back into the sea. Three battalions were to advance in line. But there was a battalion missing and the Japs walked through the gap. The missing battalion was an Australian battalion. It reappeared in Singapore town, preferring drink and rape to doing its duty.'
>
> 'The Australians were known as daffodils – beautiful to look at but yellow all through.'
>
> 'They were soon lying all over the streets of Singapore dead drunk. I have never seen men such absolute cowards and so thoroughly frightened. What fine soldiers. If America wants Australia, let her have it and good riddance to a damn rotten crowd.'

No one denied that earlier in the campaign the Australians had fought with conspicuous bravery, especially since they did not go into action until 14 January 1942, by which time the Japanese were already three-quarters of the way down the Malay peninsula. On Singapore Island, the Australians manned the north coast, west from the causeway linking the mainland. The full force of the Japanese assault fell on them here on 8 February and they had to pull back. But no one had expected them to hold the Japanese, who outnumbered them and were better supported. 'The general charge that the Australians were responsible for the fall of Singapore is simply wrong,' says Hank Nelson, Senior Fellow, Pacific and Asian History, at the Australian

National University. 'The disaster of Singapore was determined long before the fighting from 8 February.'

Nelson also says that most of the specific charges against the Australians are untrue or exaggerated. The story (told above) of the missing Australian battalion, for example, does not stand up. Records of all six Australian infantry battalions and the composite battalions have been examined and 'this cannot refer to any of them'.

Desertions? Nelson says, 'Less than twenty in a thousand Australians left Singapore before the official surrender and either reached Australia or were captured in Java or Sumatra. Many more were not with their units at the time of surrender, but most of these were unwitting absentees, victims of battle and chaotic communications.'

The controversy will rumble on, stimulated from time to time by new histories and Britain and Australia's inexorable parting of the ways. When the then Australian Prime Minister, Paul Keating, discussing relations between the two countries, said on 27 February 1992 that Britain had decided 'not to defend the Malaysian peninsula, not to worry about Singapore and not to give us our troops back', Britain's reponse was not long delayed. It released under the then Fifty Year Rule all the War Office material relating to the fall of Singapore. This produced in the *Canberra Times* of 12 January 1993 the headline: 'British report brands World War II Diggers cowards, rapists and looters' and battle was joined once again.

One aspect of the fall of Singapore is seldom discussed – the role of a British traitor. After the debacle in Crete in April–May 1941, the British had promised the Australians that they would never again have to fight without proper air cover. But that is exactly what they had to do in Malaya. The real question, one the British have been at pains to avoid, is: why?

The short answer is that the British forces harboured a traitor, an officer and a gentleman who was in the employ of the Japanese, and largely due to him Allied air power was virtually wiped out in the first 24 hours of the war – the RAF in north Malaya was reduced from 110 operational aircraft to 50. Rumours about this traitor have circulated since 1942 but he was not publicly identified until Elphick and Smith did so in 1993. No official files exist on his case; the Ministry of Defence believes that they were probably lost in those last terrible days before Singapore fell.

The traitor was Captain Patrick Heenan of the 2/16 Punjab Regiment, a career officer in the Indian Army who was serving with his regiment in Malaya. In March 1941 he was transferred to a new secret unit which liaised on intelligence matters between the Army and the RAF. This gave him access to all the Allies' plans for the air defence of northern Malaya. Since Heenan had been recruited by the Japanese while on six months' leave in Japan in 1938–9, it must be assumed that he passed these plans to his Japanese controller.

Elphick and Smith say, 'Certainly this would explain their remarkable success in destroying the Allied aircraft so quickly, many of them on the ground.' The Official History adds, 'As a means of defence and as a support of land operations in the forward areas, the British air effort had almost ceased to exist within twenty-four hours of the opening of hostilities.'

Although some of his fellow officers became suspicious of Heenan's behaviour, he was finally caught by accident. One of his colleagues examined a padre's field communion case which appeared to be abandoned. He found that it contained a radio receiver and transmitter. He replaced it and kept watch and saw Heenan collect it. Heenan was tried by court martial and sentenced to death by firing squad. But the sentence had to be confirmed and in the meantime the Japanese arrived at the outskirts of Singapore.

Heenan knew what was happening and tormented the military police who were guarding him. 'You'll be in the bag soon, you'll be in prison or shot or something.' So the MPs drove Patrick Heenan down to the harbour through the devastated dockyards and ordered him to stand on the edge of a wharf and look at the setting sun. Then a sergeant shot him through the back of the head and pushed his body into the water.

The story of Heenan's treachery is an appropriate one for Singapore because many Australians, Curtin among them, felt that Australia had been betrayed by Britain, that the Mother Country had deliberately left her at the mercy of Japan for dubious strategic reasons. After the fall of Singapore, how long would it be before Japan attacked Australia? The answer was, five days. On 19 February the Japanese bombed Darwin, the first time a foreign power had ever attacked the Australian mainland. For Australians it was to be a moment of national shame.

CHAPTER 12

THE WAR YEARS: UNCLE SAM

A drunken GI threw open the doors of a Sydney bar crammed with Australian soldiers and shouted: 'Here I am you Aussie bastards. Come to save you from the Japs.'

The bombing of Darwin came as a complete surprise to most Australians. The Marshal of the British Royal Air Force, Sir Edward Ellington, had told Australia's military planners in 1936 that an air attack on any Australian city was 'so remote it can be almost disregarded', so the town was virtually undefended. Even when the 188 Japanese bombers were actually overhead – an impressive sight – the Australian military authorities were reluctant to believe it.

A sergeant who telephoned his brigade headquarters to report what he had seen in the sky was told, 'Mac, I'm busy. Don't play games with me. How do you know they're Japs?' And the sergeant shouted back, 'Because they've got bloody great red spots on them!'

The Japanese flight commander, Mitsuo Fuchida, saw below him a target almost as juicy as Pearl Harbor, where he had distinguished himself. Forty-five ships, including the American heavy cruiser USS *Houston* and the destroyer USS *Peary*, were moored in the harbour or at anchor. The town and its citizens, lazily going about their business in the tropical heat, seemed unaware of what was happening. There was still a large civilian population in Darwin because there had been resistance to an evacuation programme.

When the South Australian government was warned to expect about two thousand evacuees from the north, the state Premier complained to the Prime Minister that he did not want any half-caste Aboriginals. His telegram read: 'Their presence likely prove source of infection to contacts. Most undesirable further parties half-castes be sent south.' So ninety part-Aboriginal children from Croker Island, near Darwin, had to walk right across Australia, north to south, hitching lifts in cattle trucks and with military convoys, living on goannas and lizards and whatever they could scrounge, until two months after setting out they arrived in Adelaide.

For the Japanese, bombing Darwin was a textbook operation. In 45 minutes, they sank eight ships, set eleven on fire and forced four to beach. A total of 243 people were killed, 160 of whom were on ships. The Japanese lost five aircraft. Jack Mulholland, an RAAF anti-aircraft gunner, remembers, 'All hell broke loose. The raid was planned to the minute and brilliantly executed. The *Neptuna*, a merchantman, was tied up at the wharf discharging her cargo, which contained a large supply of ammunition. Several bombs hit her and she exploded. The Lewis gunners on the oil tanks nearby saw some strange objects flying through the air, including a whole locomotive which had been deck cargo.'

But it was the aftermath that turned the attack into Australia's day of shame. Panic swept the city. Everyone thought the raid was a prelude to an invasion and that Japanese troops, portrayed in Australian war propaganda as sub-human, would soon be wading ashore intent on rape and murder. There was a stampede to get away. One Greek businessman, however, kept his head. As civilians from town streamed south, the Greek stood on the roadway with a wad of cash offering to buy the house of anyone prepared to accept his ludicrously-low offer. Many did.

Some officers behaved the worst. They commandeered cars, trucks and jeeps and headed south, turning up days later in Alice Springs, Adelaide, Sydney, Melbourne and Brisbane. Their men did their best to follow their example. When the army arranged a train to evacuate women and children, able-bodied men tried to force their way on board. Then an air-raid alarm sounded, so everyone got off to take shelter and to allow the train crew to drive the train out of the station to avoid being bombed. The alarm turned out to be false but the train failed to come back to collect its passengers – the crew kept it going until they were a safe 300 miles away.

Some of the military personnel in Darwin were under orders not to resist a Japanese landing. Frank Beale, who was with a RAAF construction unit, remembered in 1993: 'We were told that if the Japs invaded we were to drop everything, take to the bush, making our way to the Army front line at the Adelaide river, and not try to stop the Japs. Fifty chaps with sixteen rifles and less than two hundred bullets. One of the big jokes of the war. One wag wanted to know about how we used sticks.'

Those still in Darwin, both military personnel and civilians, now systematically looted the place. A group of provost soldiers raided the freight office of Qantas Airways and while they carried off their spoils, one of the soldiers held the airline staff at bay with a broken bottle. Government House was stripped bare. Many servicemen took their loot down to the harbour, found a freighter and sent to their friends and families in the south of the country crates of household possessions that belonged to people who had left Darwin believing that their property would be safeguarded by the Army. Nurses went out to the military hospital and worked day and night looking after the wounded, only to return to their quarters for a brief break to find that all their possessions had been stolen.

The looting went on for months, but it never became a public issue because the government, determined to play down the raid so as not to shatter national morale, had imposed almost total censorship. In fact, the government itself may not have known the true situation, because the Administrator, Charles Abbott, kept sending it telegrams saying that everything was 'back to normal' when there was hardly a building, including his own house, that had not been stripped bare. Abbott explained himself later by claiming that he did not want news of the looting 'to leak out'.

Casualties were given as fifteen killed and losses to shipping 'comparatively small'. But many Americans had died, especially on the *Peary*, and when the American press carried reports of the raid that were close to the truth, Prime Minister Curtin was forced to lie. He issued a statement saying that what he had said after the raid was accurate and 'nothing has been hidden . . . the Government has told you the truth'. The truth was that Darwin was not better defended because the government had decided it could not defend it at all.

Curtin had been fighting a bitter running battle with Churchill over what troops Australia needed in order to have at least some chance of resisting a Japanese invasion. After the fall of Singapore, Britain had ordered a fall-back to what it termed 'essential bases' – Burma, India, Ceylon and Australia. Churchill immediately made it clear to Curtin that of these four, Australia would be the last priority. Further, he wanted to use the three Australian divisions in the Middle East to reinforce Burma rather than return to Australia. Realizing that Curtin would resist this, he enlisted President Roosevelt's help to put pressure on Curtin to agree.

The exchanges between the two wartime leaders mark the beginning of the end of the special relationship between Australia and the Mother Country. Churchill was at his bullying worst. He again accused Curtin of bearing a heavy share of the responsibility for the loss of the British 18th Division in Singapore and then threatened that if Curtin did not agree to divert that Australian division, already on the high seas, to Burma he would use his influence with Roosevelt to get the United States to withhold further support for Australia.

Churchill's cable demanded an immediate answer, 'as the leading ships of the convoy will soon be steaming in the opposite direction from Rangoon'. In fact, the British Admiralty had secretly already ordered the convoy to change direction away from Australia and head for Burma, despite a decision by the Chiefs of Staff two days earlier that the convoy would not be ordered to divert until Australia replied to Churchill's cable.

Further, the Resident Australian Minister in London, Sir Earle Page, knew of this decision by the Chiefs of Staff, and had cabled Curtin assuring him that no orders had been sent to divert the convoy. When the Admiralty then went ahead and did indeed issue just such an order, the British Prime Minister's office did not bother to inform Page. Why? General Ismay, the British Chief of Staff, in an offhand note to Churchill justifying this amazing omission, said simply that things were busy that day and 'the matter did not seem of any great consequence' because the Australian government's reply would arrive before any complications ensued.

It did not. When Curtin's cable arrived in London on Sunday morning, 22 February, refusing to allow the Australian divisions to go to Burma, the troopships carrying the men had already steamed too far on a course for Rangoon to turn back for Australia without first refuelling. Churchill now had to admit to Curtin that he had diverted the convoy without consulting him. He said he had done this because he had never contemplated that Australia would refuse Britain's request, backed as it was by President Roosevelt. Now that it had, the convoy would have to refuel in Colombo. This would take a few days which would give Australia time to change its mind.

Cables of accusation and recrimination now flew between Curtin and Churchill. Curtin denounced Churchill's high-handed treatment of Australia and said he would hold him responsible if anything should happen to the troopships – there had been reports of Japanese submarines off Colombo. Churchill blamed the problem on the time Curtin had taken to reply and said he not only accepted responsibility for what he had done but would defend it publicly if it ever became possible to do so. To the end of his days he blamed Curtin's refusal to divert the Australian troops to Rangoon for the loss of Burma.

Curtin was more worried about the possible loss of Australia. Of the eleven divisions of the Australian forces, less than two divisions were effective troops, one was approaching combat condition, and the rest were composed of militia. In those early months of 1942, the 6th and 7th Divisions were at sea in the Indian Ocean between Suez and Fremantle in poorly protected convoys. A substantial part of the 6th Division would be offloaded in Colombo and the 9th Division was still in Palestine. The Royal Australian Air Force was equipped with almost obsolete aircraft and was short of engines and spare parts. The Royal Australian Navy did not have a single aircraft carrier. True, American troops had started to arrive and the Supreme Commander of Allied Forces in the south-west Pacific, General Douglas MacArthur, had turned up in Australia unannounced (for 'security reasons' the Australian government had not been advised of his impending arrival).

The *Chicago Sun*'s correspondent, H. R. Knickerbocker, reported, 'MacArthur's appointment and the knowledge that American troops are here in considerable strength have brought the first sure hope of victory.' This was a curious assessment because, as David Day has observed, 'MacArthur's generalship in the Philippines had been little better than Percival's conduct in Singapore.' And Knickerbocker would surely have been less confident if he had been party to a little scene in the train carrying MacArthur to Melbourne. His staff had just told him that the Australian defence plan envisaged withdrawing south and leaving the northern ports open to the Japanese. Historian Tony Griffiths describes the scene: 'When he heard this news, MacArthur was stunned. He turned deathly white, his knees buckled, his lips twitched and he whispered, "God have mercy upon us".'

This Australian 'plan', the infamous 'Brisbane Line', remains a controversial topic to this day. Writing about it in 1975, using as my source *The Reports of General MacArthur* (US Government Printing Office, 1966) I described it thus:

> Western Australia and everything north of Townsville would have to be surrendered. Perth and Darwin would have to be left to their fate, because there were simply not enough troops to hold them against a determined Japanese assault. The First Army was to defend the east coast from Brisbane to the Victorian border; the second was to provide local protection for Melbourne. Since civilians in those areas to be abandoned to the Japanese might have shown resentment about this plan if news of it leaked out, care was taken that it did not, even to the extent of deliberately not withdrawing those troops who were already there.

The suggestion that the Australian government under John Curtin had been prepared to abandon Australian citizens to the mercy of a Japanese invading force caused immediate outrage. I was told that there was no such thing as the 'Brisbane Line', it was a wartime myth. I had confused a 'military appreciation' (something that might have to be done if certain events occurred) with a 'plan' (a fixed course of action). If the Japanese had invaded, the end result – appreciation or plan – would, of course, have been the same.

In recent years more information about the 'Brisbane Line' has come to light. The phrase itself turns out to have been MacArthur's own. At a press conference on 17 March 1943, to mark the first anniversary of his arrival in Australia, MacArthur said that when he arrived a year earlier, 'It was the intention of Australia to defend along a line somewhere near the Tropic of Capricorn, which would be known as the Brisbane Line.'

In *Victory in Papua*, Samuel Milner, an official historian of the US Army in the Second World War, writes: 'General MacArthur arrived in Australia to find that the Australian Chiefs of Staff . . . had based their strategy for the defense of Australia on continental defense.' Milner says that MacArthur then changed this to a plan to defend Australia in New Guinea.

But where was this strategy for defending Australia on its own mainland set out, and was it the same as the Brisbane Line? On 18 February 1942, the day Singapore fell, the Minister for the Army, Frank Forde, submitted to the Australian War Cabinet an 'appreciation' prepared by Sir Iven Mackay, GOC Home Forces, of the danger facing Australia should the Japanese invade. Mackay was an experienced soldier – he had commanded the 6th Division when they had captured Bardia, Tobruk and Benghazi in the Middle East. (In fact, his record in 1942 was better than MacArthur's.) His 'appreciation' made gloomy reading. Australia did not have sufficient soldiers. Those it did have were mostly militia men and only partly trained and poorly equipped. In what was now a national emergency, the dispersal of these forces would lead to their piecemeal defeat.

Therefore, the government should decide which were Australia's vital areas and defend only these. He suggested that the vital areas were in the coastal strip from Brisbane to Melbourne. Military posts outside this strip should be defended if attacked but not reinforced.

This 'appreciation' is clearly the origin of the Brisbane Line, what MacArthur heard from his staff in the train on the way to Melbourne, and what he was talking about at his press conference a year later. It also virtually mirrors what I learnt in 1975 from *The Reports of General MacArthur* – 'military posts outside [the Brisbane Line] should be defended but not reinforced' does not differ substantially from 'not withdrawing those troops who were already there'.

So what we are left with is a decision as to whether the Brisbane Line 'appreciation' ever stood any chance of being implemented, even as a worst-case scenario. MacArthur clearly thought it did. Writing to Curtin in November 1943 he said, 'It was never my intention to defend Australia on the mainland of Australia. *That was the plan when I arrived* (author's emphasis), but to which I never subscribed and which I immediately changed to a plan to defend Australia in New Guinea.'

But MacArthur would say that, wouldn't he? The ultimate public relations general with a large staff of skilled press officers and a personal photographer who knew his chief's best profile, he devoted himself to polishing his image. Fresh from the loss of the Philippines, the worst defeat American arms had ever suffered, he set about restoring his reputation as a brilliant general.

At his first public appearance in Australia he wore thirty-six medal ribbons – nine rows of decorations – and when he addressed Australia's leaders in

Canberra he told them, 'We shall win or we shall die. To this end I pledge you the full resources and all the mighty power, all the blood of my countrymen.' It was clearly in MacArthur's best interests to suggest that he arrived in Australia to find its leaders sunk in defeatism, with a plan – the Brisbane Line – that involved sacrificing a large part of the country to the Japanese, and that he had immediately vetoed such a plan and took the fight north to the enemy.

But there is plenty of evidence that the Brisbane Line was no more than a fleeting thought by a distinguished Australian general trying to consider every worst-case scenario, that it had been rejected by *Australian* leaders before MacArthur had even arrived in Australia, and that it was the *Australian* leaders rather than MacArthur who developed the plan to fight the Japanese in New Guinea.

The minutes of a meeting in Canberra of the Advisory War Council (AWC) held on 23 February 1942 are headed, 'Future Policy and Strategy for Conduct of War in the Pacific'. Paragraph 2 records that the Australian Chiefs of Staff 'are to submit a *fresh* (author's emphasis) appreciation on the subject of the defence of Australia'. Although Curtin later defended General Mackay's Brisbane Line appreciation as an appropriate one at the time he had written it and refused to allow him 'to be the scapegoat of any controversy about what is called the Brisbane Line', the AWC felt that circumstances had changed and now wanted the country's military leaders to reconsider how best to defend Australia.

John Dedman, who was a member of the Australian War Cabinet at the time, has pointed out that Mackay's Brisbane Line appreciation could never have been a plan or a strategy for the defence of Australia because it was so limited – it dealt only with the role of the army component of the armed services. In its call for a fresh appreciation, the AWC specified that it wanted the Chiefs of Staff to take into account the return of the AIF divisions from overseas, reinforcement of Australian land forces by American troops and the fact that the United States would probably want to use Australia as a base for its operations against Japan.

Pending the fresh appreciation, the Federal government held consultations with ministers and service chiefs from New Zealand to work out a joint strategy for fighting Japan. The conclusions were approved by the Australian War Cabinet on 28 February, seventeen days before General MacArthur had even arrived in Australia and seven weeks before he became Supreme Commander of the South West Pacific Area. The two main points of the strategic plan were to secure the lines of communication between the United States and Australia and New Zealand, and 'to prevent the further southward movement of the enemy into Australia . . . via New Guinea'.

The new appreciation from the Australian Chiefs of Staff was ready by 5

March and covered the role of all three services. It was optimistic in tone, one paragraph going so far as to say, 'If there were adequate naval and air forces for the defence of Australia, an army of the numbers required could nearly be met from Australia itself – although a great deal of equipment would be required from overseas.' The appreciation stressed the importance of holding Port Moresby.

To sum up: there *was* a Brisbane Line (although its author never called it that) but it was part of a whole series of suggested defensive positions which might have been used if the worst happened, the Japanese invaded, and began to drive southwards. It was quickly superseded by a strategic plan to confront the Japanese in New Guinea and prevent them ever entering Australia. This plan was devised by the Australian government and not General MacArthur. The controversy over the Brisbane Line would never have arisen had not MacArthur mentioned it a year after his arrival as a means of dramatizing the change in plans for Australia's defence which he claimed to have made on his arrival – a claim which was clearly false.

In the event the first victory against the Japanese came not on the Australian mainland nor on any land at all, but at sea. On 5–8 May 1942, an American naval task force, supported by a small Australian contingent, joined battle with a Japanese convoy en route from Rabaul to invade Port Moresby in New Guinea. The interception was no accident. The Americans had broken some of the Japanese naval codes and so were able to send sufficient forces to the largely undefended seas of the south-west Pacific and intercept the Japanese, who had already reached the tip of New Guinea. The Battle of the Coral Sea went down in history as the first naval action in which surface ships did not exchange a single shot – indeed they never came within sight of each other.

The battle was fought entirely in the air, a fact noted by the Australian-born correspondent for the *Chicago Tribune*, Stanley Johnston, who was on board the American aircraft carrier *Lexington* until she sank. Johnston wrote that the battle showed how completely the carrier had displaced the battleship in importance in modern war, but his considered conclusion was lost in the lies that both sides told about the engagement.

The Americans claimed to have sunk between 17 and 22 ships, while the Japanese, describing the battle as a great victory, said 'the US is panic-stricken'. Curtin was cautious, telling Australians that no one could tell what the effect of the battle would be and that 'invasion is still a menace capable hourly of becoming an actuality'.

Was it – as later claimed – the battle which saved Australia? The American losses were greater than the Japanese, but the Japanese force turned back to Rabaul and therefore did not achieve its objective – the capture of Port

Moresby. The American historian S. E. Morrison gets it right in the official *History of the United States Naval Operations in World War II*: 'It was a tactical victory for the Japanese but a strategic victory for the United States.'

Either way, the Battle of the Coral Sea did not save Australia because at the time Japan had no intention of invading Australia. She was incapable of doing so. The Japanese Navy had wanted to occupy such key areas as Sydney, Brisbane, Townsville and Darwin. But the Japanese Army had complained that this would need ten to twelve divisions, and that it could not risk moving so many troops from Manchuria and other occupied areas at that time.

Whether the Japanese *ever* hoped to mount an invasion is debatable. They would certainly have taken New Guinea if they could have and they hoped to cut Australia off from the United States by taking New Caledonia, Fiji, Samoa and Palmyra and, ultimately, the Hawaiian Islands. But recent studies (see American Historical Publications, New York) have uncovered a shocking fact about Japan's war in the Pacific in the early 1940s – she had no firm, long-range strategic objectives. After Pearl Harbor, conquest took on a life of its own and the Japanese were quite unprepared for the extent of their successes, and were often at a loss as to what to do next.

All this occupied little thinking time for most Australians. They were more interested in what was happening on their doorstep – the war had come to Sydney Harbor. On Sunday night, 31 May, three Japanese midget submarines, each manned by two men, penetrated Sydney's defences, and although they missed two heavy cruisers in port at the time, they sank a depot ship killing 19 sailors, mostly Australian.

The midget subs had come from the 'Sydney Attack Force', five large I-class submarines which had assembled a mere 35 miles north-east of Sydney Heads. Three of the submarines had midgets clamped to their decks and the other two carried float planes in waterproof hangars. One of these had taken off on 23 May and had actually flown over Sydney Harbor in broad daylight on a reconnaissance mission. No one spotted it.

The pilot and observer reported seeing two cruisers and several other warships. The cruisers were the USS *Chicago*, fresh from the Battle of the Coral Sea, and HMAS *Canberra*, refitting in Sydney. There were also the American destroyer *Perkins*, the tender *Dobbin*, the Dutch submarine *K-9*, HM Indian Ship *Bombay*, a minesweeper, the Australian auxiliary cruisers *Kanimbla* and *Westralia*, the minelayer HMAS *Bungaree*, and the corvettes *Whyalla* and *Geelong*. These were enough juicy targets for any submarine attack, but the Japanese officer in charge of the operation, Captain Hankyu Sasaki, thought that the reconnaissance crew were wrong – they had somehow missed the British battleship *Warspite* – and he was determined to get her.

With insolent confidence, the Japanese made another reconnaissance mission at 2.30 a.m. on 30 May, flying down the Harbor at about 600 feet – low enough for the pilot, Second Lieutenant Susumo Ito, to see the welding torch flashes from the night shift at the dockyard. After the war Susumo Ito recalled: 'Then I could not find my way, and tried to see Mascot aerodrome. The map was wrong somehow. Suddenly Mascot switched on its landing lights. I don't know whether they expected a friendly plane, or mistook me for one. But it was very helpful.

'I flew back towards the Harbor, over the island where the cruisers were, and climbed into the clouds to miss some more searchlights which were looking for me. We flew out over North Head which showed up in the faint moonlight that came through the clouds. The submarine switched on a searchlight for a second to show me where to land. But it was not a good landing. The plane capsized and I nearly drowned. My observer and I managed to scramble clear and then swim to the submarine.'

The two Japanese had brought back with them sketches of ship positions and the opening to the anti-submarine net boom being built between Watsons Bay and Taylors Bay, and the crews of the midget submarines spent hours that morning studying them. They were not, as has been often said, on a suicide mission. Admiral Isoroku Yamamoto, who had planned the attack on Pearl Harbor, had personally briefed Captain Hankyu Sasaki for the Sydney attack and had ordered him, 'Be sure to recover the crews.' A rendezvous point for their pick-up had been fixed off Port Hacking, some 20 miles south of Sydney.

The midgets had no difficulty in getting as far as the boom. Sydney's coastal suburbs were only half blacked out – the 'brown out' – and further inland the street lights were still on. But the first submarine got itself helplessly entangled in the boom and its bow broke the surface. A night watchman at the Maritime Services Board spotted it, rowed across to investigate and, uncertain what it was, called out HMAS *Yarroma*, a converted pleasure launch armed with a machine-gun and small depth charges. But before *Yarroma* could go into action, the Japanese submarine blew itself up with its own demolition charge. The explosion was heard all over Sydney but since only the *Yarroma* knew what had caused it, all the other ships went to 'air action' stations.

Canberra put up a spectacular anti-aircraft barrage until she realized the real direction of the threat, and then joined other ships firing at a periscope that had been spotted near the USS *Chicago*, which was frantically trying to slip her moorings and get underway. The periscope belonged to the midget submarine commanded by Second Lieutenant Katsuhisha Ban. It submerged when the firing began, waited until it had subsided, and then fired both its torpedoes at the *Chicago* at point-blank range.

But they had been set too deep and both passed under the cruiser. One beached itself near Garden Island gun wharf and lay there, its propellors thrashing away in the air until they ran out of power. The other went under the Dutch submarine *K-9* as well as the *Chicago* and then hit the RAN depot ship *Kuttabul*, an old ferry that was being used for accommodation. The explosion lifted it right out of the water, debris flying in all directions. Then she quickly began to go down. Most of the 19 men who lost their lives in her were drowned in their hammocks.

Able Seaman Colin Whitfield had slept through the earlier battle – 'Guns don't wake me' – and was innocently walking along *Kuttabul*'s upper deck on his way to go on watch at midnight when the torpedo hit. 'I remembered that a table was supposed to be a good air raid shelter, so I ducked under one. Then the ferry began canting over and I thought it was time to make a move.'

Katsuhisha Ban and his submarine vanished. He must have found his way down the harbour again and out through the boom because extensive searches within the Sydney Harbor located nothing. But he did not make it to the rendezvous point and nothing was ever heard from him again.

The third submarine followed the *Chicago* out through the boom while the gate was still open and was spotted at dawn on the surface in Taylors Bay. It crash-dived but was depth-charged until big bubbles of oil and air began to break the surface. When it was raised soon afterwards, the two crew members were found dead. They had shot themselves in the head. Their bodies and those from the submarine that blew itself up were cremated with full naval honours and the Australian government arranged for their ashes to be returned to their relatives in Japan.

Seven nights later, one of the mother submarines, I-24, fired ten shells at Sydney from five miles off Bondi. They all landed in a triangle between Rose Bay, Bondi and Bellevue Hill, a densely populated area, but half of them did not explode because they were meant for use against steel ships, not bricks and mortar. No one was killed but one shell did confer on Ernest Hirsch, a 35-year-old electrical engineer, the distinction of being Sydney's first civilian casualty. Hirsch, who had arrived in Australia five years earlier as a refugee from Nazi Germany, was asleep in his bed in his top-floor flat in Rose Bay when when one of the shells came in through the flat's outer wall, crossed the room in which Hirsch's mother was sleeping, and then went out into the stairwell where it stopped without exploding.

Hirsch was buried under a pile of dust and masonry. His wife first made certain that their 18-month-old baby was not injured – in fact she was still asleep – then ran downstairs to the local air raid warden post. The wardens cleared the rubble away from Hirsch and carried him out into the street to await the ambulance. The Hirsches behaved as every Sydneysider liked to think they would have themselves in similar circumstances. Ernest smoked a casual cigarette until the ambulance came, and then his wife went back into the flat to sleep among the debris. 'The baby hadn't woken up and it was still our home,' she said.

The next day there was a row over whose job it should have been to declare an alert. The authorities announced a complete blackout, people queued to see the interiors of the shell-damaged houses, and there was a wild rush to take out war damage insurance. Then as it sank in that this might not be the only attack on Sydney, people began to move out of the eastern suburbs. Some locked their doors and went to live in rented accommodation or with friends in the Blue Mountains, out of reach of Japanese naval guns.

Others sold their houses and flats at knock-down prices. Most of the buyers were European refugees for whom a few shells or the prospect of a bombing raid did not hold the same terror as it obviously did for some Australians. Their gamble paid off. There were no more shellings or raids on Sydney and after the war, when the eastern suburbs became one of the most desirable residential areas in the city, property values soared and these far-sighted European migrants found themselves deservedly rich.

Despite these two attacks on Sydney, for most of us the war still seemed far away. The only rationing that impinged on our lifestyle was that of petrol and we seemed to manage, diluting our ration with kerosene. An ingenious neighbour even converted his car to run on coal gas and drove around with a large rubber gas bag secured to the roof rack.

At school, the playground was crisscrossed with slit trenches that always seemed to be filled with rainwater, and the windows of classrooms were coated with anti-blast tape. Horse racing continued in some states, as did many other sporting fixtures, and if the beer was sometimes in short supply in hotels, there was plenty of black-market liquor around for those who could afford it.

The fact that Australians were continuing their hedonistic lifestyle despite the war worried the Prime Minister, servicemen returning from the Middle East, and the keepers of the nation's sexual morals. Mr Curtin complained that when Australians enjoyed their high lifestyle they took up valuable resources necessary for the war effort. No one took any notice of him.

When the 9th Division, back at last from the North African campaign and ready to fight the Japanese, marched through the streets of Melbourne they got a great reception. But when they had been back a little while, they began to complain. 'Everyone seems to have plenty of money,' said a Brisbane-born private. 'But there's nothing to spend it on.' He suggested that they should invest in the War Loan which had been consistently under-subscribed. 'Some people here still don't seem to know there's a war on.' No one took any notice.

Police in all cities complained that there was too much drunkenness, particularly among young women who were accompanied by servicemen with plenty of money, and the Tramwaymen's Union said its members deserved double-time rates for coping with abusive and aggressive servicemen. 'The war may well leave a generation of Australia's willing young women corrupted by their exploits with servicemen,' said one clergyman. No one took any notice.

All this was coded criticism for the fact that the Yanks had arrived in Australia in large numbers – nearly 180,000 by mid-1943 – and were stealing all the girls. The GIs looked down on the Australians for their low pay, shabby uniforms, and lack of style. Australians resented the Americans for their high pay, smart uniforms and brash and successful manner with girls.

Ill feeling between the troops of the two countries occasionally reached such a pitch that in the so-called Battle of Brisbane in 1942, an American MP shot and killed one Australian soldier and wounded eight others. Censorship of what happened made matters worse because rumours exaggerated the casualties, as they did with the 'Battle of the Trains' in Queensland. The story is that a train load of Australian commandos on their way to the battle zone fought a pitched battle with a train load of GIs returning south on leave, after the GIs boasted of what they were going to do with 'your wives and sweethearts' in Brisbane and Sydney. But censorship at the time and disputes since over whether this happened at Rockhampton or Townsville, or somewhere in between, raise doubts as to whether it happened at all.

There were fights between Australian and American servicemen, usually fuelled by drink and boasting – the reader can imagine what happened to a drunken GI who threw open the doors of a Sydney bar crammed with Australian soldiers and shouted: 'Here I am you Aussie bastards. Come to save you from the Japs.' But many of the fights were between Australian and American soldiers on the one side, and the notoriously brutal American military police on the other. These usually started when the military police were clubbing a GI and the Australians intervened, demanding in typical Australian fashion that the GI be given 'a fair go'.

There was another aspect of the relationship between Australians and their American saviours – a racial one. There were more than 5,000 black Americans in labour units in the American forces, 2,000 of them in Townsville, busy enlarging the aerodrome there for American bombers. The local white Australians watched this work with growing unease and eventually the local Trades and Labour Council asked the Federal government to issue a statement about this use of black labour to 'allay any doubts which might exist in the minds of the workers as to the future of their hard won conditions'.

The government refused, pointing out that it had agreed to admit black construction units to Australia and any public statement on the subject 'might only serve to emphasize this fact and draw attention to their presence'. In other words, let's keep quiet about these black Americans and hope that no one notices them. From the official point of view this was a wise precaution. In 1928 Sonny Clay brought a jazz band of black Americans on a tour of Australia. The press, aided by the police, persecuted them, following them everywhere and watching their parties through windows – 'The niggers and girls partly discarded their clothes and glistening black arms wound round white shoulders,' said *Truth*.

Some politicians supported this persecution. A former Prime Minister, Billy Hughes, lamented the fact that the black musicians had not been lynched, and lashed out at America's claim to be the melting pot of Europe. 'She got the people there but they did not melt. Jew and gentile, the scum of Europe.' He worried about the future of Australia. 'Are we to be subservient to the dago? We believe in a White Australia Policy and a British White Australia Policy at that.'

In keeping with this racial sentiment, when the American army proposed to billet some of the black American workers on a Sydney sports oval, the local member of Parliament protested and another site had to be found – in a railway goods yard. Britain considered that Australia had handled this problem of black American servicemen so well that it asked for its advice on how to restrict their number in Britain without upsetting the American government.

It need not have worried. Race was an equally sensitive issue in the United States, because although black Americans were encouraged to fight for their country, they were not to get the idea that this would entitle them to civil rights in peacetime. So blacks in the army served in strictly segregated units and in the navy did all the menial work. The all-white Marine Corps refused to have any blacks at all. Any media material showing blacks mixing with white soldiers, or relaxing off duty with white girls, was banned.

The propaganda films, *Hollywood Canteen* and *Stagedoor Canteen*, defined democracy as a system where all worked together – black and white, short and tall, people from many different religious and national backgrounds. But neither film, nor any other American wartime film, showed racially mixed dancing at these social centres.

The Battle of Midway in the first week of June 1942 took the pressure off any early invasion of Australia. Japan intended to destroy what was left of American naval power in the Pacific, invade Hawaii in December 1942 and then hope for a compromise peace whereby she would be allowed to retain what she had seized of the British and Dutch empires in south-east Asia.

Although Midway was an uneven confrontation – 200 Japanese ships against

76 American – the Americans sank four of the Japanese carriers and forced Japan on to the defensive for the rest of the war. When the news of this victory arrived, Australians, the eternal optimists, immediately set up a committee to look at the best ways of implementing demobilization.

Australia had had nothing to do with the Midway victory – it was a purely American affair – but by now it was heavily engaged in a jungle war against the Japanese in New Guinea, a dirty, debilitating battle against a determined, battle-hardened foe. For a long while the facts of this war were obscured by allegations of incompetence and cowardice on the part of the Australian soldiers. But today they are credited with being the men who saved Australia.

CHAPTER 13

THE WAR YEARS: THE BOYS WHO SAVED AUSTRALIA

'We'd survived. We didn't blame anyone. I didn't know anything about war and I don't think any of the others did, either. We thought all wars were like that. Badly managed.'

– Private Jack Manol, 39th Militia

On 9 November 1942, in the steamy heat of a New Guinea afternoon, the ragged troops of the 21st Brigade AIF were paraded before their commander, General Blamey, near a village called Koitaki, a rest camp near Port Moresby. Throughout the four previous months the Australian soldiers had been been fighting a gruelling jungle war against the Japanese along a mountain track known as Kokoda.

The Japanese South Seas Force, some 13,000 strong and well supported by mountain guns, had planned to advance from the northern beaches of the Buna-Gona area of New Guinea, up over the Owen Stanley Range and down to take Port Moresby, the gateway to Australia.

The battle to stop them had taken its toll. The Australians' skin was waxen from exposure to the tropical climate. They had not had a proper meal in months. They were exhausted, their eyes sunken, their feet torn and bloody. They were dirty, scruffy and unshaven and many were suffering from diarrhoea or dysentery.

But they were quietly proud of themselves. For one thing, they had survived. And during their long withdrawal along the Kokoda Track they had worn down the superior forces of the advancing Japanese, maintaining themselves as a threatening fighting force between the enemy and his goal.

Other Australians who had taken part in the battles had been praised for their fine work on the Kokoda Track. The 39th Militia battalion had been congratulated by Colonel Ralph Honner, one of Australia's great battalion commanders, at an earlier parade at Menari. Honner had praised his troops and had emphasized that they should not blame their fellow 53rd Battalion that had broken and left their flank exposed. In 1995 Sergeant Jack Sim, who was there when Honner spoke, recalled what an impression it had made on him. 'I'll always remember what he meant and what he implied when he said, "You're all Australians and some things you've just been through you must forget. Some of the men who were with you, you feel have let you down. But they didn't. Given different circumstances they'd be just the same as you. The fact that their leaders may have failed them, and yours didn't, doesn't mean they're any worse than you are."'

Now the men of the 21st were in for a bitter shock. General Blamey climbed on to a small platform and began to address his troops. They could hardly believe the words they were hearing. They had been defeated, Blamey said, and he had been defeated, and Australia had been defeated. This was simply not good enough. Every soldier there had to remember that he was worth three Japanese. In future he expected no further retirements but advance at all costs. 'Remember it's not the man with the gun that gets shot,' Blamey said. 'It's the rabbit that is running away.'

As the men realized that Blamey, intentionally or not, had implied that they were rabbits, a murmur of protest ran through the ranks. An army doctor recalled that the entire parade, officers and men, were 'almost molten with rage and indignation'. One man, his hand on his side arm, was moving towards the dais when he was hauled back by his fellows. In 1995, Sergeant John Burns said he considered that Blamey was 'lucky to get out of it alive that night'. There were further protests as word of Blamey's attack on his men spread throughout New Guinea and then to the Australian mainland.

Later that year, when Padre Fred Burt began telling public meetings in Perth what had happened, he was hauled before the Adjutant General of the army, Major General C. E. M. Lloyd, who tried to excuse Blamey, arguing that mistakes are inevitable in any campaign. Padre Burt was having none of this. 'On this occasion, the men who saved Australia in spite of your mistakes are the men who are blamed,' he said. Then without waiting for Lloyd's reply, Burt walked out.

Blamey's behaviour at the Koitaki Parade arouses anger in many Australians, even today, because behind it lies a story of bravery and dedication, frustrated by the apparent determination of Australia's leaders to play a subservient role to the United States, to replace Mother Britain with Uncle Sam.

When it had looked as if a Japanese invasion of Australia was imminent, the government rushed troops to New Guinea to meet any Japanese attack there. Most of Australia's experienced soldiers were either on the high seas on their way back from the Middle East or, like the 9th Division, still there. So the bulk of the New Guinea force had to come from the Militia, who were conscripts, known as 'Chocolate Soldiers', or 'Chocos' for short.

They had had little or no training and were thought not to be highly motivated. The conditions of their conscription specified that they would not be be asked to serve abroad, like the volunteers of the AIF, but only within Australia. The government felt that they would be resentful at being told that they could be required to fight in New Guinea even though this was legally Australian territory.

The government was right. Some of the men had to be press-ganged into service. Private Jack Boland of the 39th Militia, remembered in 1995 what happened when his troopship was about to sail from Sydney to New Guinea. 'They marched a bunch of boys on board under guard. They'd been press-ganged. They'd picked them up at home that morning, took them to a camp and brought them straight to the boat. There were women – mothers and sisters – screaming and everything.'

But these were the young men – average age only twenty-one – who were among those about to inflict on the Japanese army its first defeat on land in the war, the ones who showed that the Japanese soldier was not invincible. Their battles were fought largely alone – the Americans did not arrive until much later – and even today most Australians know little of their achievements. Gallipoli, a defeat, remains better known in Australia than the victories in New Guinea.

They fought in what the film-maker Chris Masters, echoing the Australian military historian, Peter Brune, calls in his masterly documentary *The Men Who Saved Australia*, 'a battlefield designed by the Devil'. The rain, the mud, the undergrowth, the disease, the lack of supplies and proper jungle-fighting equipment, the mistakes in leadership, the difficulty in getting medical aid for the wounded, the implacable nature of the enemy, all combined to turn them into an army of deadly scarecrows.

The condition of the men is best described by an AIF captain, Keith Norrish. Crouched in the undergrowth, rain streaming down, visibility limited, feverish and hungry, his feet bloody and diseased, he suddenly heard a mysterious, soft moaning sound. Was it a Japanese? 'I couldn't work out where it was coming from,' he remembered in 1995. 'I listened a bit harder and then I suddenly realized it was me.'

The Chocos soon learnt what war was about. Jack Manol, from Sydney, a private in the 39th Militia, recalled in a 1995 television interview what happened as he was pushing his way through the undergrowth. 'Suddenly I came face to face with this Japanese officer. He was just as bloody scared as I was. Luckily I pulled the trigger first. It was an experience I never want to have again. It's haunted me for years. When I went through his equipment – that was my job – I found a photograph of his wife and three little kids.' At this point in the interview Manol broke down, wept, and could not continue.

In mid-August the 'Choco' 39th Battalion from Victoria dug in at Isurava on the Kokoda Track to resist the Japanese advance. Earlier they had made two unsuccessful attempts to hold the Kokoda airstrip before withdrawing to Deniki and then Isurava. During this fighting their commanding officer had been killed and replaced, just before Isurava, by Ralph Honner. Behind the 39th were elements of another 'Choco' battalion, the 53rd from New South Wales, poorly armed and inexperienced. Their biggest weapon was an obsolete First World War machine gun and some of the soldiers in the 53rd could not even assemble their light machine guns.

By now some 10,000 troops of the Japanese South Seas Force were about to concentrate against the 39th. When the Japanese came, their bayonet charges, heralded by a bugle call, were terrifying. Yet the Australians held on until the 2/14 battalion of the AIF 21st Brigade arrived to reinforce them. Fresh from the Middle East, fit and tanned, they appeared to the exhausted 'Chocos', as one said, 'like Gods'. Outnumbered and out-gunned, the 2/14, supported by the 39th Brigade, fought the Japanese for four crucial days, fatally disrupting the Japanese 10-day timetable for crossing the Kokoda Track and taking Port Moresby, from where they could dominate the Coral Sea. Ralph Honner's son summed it up at his father's funeral over fifty years later: 'The Japanese were doing pretty well until they ran into my father and five hundred Australian Diggers.'

The Australian resistance astonished the Japanese. In 1995, Shigenori Doi, formerly a private in the 144th regiment, recalled, 'When we charged we saw, about two hundred metres away, a young Australian soldier wearing only a pair of shorts. He was holding a grenade in his hand and he attacked us. He attacked us and threw the grenade. Even a Japanese soldier would not have had such courage.'

But the Australians, short of promised supplies, could not hold on forever and they retreated down the Kokoda Track back towards Port Moresby. Those wounded soldiers able to walk did so (one with four bullets in his chest walked for five days) and the others were carried by their fellow soldiers and by New Guinea natives, the 'Fuzzy Wuzzy Angels' of Second World War legend. It is fascinating to hear what these Australians remember of that retreat – the uncomplaining endurance of their comrades, the caring compassion of the native bearers, and simple events, often to do with food.

One soldier remembered in 1995: 'I could see something glistening in the dark a bit further down the track, but I couldn't make out what it was. Finally I got up and went down to have a look. It was a tin of IXL blackberry jam.

We levered it open with a bayonet and spooned it out. It was the most beautiful thing I've ever tasted.'

There were unexpected surprises. 'We came around a corner on the trail and there was a Salvation Army post. A boiling hot cup of tea, some Arnott's milk coffee biscuits, a packet of three threes [State Express 333 cigarettes], and a piece of chocolate. Unforgettable. It's marvellous the pleasure you get out of simple things when you're desperate. Thanks to the Salvation Army. That's why today they never have any trouble getting a quid out of me, those blokes.'

But back in Australia, this fighting withdrawal which had worn down the Japanese strength by half and was about to make the Japanese high command decide to abandon the attempt to take Port Moresby, was seen as a defeat. Panic set in, especially when General MacArthur announced that it was all Australia's fault. In truth he was the one who had failed to appreciate Japan's determination to take Port Moresby – he had assured Curtin that the Japanese could not attack the town with any strength over the Owen Stanleys – and therefore had not deployed sufficient forces early enough. Worried that his position might now be under threat from his enemies in Washington, who were many, MacArthur decided to make the Australians the scapegoats for New Guinea and General Blamey the biggest one of all.

First he planted the suspicion in Washington that the Australians might not be up to the challenge. He signalled to the US Army Chief of Staff, General George Marshall, that the Australians had 'proven themselves unable to match the enemy in jungle fighting. Aggressive leadership is lacking.' Then he began to undermine Blamey and the Australian high command, cleverly mixing flattery with devastating criticism. Blamey was 'a tough commander, likely to shine like a power-light in an emergency'. But he was also 'sensual, slothful and of doubtful moral character'.

Finally, he went to Curtin and recommended that Curtin should order Blamey to go to New Guinea to take personal command. If the Japanese now took Port Moresby, Blamey could take the blame. Or, as one of the Australian ministers put it, 'Moresby is going to fall. Send Blamey up there and let him fall with it.'

What was really at stake here was who was commanding the Australian Army – MacArthur or Blamey? By persuading Curtin to order Blamey to New Guinea, MacArthur had effectively removed him from command of the Allied Land Forces in Australia. Blamey knew this and was determined to seize control of the New Guinea operations and save his military career. He did not care who or what stood in his way. This put him in conflict with the press, politicians and his fellow officers, and the fact that these battles behind the lines went on while men were dying at the front constitutes one of the more remarkable episodes in Australia's military history. It is a story of such rampant ambition, betrayal, backstabbing and intrigue that ancient Rome would be envious. One Australian officer said that he was more afraid of being knifed in the back by his own headquarters than he was of the Japanese.

If Blamey was to run the campaign in New Guinea, then the current commander, Lieutenant General Sydney Rowell, was superfluous. At first Blamey tried to convince Rowell that this was not so, asking him not to think that 'my coming implies any lack of confidence in yourself'. But Rowell saw through this. In the assessment of Blamey's biographer, Dr David Horner, of the Australian National University, 'Rowell thought that Blamey was coming to Port Moresby "when the tide is on the turn and all is likely to be well. He cannot influence the local situation in any way, but he will get the kudos . . ."'

It was just a matter of time before Blamey sacked Rowell, and not long after Rowell had nearly thrown a boot at Blamey – as he confessed later – he was relieved of his command. Rowell was furious. In letters and conversations he referred to Blamey as 'an evil cancer in the body of the public', spoke of Blamey's 'rotten influences', and described him as being like 'all crafty gangsters'.

Rowell had admirers, one of whom, Chester Wilmot, the ABC war correspondent, thought that Rowell was a fine fighting soldier while Blamey was unfit to be commander-in-chief of the Australian forces. Rowell had sent Wilmot and two other correspondents – Osmar White of the *Herald and Weekly Times* group in Melbourne, Damien Parer, the official war cameraman, who was shooting both film and stills – up the Kokoda Track so that they could see for themselves the conditions under which Australian troops were fighting.

This was where Parer shot the footage for *Kokoda Frontline* which, when edited and produced by the Cinesound legend Ken G. Hall, won Australia's first Oscar for best documentary of 1942. Parer's film and Wilmot's radio dispatches told the story of the stubborn resistance of troops who lacked supplies, jungle training, proper uniforms and equipment against an enemy with all these advantages.

Since Wilmot was reporting for radio, his dispatches had the attraction of being immediate and therefore had the capacity to improve matters. But Blamey's censors refused to pass a word of what Wilmot had written. Wilmot had refused to accept this and had gone straight to the American censors at MacArthur's headquarters. Since Wilmot's sentiments echoed MacArthur's criticisms of the Australian high command, they passed most of his dispatches *in toto*, to Blamey's fury.

So when Wilmot was back on a visit to the Australian mainland he went to see Prime Minister Curtin, told him what he thought had gone wrong with the New Guinea campaign and urged him to overrule Blamey and reinstate Rowell who, he said, was being victimized. Wilmot said later that Curtin had listened to him carefully and had then told him that he was well aware of Blamey's shortcomings, but that Blamey had been spot-on with his predictions of the Japanese Army's moves. In any case it would be impossible to dismiss the commander-in-chief when the nation was in such dire peril.

It did not take Blamey long to learn of Wilmot's action. It provided the excuse he had been long awaiting. On 1 November 1942 he summoned Wilmot to his headquarters, cancelled his accreditation and ordered him to return to Australia. He said Wilmot had been guilty of trying to 'undermine my authority as commander-in-chief'. This was true, but there was more to it than that.

Wilmot had been intermittently pursuing the allegation that Blamey had received cash 'kickbacks' from the contractor who exhibited films for the troops in the Middle East (see Chapter 11) and had been seeking further information in Melbourne from other Australian army officers. Blamey had heard of this and now felt it was time to get rid of Wilmot for good.

But Wilmot had friends. Charles Moses, the general manager of the ABC, considered Wilmot a brilliant correspondent with a bright future. Moses had joined up in 1941, but after serving in Malaya, where he had organized Gordon Bennett's escape, he returned to the ABC at the request of Prime Minister Curtin, who was worried about a drop in standards during Moses's absence. Still connected to the army, he therefore had a foot in both camps. He worked out a deal to get Wilmot back to the front. Wilmot would swallow his pride and write a letter of apology to Blamey, if not grovelling then certainly contrite. Blamey would graciously accept it and then restore Wilmot's accreditation.

It all went dreadfully wrong. Wilmot wrote the letter, Moses made certain that Blamey received it, and then Blamey double-crossed them both and refused to reinstate Wilmot. The ABC was unwilling to confront Blamey openly, so although it continued to make representations on Wilmot's behalf, it was all done behind the scenes and to no avail.

For the next 17 months, thanks to Blamey, Australia's best war correspondent was unable to report the war. Then Moses heard gossip that Blamey had not finished with Wilmot and was about to use the Manpower Act to conscript him into the army as a private. 'A latrine unit was waiting,' Moses recalled. Remembering that the BBC had been impressed with Wilmot's broadcasts from Tobruk, Moses cabled London: 'Wilmot available.' The BBC promptly offered Wilmot a job as one of a team of broadcasters to cover the Allied invasion of Europe for what became known as 'War Report'.

When Blamey learned that Wilmot was about to escape from him he wrote to the British Army advising it to withdraw Wilmot's accreditation. It referred Blamey's letter to General Rowell, who was in Britain on secondment, working on the planning for D-Day. Rowell was only too pleased to give Wilmot a glowing report and tell the British what he thought of Blamey. When later someone thought they had better get General Montgomery's views on Blamey, he replied in his laconic fashion, 'There was the drink and then the women.'

Wilmot landed in a glider on D-Day and went on to become a famous war correspondent in Europe, a war historian (*The Struggle for Europe*) and a broadcaster of renown. He died in the Comet jet crash in the Mediterranean in 1954. (Blamey, who became Australia's only Field Marshal, had died three years earlier.)

The Koitaki Parade, where Blamey berated many of the men who had helped save Australia had a grim legacy. As historian Peter Brune discovered, it planted in the mind of the new commander of the 21st Brigade a seed of doubt as to the men's worth and Blamey's sacking of Rowell and other commanders so that Blamey would appear to be 'energizing the situation' put his officers under pressure to deliver quick victories.

So at Gona, over the protests of the battalion commanders, the Australians were ordered to make a series of futile frontal attacks against the Japanese without even scouting the ground. It was the First World War all over again, only this time it was Australian commanders who were wasting their own soldiers' lives. Gona was taken only when the 39th Battalion arrived, and Ralph Honner ignored his orders and attacked where he knew his battle-hardened troops could break through. Photographer George Silk arrived in time to photograph the piles of Japanese corpses.

MacArthur received his comeuppance when raw American troops broke and ran outside Buna. The position was then taken by a single Australian brigade. When MacArthur promised further American reinforcements, Blamey twisted the knife for all his countrymen by replying, 'I'd rather have Australians. At least I know they'll fight."

Historian David Day says that from the time the Japanese first landed at Buna on 21 July there had been a lot of pressure on MacArthur and Blamey but 'they had survived by ruthlessly watching out for their own interests'. MacArthur set the pace. Blamey followed him, but in so doing lost the respect of the ordinary Australian troops. One of them, a 'Choco', Private Manol, deserves the last word. 'We'd survived. We didn't blame anyone. I didn't know anything about war and I don't think any of the others did, either. We thought all wars were like that. Badly managed.' And the 6th Division veteran Bill Broes, who had seen action in the Middle East, Greece, Crete and New Guinea, summed up the emotions of many an AIF soldier at the end of the war: 'We disembarked at Sydney without fuss or ceremony. There were no flags or cheering crowds. None was expected. We were just glad to be back in one piece.'

Once the Japanese had abandoned their attempt to take Port Moresby and were defeated at Buna-Gona, the war in New Guinea swung in the Allies' favour. Not recognized at the time – although fully appreciated by the British who fought the Japanese in Burma – the Australian forces had repelled a Japanese invasion force and inflicted their first land defeat of the war at Milne Bay, victories that could be seen as the beginnings of the Japanese defeat in the Pacific. In Peter Brune's words on the fighting along the Kokoda Track: 'The soldiers of "Maroubra Force" took on an enemy that outnumbered them five to one and held them until the Japanese gave up and turned back. These were the ragged bloody heroes who saved Australia.'

As the war moved away from Australia and it became clear even to the dour pessimists that their country was not going to be invaded, life on the home front returned to its pre-war hedonism – with the added spice and excitement of the presence of so many Americans.

Pubs and nightclubs were packed. The black market – especially in alcohol – thrived. Many a postwar fortune had its origins in the sale of beer and whisky to thirsty American servicemen. I still went to the cinema once a week but most of the films were about the war in Europe and featured stiff-upper lip Englishmen and brave working-class heroes – *San Demetrio, London*; *One of Our Aircraft is Missing*; *In Which We Serve*. Then the Hollywood wartime blockbusters began to arrive – *Stage Door Canteen*; *This is the Army*. Most of us preferred the American version of the war, and at my school, Canterbury Boys' High, we became experts in identifying American aircraft and following MacArthur's island-hopping progress towards the Japanese mainland. And we longed for school holidays and an end to examinations so that we could go into the city and wander around and maybe, just maybe, find a girl who would take our arm and walk and laugh and joke with us like they did with Americans.

By now reports had begun to filter back about the treatment of Australian prisoners of war by their Japanese captors, but it was not until the war was over that we learned what it was really like – malnutrition, beriberi, malaria, dysentery, tropical ulcers, cholera, rotting clothes, forced marches, beatings and humiliation. For the first time white Australians were caught up in the expression of man's inhumanity to man, something they thought only happened to others in far-away places.

But even then, in the direst of circumstances, some special Australian collectivism broke through. In 1998 Tom Uren, a distinguished Parliamentarian, remembered his time in Hintock Mountain Camp near Hellfire Pass on the notorious Burma–Thailand railway, where he served with Dr Weary Dunlop, a true Australian hero. 'We had two other quite remarkable doctors, Arthur Moon and Ewan Corlette. They combined medical ingenuity with leadership and comradeship. They acquired medical supplies and drugs by trading on the black markets with the Thai and Chinese. But like all of us they worked as a part of a team. We were living by the principle of the fit looking after the sick, the young looking after the old and the rich looking after the poor.'

Uren contrasted this Australian way with the British way. 'About 400 men from British "H" force arrived at the camp after us. The officers selected the best tents, the non-commissioned officers the next best and the men got the dregs. Soon after they arrived, the wet season set in, bringing with it cholera and dysentery. Six weeks later only about 50 men marched out of the British camp, and of that number less than half survived. Only a creek separated our two camps but on one side the law of the jungle prevailed and on the other our collectivism.'

Uren had other memories of the ways the Australians struggled to maintain their collective spirit, like their 'get-togethers'. At one of these gatherings, before 800 men, Major Alan Wood recited Banjo Paterson's heroic poem of bushland myth, *The Man from Snowy River*. 'Woodsy recited it in the logfire-lit night, a slouch hat on his head, his face proud and expressive, his jaw jutting.

'"There was movement at the station for the word had passed around . . ."

'All you could hear apart from Woodsy's voice was the crackle of the fire. This great courageous Aussie major captivated his audience. Never before or since have I heard *The Man from Snowy River* recited so powerfully. It enhanced my love of our country and made me proud to be an Australian.'

The Labor government in Australia, determined to make a country fit for its heroes when the war was over, pushed ahead with all sorts of legislation for social reform, even when the outlook was at its blackest. These laws indicate the sort of Australia these Labor leaders envisaged. A National Welfare Scheme provided benefits for sickness, unemployment, old and invalid pensions and maternity payments. The Minister for Labor, E. J. Holloway, said that the government regarded the scheme as 'a piece of bridge building to carry people over those economic gaps which occur from time to time. The government wishes to ensure that people during periods of unemployment shall remain a social asset by retaining some spending power.'

And although the war was on, this was a good moment to introduce such legislation. 'At the present time, unemployment is practically confined to the unemployable. Moneys set aside now for the purpose of paying benefits will establish a reserve which will be available should unemployment increase in the future.'

But the old battle between bosses and workers also continued despite the war. A waterside workers' strike in Sydney in April 1943 was broken by Australian and American servicemen. Tram and bus workers went on strike the following year in protest at the government's refusal to release servicemen to the transport industry. Journalists, printers and other newspaper workers went on strike in October 1944 to press for a shorter working week and longer paid annual holidays. There was a miners' strike in northern New South Wales in August the same year, with 4,000 miners demanding increased wages and improved working conditions, and the waterside workers were out again just after the war's end in support of the Indonesian independence movement. In the field of Australian industrial relations, nothing much had changed.

It is difficult even today to grasp the scale of the Second World War and the influence it had on our civilization. From the skies over Britain to the icy vastness of the Russian plains, from the deserts of North Africa to the jungles of Burma and the Pacific Islands, armies the size and firepower of which the world had never seen before fought terrible battles of survival. Before it was over, every major power in the world had been engulfed and some sixty million people were dead, on average more than 20,000 for each day of the war.

What was it all about? Richard Overy, Professor of Modern History at King's College, London, points out that so much of what we believed during the war was propaganda and moonshine. It was not about the defeat of tyranny by democracy. In 1939 democracy barely existed outside the United States, France, Britain and her white empire (Australia, New Zealand, Canada.) It was about 'the very survival of democracy in its besieged heartlands' and victory 'made the world safe for Communism, which was as embattled as democracy in the 1930s and close to eclipse by 1942'.

And for Churchill to say on VE day, 'We were the first, in this ancient island, to draw the sword against tyranny; after a while we were left alone against the most tremendous military power that has been seen,' was pure nonsense. Britain entered the war with the French and after their defeat was not alone. What about Australia, Canada, New Zealand, South Africa and India, countries which provided, among other things, nearly half of the RAF pilots during the war?

Compared with, say, the Soviet Union, without whose tenacity the Allies could have defeated the Germans only by 'sitting tight and waiting until atomic weapons had been developed', or the United States, which lost the flower of her young men on the beaches of Normandy and in the hills of Okinawa, or Britain, who lost 30,000 merchant seamen alone – volunteers at £9 a month and two and sixpence a day danger money – Australia's contribution to victory was minor. But like every other country, its world too was irrevocably changed.

The sun had set on several empires, including Britain's. Australia's ties with the Mother Country, when it came to the crunch, had frayed all too easily. Total war had turned out to mean exactly that – everyone was a warrior, no one was innocent. Ordinary men and women learnt that, given the right circumstances, they could carry out acts from which they would normally recoil in horror. The illusion that there were limits to man's inhumanity to man was shattered – the Holocaust and the atom bomb put an end to that idea.

Perhaps there were two wars – the one experienced by the soldiers and those civilians unlucky enough to be caught in the firing line, and the one experienced by those at home in the two lucky countries, Australia and the United States. For the first group, it was a horrible war. People were shot, bombed, incinerated, hanged, drowned, starved, gassed, worked to death, used for medical experiments, buried alive and beheaded.

There were incredible acts of bravery and defiance, especially among those defending their own soil. On 29 November 1941, in the village of Petriščevo, in the Soviet Union, the Germans led a civilian, an 18-year-old girl called Zoia Kosmodemianskaya, to a scaffold they had erected in the square. Around her neck they had tied a placard reading, 'She set fire to houses.' As they put the noose over her head she spoke her last words to one of the young German soldiers at her side: 'You can't hang a hundred and ninety million of us.'

In the killing zone, soldiers wet and soiled their pants in fear, vomited, screamed, wept and had mental breakdowns – one million American servicemen alone suffered psychiatric symptoms that put them out of action for some period. Normal human emotions kept overcoming the demonization of the enemy that was necessary to arouse the soldiers' fighting fervour. An Australian infantryman recalls a Dyak warrior presenting him with a bloody, dripping Japanese head, still wearing its spectacles. 'Somehow it brought home to me as nothing else that the enemy were fellow human beings. I was very grateful that the war ended soon afterwards.'

But for the second group, safe on mainlands that escaped invasion and serious bombing, life went on much the same as usual. If anything it was more exciting, and for some Australian civilians the war was the most thrilling and productive period of their lives. The Depression had shaken everyone, but between 1941 and 1945 business was good, wages went up, fortunes were founded and there was an exhilarating sense of joint effort. Women took a huge leap towards liberation and Australians who had been marginalized in peacetime suddenly felt that they had a role to play.

Secure in the belief that they could rely on their newly found uncle, the United States, to lead them to victory, far away and shielded from the reality of combat, watching war films in which the goodies always won, their emotions unnaturally heightened, is it any wonder that for many young Australians the war years were the best of their lives?

An American historian, George H. Roeder, Jr, has argued that to understand the Second World War you have to see it in Hollywood terms. The war was a big, blockbuster movie produced by the Allied governments with a script by the news and entertainment media. It had a compelling story line, a huge cast, easily identified goodies and baddies, and an outcome that the audience could anticipate but not take for granted. And with the dropping of the atomic bomb on Hiroshima, it had a climax as dramatic as the opening scene, the bombing of Pearl Harbor. Moreover, since the Allies had declared it to be a total war and the main propaganda theme was that everyone had a part to play, it was the first movie *everyone* could be in.

Australia's wartime leader, John Curtin, did not live to see victory against Japan. Worn out by the long hours and the strain of office, he died on 5 July 1945. He had predicted two years earlier that he might not see the war out. He had stood up for Australia against Churchill and started the unravelling of Australia's ties with the Mother Country with his historic speech – 'Australia looks to America, free of any pangs as to our traditional links or kinship with the United Kingdom.' But then in handing over almost complete control of Australia's military resources to General MacArthur, he made it obvious that Australia was unable to stand alone. True independence was still a distant dream.

CHAPTER 14

THE NEW SETTLERS

Some looked around for a place as far away from Europe as they could get and decided on Australia. 'The moon would have been my first choice,' said one. 'Australia came second.'

The war was over, the Allies had won and Australia had escaped invasion. But her population was only seven and a half million and the old worry about envious Asian eyes looking south at a large land with so few people quickly resurfaced. 'Armies recruited from the teeming millions of Japan threatened to overrun our cities and our broad hinterland,' said Arthur Calwell, the newly-appointed Minister for Immigration. 'They were so many, we were so few.'

The Labor government under the new Prime Minister, Ben Chifley, said Australia would have to 'populate or perish'. But where would the extra people come from? No one seriously thought that an increased birth rate would supply them. How about Britain, the traditional source? Many would certainly come to Australia – especially when the Australian government offered to bribe them by paying 40 per cent of the fare – but not in the numbers the government wanted: it was thinking of achieving a target population of 20 million people as soon as possible.

The answer was, of course, staring Australia in the face. The war had left millions of people in Europe homeless and destitute. They were living in displaced persons camps, their future bleak. As well, many people from countries that had been occupied by Germany wanted to get away from Europe and start a new life somewhere free from fear, free from a history in which war loomed large. Some looked around for a place as far away from Europe as they could get and decided on Australia. 'The moon would have been my first choice,' said one. 'Australia came second.'

There were, however, tricky problems. The Australian trade union movement had always been against immigration unless it could be shown not to cost Australian jobs and not to threaten wages. But now the unions were convinced that this would not be the case. Returning servicemen had been quickly reabsorbed into the workforce, many employers sacking the women who had replaced men during the war – 1940s and 1950s Australia still believed that a woman's place was in the home. (Although a tall, young girl called Germaine about to join Sydney University would later change all that.) If anything there was a shortage of labour, and competition for it was forcing up wages. The government and the unions saw no incompatibility between a vigorous immigration programme and its main plank, full employment.

But how to sell such a scheme to a population traditionally wary of outsiders? Calwell and his department thought that they had the answers. There would be no watering down of the White Australia Policy and to prove this the government began deporting the five thousand Asians and Pacific Islanders who had sought refuge in Australia during the war. Some had married Australians but this did not touch Calwell's heart.

'We can have a white Australia, we can have a black Australia, but a mongrel Australia is impossible and I shall not take the first steps to establish the precedents which will allow the floodgates to open,' he said. Then he deported the O'Keefe family – an Indonesian woman and her eight children who had married an Australian after her Indonesian husband had died. A Chinese family got an even harsher response. There were many Wongs in the Chinese community, Calwell told Parliament, 'But two Wongs don't make a white.' There was to be no exemption even for those Japanese women who had married Australian soldiers in the Occupation Force in Japan; or for the children of such marriages. 'No Japanese women, or any half-castes either, will be admitted to Australia,' said an Immigration Department instruction issued by Arthur Calwell in 1948. 'They are simply not wanted and are permanently undesirable.' (Eventually Australia relented and about 600 Japanese wives were admitted from 1952 onwards.)

Even those migrants coming to Australia from Europe should not look too different, the government decided, so the first wave were recruited from northern Europe rather than southern and had an Aryan or Caucasian look. The Baltic countries, home of 'the beautiful Balts', were an ideal recruiting ground. As Calwell wrote later, 'Many were red-headed and blue-eyed. There were also a number of natural platinum blondes of both sexes. The men were handsome and the women beautiful. It was not hard to sell immigration to the Australian people once the press published photographs of that group.' (The photographs had, of course, been published at the instigation of the Immigration Department.)

Some Jews could come to Australia but not too many. (The government was ahead of an opinion poll in 1948 that revealed that only 17 per cent of Australians were in favour of Jewish migration.) Calwell stipulated that Jewish immigrants must not exceed 25 per cent of the passengers on any migrant ship coming to Australia. Later he increased this quota to 50 per cent, but limited the total number to 3,000 a year.

There were two hidden agendas in this early postwar immigration policy, one to do with workers and one to do with politics. In return for being brought to Australia, the migrants had to enter into a contract with the Australian government under which they agreed to work wherever directed for a period of two years. Thus Australia created a pool of docile labour which it could move anywhere it wanted – just like in the convict era – and which was not in competition with Australians and therefore of no interest to the unions. Employers loved the idea.

But the Soviet government called it 'forced labour' and warned displaced persons emigrating to Australia that they were doomed to conditions of slavery. It also claimed that the German authorities were coercing displaced persons into going to Australia so as to reduce the problems of homelessness in Europe. Calwell angrily denied the Soviet charges but there was some truth in them. Although migrants received the same wages and conditions as Australian workers, they were not able to move freely and had to do whatever job the Australian authorities decided, whether they liked it or were suited for it or not. So doctors, dentists, nurses, musicians and teachers found themselves working in road gangs, in fruit-canning factories or as labourers in the construction industry.

High-up in the Australian Alps where forty-two different nationalities were working on the Snowy Mountains River Scheme, one of the world's greatest engineering projects (of which more later), a Czech ambulance driver asked to be allowed to watch an Australian surgeon performing a tracheotomy. As the doctor went to make the first incision, the driver yelled, 'No. No! Not there. Cut here!' The doctor brandished a manual at him and yelled back, 'Listen you mug, I happen to be following the instructions of a world authority on this surgical procedure.' And the Czech driver said, 'Yes, yes, but there've been developments since I wrote that.'

There were other criticisms apart from the Soviet one. Australia was accused of stripping Europe of all her young, strong and healthy people – the age limits were 45 for single men, 35 for single women and 50 for married persons with children – and leaving behind the maimed, ill and incapable. And there was the question of political screening. The government was worried that Communists had infiltrated the ranks of the displaced persons, so immigration officers were ordered to probe the past of all applicants to make certain they were not

Communist Party members, nor held Communist views. They were less concerned about Nazis. The chairman of the Immigration Planning Council, Sir John Storey, said at the time, 'Ex-members of the Nazi Party should not necessarily be excluded in the selection of migrants from Germany.'

As a result, not only did Nazi sympathizers manage to get into Australia but also a number of men wanted for war crimes. I saw an early suggestion myself that this might be so. At Bonegilla, on the border of New South Wales and Victoria, the government had set up a hostel where migrants could live while they were processed and allocated jobs. When their ships landed in Melbourne, the migrants were taken by train direct to the camp and all the immigration and customs formalities took place there.

I visited the camp to write an article about it and on my second day there a customs officer took me aside and said he might have a story for me. He took me into a storeroom behind the customs examination hall, unlocked the door and ushered me inside. It was like entering a Nazi shrine. Two tables were covered with Nazi memorabilia – busts and photographs of Hitler, Goering and Himmler, SS daggers, Nazi flags, copies of *Mein Kampf*, photographs of officers in SS uniform, Nazi party cards and pamphlets. The customs officer said that all this had been seized from migrants' luggage.

I could not find anyone prepared to publish my article about this, and it was not until 1987 that the Australian government, after fourteen months of parliamentary debate, decided to set up a Special Investigations Unit to root out any war criminals who might have come to Australia in that postwar mass migration. The unit went to the Ukraine to examine a Jewish mass grave and was preparing to prosecute a Ukrainian forest worker who had lived in Australia for forty years, when he shot himself. He survived his suicide attempt but the government abandoned the case and closed down the Special Investigations Unit soon afterwards. The Simon Wiesenthal Centre says, 'Australia is the only major Western country which admitted large numbers of Nazi war criminals and has failed to successfully convict a single one.'

Bonegilla was a former army camp and the migrants led a spartan existence there. Eugenia Bakaitis, a Lithuanian, came to Australia in 1949 'because it was as far away from the Communists as we could get'. In 1995 she remembered: 'Bonegilla was in the middle of nowhere. There was not a tree, not a flower, just an army barracks in an empty, hot, dusty place. There was barbed wire all around it like a German concentration camp. And the heat. The supervisor used to tell us to bring bucket after bucket of cold water and pour it on the floor and then we would all sit in it. We felt this enormous sense of deepening isolation. We didn't know where we were, we just knew that there was no way back.'

If you were a Holocaust survivor – 15,000 of them arrived in Australia between 1946 and 1951 – the sheer joy of still being alive overrode everything else. Regina Zielinski was born in Poland and lost all her family in the Sobibor death camp, surviving herself because the camp staff needed girls to knit socks and do their washing. In 1943 she took part in the only mass escape ever staged from a German death camp and saw out the war with false papers given to her by a school friend and which identified her as a Roman Catholic. When she came down the gangplank of her migrant ship on to the wharf in Sydney in 1949, she knelt down and kissed the ground. 'I couldn't help myself,' she remembered in 1998. 'After what I'd been through Australia seemed like sacred soil.'

Halina Kalowski and her husband Michael came to Australia in *Derna*, a rusting Panamanian freighter built in 1917, converted and pressed into service on the migrant run because of a desperate shortage of shipping. It was a voyage from hell. The captain ordered all pregnant women off the ship because there was no doctor or medical facilities on board. Mrs Kalowski, who was nearly five months pregnant, wore a loose dress and kept quiet about her condition.

The chef died two days out from port and the *Derna* had to return. Then the engines failed and it drifted in the Indian Ocean for several days before the crew could could get them working again. When a fire broke out the hoses proved to be rotten. The refrigeration system broke down and all the chilled meat went rotten and had to be thrown overboard. For the rest of the voyage the migrants lived on pickled vegetables, pea soup and olives.

After ten weeks at sea and only two days away from Melbourne, Mrs Kalowski went into labour two months early. Thirty men had to vacate the cabin they shared so that she would have somewhere to give birth. A medical student who had never delivered a baby before was about to try when one of the other migrants, Dr Herman Frant, who was escorting a group of 61 Jewish orphans, announced that he was an obstetrician.

In 1998, Mrs Kalowski remembered, 'My baby was like a little wounded bird and as soon as we landed we rushed her to hospital. I couldn't speak a word of English and when the nurse took her away and later brought me her clothes, I panicked and started screaming. I thought she must have died.' (She survived and made a home and career in Australia.)

Most of the Holocaust survivors arrived in 1948 and fifty years on, at an exhibition at the National Maritime Museum, some of them recalled how they coped with a strange country where they knew no one, did not speak the language, had no money and no access to social services. Mendel Gelberman, who was an orphan from Czechoslovakia, said, 'From the moment our ship docked my friends and I made a pact not to speak anything but English no matter how badly we spoke it.' Another migrant, Sabina Lis, recalled, 'Coming here wasn't like choosing a dress that you could take back to the shop. I was determined to make a go of it. Life was hard at first but Australia has been good to me.'

Dr Suzanne Rutland, of the department of Semitic Studies at Sydney University, said in 1998 that in many ways the Holocaust survivors made ideal migrants. 'They had enormous willpower, determination and resilience to survive, and those qualities helped them to build new lives and to enrich Australian society.' Or as Mendel Gelberman put it, 'We're very proud of our adopted country, but our country should also be very proud of us.'

Some Australians went out of their way to make migrants welcome, knowing that they would need a positive experience to balance the negative ones that would surely be forthcoming. Nuns at the migrant centre near Parkes, in western New South Wales, brought boxes of clothes for each newborn baby. 'When my husband was sick and couldn't work, Australians left chickens, vegetables and fruit,' recalled Aurelia in 1988. 'And they did all this anonymously so as not to embarrass us. How could you feel hostile to a country whose people did this?' And Elizabeth recalled: 'When I got ill the local doctor treated me for nothing. He explained Australia to us and said not to take any notice of people who didn't like us because we were strangers. He said, "They don't know how badly Australia needs you."'

It was not easy on either side. For the migrants, the climate, the easy pace of life, the sense of security, the acts of kindness they found in Australia did not always compensate for the strange food, customs, and language and the alienation they felt in this far southern land. And it was so big. 'I can still remember being frightened every time I looked at the horizon,' recalls Helmut Bakaitis. 'I had never seen such flat country, and the sky seemed lower. I expected it to fall in.' It was also dangerous – snakes, poisonous spiders even in the suburbs of the cities, vicious mosquitoes, annoying flies, and powerful sunshine. For Europeans used to placid lakes and beaches, even a swim in Australia could be life-threatening.

And then there were the Australians themselves. Who were these strange people? They looked like Europeans and they spoke a European language, but they were very different. They had peculiar eating habits. They ate spaghetti out of a tin. They loved cold spaghetti sandwiches. They had their main meal of the day not at 1 p.m. or 8 p.m. but at 6 p.m. and they called it 'tea'. But tea was also what they drank, not coffee. They knew what coffee was but had no idea how to make it. They had no knowledge of croissants or pastries and certainly would not consider eating them for breakfast. For breakfast they liked chops, or whatever was left over from the previous night's tea, all mixed together and then fried in dripping and called – for reasons no one could explain – 'bubble and squeak'. They saved up stale cakes, poured hot chocolate over them, sprinkled them with dessicated coconut, called them lamingtons and sold them to each other at school fetes. They ate oysters but not mussels –

mussels were dirty. They scorned rich European sausage like salami but loved saveloys made from slaughterhouse scraps. They had restaurants but the food was bad, the service was appalling and many of them closed at the very time people wanted to use them – at weekends. In some parts of the country you could eat in a restaurant or drink alcohol in a pub, but you could not do both on the same premises without breaking the law.

The men drank large quantities of beer during the day but stopped promptly at 6 p.m. Many were then sick in the gutter. Wine was available but it was mostly sweet and fortified and drinking it was considered decadent – except at Christmas when old ladies were allowed a small glass or two. Some adventurous young people drank sparkling burgundy which gave them dreadful headaches. At parties and dances, the men stood at one end of the room and drank beer out of a barrel – the only time they were allowed to do it after 6 p.m. – and talked about sport. The women stood or sat at the other end of the room, talked about babies and only spoke to the men to tell them it was time to go home. A woman who joined the men's group was considered to have loose morals. A man who joined the women's group was considered to be effeminate, probably a homosexual, or a 'poofter', whatever that was.

The national language was English but Australians used many peculiar words and expressions not found in dictionaries and they often failed to finish their sentences – 'My name's Bill anyway' and 'We had a terrific time on Saturday but.' They did not address each other by their correct Christian names but abbreviated them and twisted them. William became Willie or Bill; Barry became Bazza, Margaret became Meg – but all the women also seemed to have been christened Sheila. Nicknames were even more difficult to fathom because Australians often reversed the meaning: a tall man was called 'Shorty', a man with red hair was known as 'Bluey', a fat man was called 'Slim', a bald man 'Curly'. An elderly migrant whose sex life was only a memory went home and told his wife with great pride that an Australian had called him 'a fucking Greek'.

A common complaint was that Australians had no 'spiritual life'. They were 'not cultured' – Sydney did not even have an opera house and few people went to concerts. No one told them about Henry Handel Richardson, Australia's famous writer, whose best novel, *Maurice Guest* (1908), is about the lives of music students in Leipzig, where Richardson herself had studied. So they believed that all Australians were interested only in sun, surf, racing, football, cricket, fishing and enjoying themselves. Work was something that interrupted pleasure, ambition was considered unAustralian and anyone who showed it would be punished. Australians did not hold elegant dinner parties at which guests debated issues of intellectual importance. Australians had no understanding of history, even their own, and their geography was sketchy.

All blond migrants were 'bloody Balts', all with dark hair and olive skin were 'bloody dagoes' and collectively they were 'bloody reffos' (refugees).

Australians for their part, sheltered for years from contact with outsiders, overwhelmed by American servicemen but at least assured that they would eventually go home, found the sudden influx of European refugees very unsettling. They were here to stay but they would never be real Australians. They were too reserved, too formal, too quiet. They gave themselves airs, pretended to be better than they were, looked down on Australians as uncivilized, preferred their own company, and were not grateful enough for what Australia was doing for them. Above all, they were *different*.

'There was an Italian girl in my class and we got friendly,' one Australian girl remembered. 'She invited me home after school one day and all the time I was there I could hear mousetraps going off in the garden. Her father was catching sparrows for soup. She told me that the family went to the beach each weekend, not to swim or build sandcastles, but to collect shellfish and mussels. At night her Dad used to walk around the back streets of South Melbourne and go through the factory rubbish bins looking for anything useful for the house. She said he once came home with a roll of towelling big enough to keep the family in underwear for years.'

Perhaps behind this Australian resentment of migrants were two thoughts. The first was that old Australians had been given no say in the new migration policy. Governments that had previously been prepared to hold referenda on all sorts of issues did not consult its citizens about remaking the country by bringing in all these new people. (This resentment was – as we shall see – to find a public voice only in the late 1990s.) The second thought was that these migrants had not really *chosen* Australia as their new home – they had been *forced* to leave Europe and the Australian government was prepared to have them.

Pino Saccaro, a Melbourne writer, said in 1998 that his father had come to Australia to 'look for America'. America was *any* place Italians migrated to, 'any place where you could escape poverty and find your dreams; hope for the hopeless and riches for the poor. My father applied to Australia because that was where his friend Rosario had migrated to. Australia could well have been the US, Argentina, Brazil or any other country.' Harry Roberts came to Australia from Poland via Palestine because his father applied to the first two countries on an alphabetical list and Australia answered before Argentina.

Those who did consciously choose Australia were often hard-headed, ambitious people who would have made a success of their lives anywhere in the world. Sir Arvi Parbo, chairman of the giant WMC Limited, an example of the immigrant who made good, chose Australia because 'I wanted to be a mining engineer and there was no future for one in postwar Germany. I knew there was a mining industry in Australia so I came to Australia. It turned out to be the right choice. There was an enormous expansion of the minerals industry going on and Australia being an open country there was no question of where you came from, just "Can you do the job?"' Meeting Sir Arvi even for a short time you left him knowing that this was a man who would have made good no matter where he went.

During the next half century these early New Australians battled to make Australia their home. Many gave up and those who could went back to Europe – and then some came back again. Just how many left Australia for good remains in dispute. For a long time the government insisted that only 6 per cent of British migrants returned home. But independent assessments put the figure as high as 14 per cent. In 1973 new figures suggested that 19 per cent of all 'settlers' who had arrived in Australia in the past 25 years had eventually left. Germans headed the list (30 per cent of those who came), the Dutch came next (25 per cent), then British and Italians (22 per cent).

Australians were most irritated by the British who wanted to go back – the archetypal 'whingeing Pom'. These were cousins who should have been delighted to be in Australia but who complained about their accommodation, the food, the weather, the government, the people, the prospects. There were reasons behind this behaviour that Australians did not appreciate. People are emotionally tied to the place where they were born and spent their early years, so migration is essentially an unnatural act. You can make it less unnatural by not going too far. Most people who contemplated leaving Britain hesitated about choosing Australia for the simple but immutable reason that it was so far away. Canada and the United States, one quarter of the distance, were much more attractive choices.

To overcome this reluctance, successive Australian governments offered bribes to lure British settlers. These ranged from free land, then land on long lease at peppercorn rents, to free and then heavily-subsidised fares. (These did not end until 1981 when the successor to the famous £10 fare was finally abandoned.) There was also a lot of pro-Australia propaganda – 'A British People in a British Land' said the caption to an Australia House brochure which showed a map of Australia coloured red with a Union Jack flag sticking up out of Alice Springs.

'They begged you to come out then,' Mary Proudlock remembered in 1997. 'They showed you film of beautiful beaches and legs of lamb – we hadn't had meat for six years. We'd had a bitterly cold winter and I had visions of Christmas on Bondi Beach. We ended up in a very basic little house in the bush. I had to carry water from outside tanks for the tin bath and one bathful had to do for everyone. I would never have left England if I'd known what was ahead, but our children married happily, had children and are content. In the end it was the right choice.'

While middle-class Englishmen served in India or migrated under their own steam to Rhodesia or Kenya, those British families who were attracted by Australian bribes came largely from the depressed industrial areas and felt that they had done Australia a great favour. When their new country failed to live up to their expectations, inflated by Australian propaganda, they did not hesitate to complain.

Over the years the 'whingeing Pom' epithet became so powerful that when an Australian TV producer was making a serious documentary about British migrants in Australia she had to use an all-British crew – the migrants would not speak freely in front of an Australian crew out of fear of being labelled a 'whingeing Pom'.

My father's closest friends, Pat and Fran Holmwood, had migrated from Sheffield in 1922 and, according to my father, had whinged for thirty years. Nothing in Australia was as good as 'back home'. In 1952, having run a successful butcher's shop in Sydney until their retirement, Pat and Fran went 'home' to Britain, planning to spend at least six months there. They were back in Australia four weeks later. 'I've never been so happy to see Sydney Harbour,' Pat said. 'And I'll never whinge again.' Others stayed in Britain. 'I just felt good,' one woman said. 'Everything was so homely. Just houses out there in Australia, not homes. Everything is clean and fresh here, after the dust and sand and dirt out there.'

Non-English migrants had the toughest times. The government had unrealistically hoped that every migrant from no matter where would be absorbed into the Australian community quickly and painlessly. The policy was assimilation. *They're a Weird Mob*, the autobiography of an Italian immigrant, Nino Culotta, suggested that this was possible. Nino worked on building sites, accepted his workmates' jibes and insults, learnt from them, and was eventually accepted as one of them. But the book turned out to have been written by an Irish-Australian, John O'Grady, and was far from the truth.

The reality was more like Susan Varga's struggle to assimilate or Pino Saccaro's long and draining experiences on the path to becoming an Australian. Both were small children when they arrived – Susan from Hungary, Pino from Italy. Both came from hardworking, ambitious families determined to make a new life in this new land.

In 1994 Susan remembered: 'It was like being born again, but without the luxury of time to grow and learn. Knowledge had to be acquired very fast and acted upon immediately. Small mistakes had disastrous consequences. But after the first awful months were over, the energy of the battle gripped us, and in many ways the early years were the best.

'My parents were relatively young and had the will to make this strange new country yield them not only a living but a life. Each year brought new achievements. In the first year they bought a radio and started a business. In the second they bought a car, an Austin A40, and dabbled in a second venture. In the third year they moved house and began to have a bit of a social life. In the fourth year they bought a Holden and went on a holiday to the Blue Mountains. By the fifth year, they were fulfilling the migrant's dream: building their own home on their own block of land.'

In many ways it was easier for the parents than for the children. 'We just wanted to stop being different in any way, in the shortest possible time. Within a year, as if by osmosis, we had picked up the language and lost all traces of an accent. We trailed our battered Globites [a popular brand of schoolcase] home from school on hot afternoons, pretending, even to ourselves, that we were not so very different from other kids. We were so intent on mastering the present that our former lives receded, half-forgotten, certainly repressed.

'But every time one of our new friends met our parents there were awful reminders. Everything about *them* – their accents, voices, clothes, laughs – marked *us* out as different.'

It was much the same with Pino – he became ashamed of where he came from and how different he was. 'My olive skin, the loaf of bread and salami for lunch, my non-English-speaking parents and tomatoes growing in the backyard.' So he copied the slang, the sport, the Australian ethos. 'I spoke English with an Australian accent, my friends were Australian, my heroes were Australian, my dreams were Australian.' Just when he thought he *was* Australian came a defining moment. 'I asked Wendy to be my girlfriend. She said yes, so we walked hand-in-hand home from school. Then one of her girlfriends saw us. "You're not going out with a wog, are you?" It hits you straight between the eyes.'

Saccaro decided that he had an inner psyche, thought processes and cultural identity tied up with the land of his birth. Instead of Australia rejecting him, he rejected it. He embraced everything Italian. He learnt Italian, played Italian songs, frequented Italian cafes and bars. But it did not work. He was not Australian, he was not Italian. 'I became angry with my parents for my identity crisis. Like so many migrants before me, I needed to resolve or at least exorcise this sense of dislocation. Something was missing.'

So he went back to his *paese*, Calatafimi, on a hilltop in north-western Sicily – 'Houses with terracotta roofs set among the wild fennel, goat bells tinkling in the distance, narrow cobblestone streets, men with dark caps, women in black.' He thought he would find everything as he had left it but he did not. 'Nothing had changed yet everything was different. Disillusioned, every time I came back I swore I would never return.'

Then in 1997 Saccaro went back to Calatafimi with his wife. All the old aunts and uncles turned out to welcome the couple. There was a feast with a cake inscribed *Benvenuti a Calatafimi*. 'I almost cried. This was the place I was born. My home. I was among my family . . . I had discovered my past and my history. I was grateful to find the missing jigsaw piece in my life.'

Then he was ready to go back to Australia. 'An orphan having searched and found his real parents returns to the adopted parents where the bonds are stronger. I am grateful to be an adopted son of Australia. I was born in Italy but Australia gave me an identity. This year, for the first time, I will celebrate Australia Day. I will rejoice that we now live in a multicultural society. How privileged we are that this is the country that the world came to live in.'

The world certainly came to live and work at the Snowy Mountains Scheme, an example of Australia's preoccupation with water that dates back to the shock the early settlers got when they realized that there was no giant, inland lake filled by crystal inland rivers and that Australia was a dry, merciless country. How dry? The driest of all the world's inhabited continents – the Danube carries more water than all Australia's rivers combined.

Part of the problem was that the country's rivers ran the wrong way. The mighty Murray River ran west of the Great Dividing Range, fed by some 26 feeder rivers, and emptied into the sea in the Great Australian Bight. On the way it watered nine-tenths of Australia's irrigated land. But there was never enough water and in some drought years it dwindled to a chain of pathetic pools. East of the Great Dividing Range, fed by snow-melt and semi-tropical coastal rains, rivers like the Snowy ran fast but wastefully the short distance into the Tasman Sea.

So with the typical belief of many Australians that everything is possible, in 1949 the Snowy Mountains River Scheme was born. They would take the Snowy River water and send it up over the mountains and down the other side to join the Murray. They would do this with 16 major dams and 134 kilometres of rock tunnels, and while the water was on the move it would power seven alpine electricity stations. This would take 25 years to build and would employ 100,000 workers, most of them migrants.

They came from 42 different countries around the world – whole new towns went up to accommodate them – and if it were possible to point to one event in the history of Australia that sowed the seeds of the country's present multicultural society, the moment when Australia began to re-make itself into its Millennium image, most Australians today would choose the Snowy Mountains Scheme – 'the day the men from the Danube, the Oder and the Dnieper joined the Man from Snowy River'. It not only introduced Australians to European migrants but European migrants to each other. In 1996, Andrew

Laycock, then the owner of the Alpine Inn at Khancoban, recalled the 1950s. 'We had survivors from the concentration camps, a tribe of European aristos including a gambling-mad Polish count who worked a jackhammer, plenty of academics and professional men. Hell of a culture shock for the Aussies.'

A former carpenter called just 'Lew' remembered, 'One night at Talbingo they showed a movie about Dunkirk. Afterwards two migrants started arguing about the film. It turned out that one was a Pom who'd been stuck on the beach and the other was a Luftwaffe pilot who'd been shooting him up. The German carpenters used to come to work in black corduroy suits with mother-of-pearl buttons and Homburg hats. The Italians did the deep tunnelling. Happy-go-lucky bastards, but brave. Used to chase the sheilas all the time. You had all these blokes making good money with hardly a woman in sight. So the girls came down from Sydney. There was a shack at Clovers Flat where they'd hang up blankets and operate four at a time, two quid a go, with the men still in their gumboots queuing half way to Towong. At Cooma they had nightclubs. Harry Belafonte sang there once.'

On a smaller scale, the culture shock of the Snowy Mountains Scheme was mirrored in nearby towns. Bruce Elder was a schoolboy at Tumut in the 1950s. 'I was surrounded by an unremitting monoculture,' he recalled in 1998. 'Beer, strongly-brewed tea, lots of fatty mutton with peas and burnt potatoes and pumpkin, loaves of white bread with butter and jam. The town was divided aggressively between the descendants of English and Scottish Protestants who went to the State school and Irish Catholics who went to the convent.

'The only non Anglo-Saxons in the town were the Greeks who ran the milk bar, a few Chinese of whom the most notable was Teddy Shai Hee, a photographer who rode around town on a bicycle, and an entrepreneurial family named Moses who, given the prevailing Anglo-Saxon orthodoxy, quietly changed their name to Manning.'

But then the Snowy Scheme got under way and Elder found his town at the cutting edge of Australia's profound postwar changes. 'By the time I was in fifth class a strange and particularly good-looking German, Horst Neumann, had arrived at Tumut Public School and was grappling with the idiosyncracies of English. By the time I was a teenager I was working at weekends in a milk bar–delicatessen where, catering for the changing needs of the population, the owner was selling strange-smelling sausages with names like liverwurst and bratwurst, and peculiar cheeses with holes in them. The days of brawn and cheddar were disappearing. It was around this time, shock, horror, that one of the local primary school teachers married one of these foreigners. Suddenly Miss Witherby became Mrs Markevics.'

Elder has lived long enough to see the outcome of the winds of change that began to blow in Tumut in the 1950s and to realize that for someone of his background old prejudices are hard to eliminate. 'The idea of Australia as an Anglo-Saxon society made up almost exclusively of the English, Scottish and Irish is deeply rooted. My generation still tend to think of non-Anglo-Saxons as outsiders, visitors, itinerant aliens. Even though we love to go out and eat Italian, Vietnamese and Thai and our preferences in liquids now tend more and more towards real coffee and good wine, when we see large numbers of Asians, or are served by Asians in restaurants, we assume that they are students rather than Aussies.

'The wrongness of this perception was brought home to me in 1993 when I was teaching a class of electrical engineers at Wollongong. At the beginning of the year, I asked the fifteen people in the class to write their names on a sheet of paper. There wasn't a single Anglo name on the list. There was a Nguyen from Vietnam, two Maltese, a Chinese, three names from the Balkans, a couple of Turks, a Filipino and a miscellany of other nationalities.

'Instantly I thought, "We're going to have some language problems here." Then I asked about their background and their education. Without exception they had all been born in Australia, had all completed 12 years of New South Wales education, and were as dinky-di as any Aussie can get. Now when I ask myself who is an Australian, I think back to the Kells and the Webbs and the other Anglos in the Tumut valley in the 1950s and I also think of Horst, Binh, Fatima and Johnny Markevics. I now live in a glorious, life-enforcing multicultural society, enriched by those who, for a million and one reasons, had the courage to travel halfway around the world and settle here.'

CHAPTER 15

The Menzies Era

As if to confirm that the Australia-Britain link was once again safe, the new Queen Elizabeth made a grand tour of Australia in the summer of 1954, the first by a reigning monarch. She was greeted with proper deference by Menzies and wild enthusiasm by the ordinary people of Australia.

The Anglo-Australians naturally felt that the mass influx of European migrants of the late 1940s and early 1950s was creating a new Australia whose ties with Britain would be weaker. They were right. The British Empire was breaking up, anyway. Respect for the Royal Family seemed to be diminishing. Why, even the great Don Bradman was criticised for having his hands in his pockets when out walking with the King. Connie Sachs, a character in the John le Carré spy novels, captures the approaching twilight of Empire better than many histories. 'Poor loves,' she says of the postwar generation of British intelligence officers. 'Trained to Empire. Trained to rule the waves. All gone. All taken away.'

There was a moment when it looked as if the jet age would save Britain. For a bankrupt country's media – four-page newspapers, black-and-white newsreels and the BBC Home Service – 'the 600 mph men carried the future of the country and empire'. 'Test pilots,' wrote one admirer, 'will make it possible for the empire once more to be strong and rich.'

Not the empire he knew. In April 1949 India announced its intention of becoming a sovereign independent republic but expressed a wish to continue as a member of the Commonwealth. This required major adjustment. How could a republic swear 'common allegiance to the Crown' as membership of the British Commonwealth required? But if India was not allowed to join, then it could lead to a future Asian coalition of countries led by India which could be critical of – perhaps hostile to – Britain. The answer was the London Declaration issued at the Commonwealth Prime Ministers' meeting that same month.

Common allegiance to the Crown would no longer be required. Instead India agreed to recognize the King as symbolic of the free association of the independent member nations and, as such, the head of the Commonwealth. The word 'British' was removed from the association's name, making it 'the Commonwealth of Nations' – known later as simply 'the Commonwealth'. When Pakistan, Sri Lanka and later Malaysia joined, and was followed by the old British colonial countries in Africa and the West Indies, the Commonwealth became fully multi-racial, and 'white Commonwealth' countries like Australia slipped to a tiny minority.

That same year, on 26 January, Australia Day, the Federal government declared the National Citizenship Act. Australians, instead of being British subjects, citizens of the United Kingdom and Colonies, became British subjects, citizens of Australia. It was a small but significant step, recognizing for the first time a distinctive identity for the people of Australia.

If the Labor Party had not lost power later that year the Republican movement that came to prominence in the 1990s might have had a much earlier birth. Australia might well have felt that with a growing migrant population – many of whom found swearing allegiance to the Queen of England curious, to say the least – the new decade of the 1950s might have seemed a good moment to follow India's lead and become a republic within the Commonwealth.

But on 10 December 1949 the Labor government of Ben Chifley, the Prime Minister who had followed Curtin, was swept from office in an electoral landslide to the conservatives. The electorate felt that it had had enough of wartime controls and four years after the end of the war wanted to be free from rationing, regulation, and the constant battles between the courts and the government over its nationalization plans. The Liberal and Country Parties, led by Robert Menzies, stood on a platform that offered to stop 'the socializing process' of the Australian way of life, reduce taxation and ban the Communist Party.

Australia's own car, the Holden, which had come off the production line a year earlier, featured in the election campaign. An anti-Labor cartoon shows Menzies, standing at the door of a new Holden marked 'Free Enterprise', and helping a woman get out of a second car that has broken down by the roadside. A left-hand drive, it has Ben Chifley at the wheel, and its number plate reads: 'Socialization, 1921 model.' Menzies could have used the Holden to hit even harder. Labor's platform of reduced working hours and higher wages meant that the Holden – which was nearly called the Canbra – was too expensive for the workers who had made it.

Chifley had embodied much of the Australian dream, an engine driver who rose to be Prime Minister, a Prime Minister whose home in Canberra was room 205 at the Kurrajong Hotel (lavatory and bath down the corridor) and who took his own home-grown onions down to the local cafe to be fried for his dinner – when dinner was not a meat pie eaten at his desk. In how many other democratic countries around the world, societies which preach equal opportunity, could an engine driver become leader of his people?

Manning Clark said of him, 'Ben Chifley drew two great prizes. He was loved. He was Prime Minister of a government which put on the statute book much of what Labor had preached, but had never been able to practise. He gave capitalism a human face, he encouraged excellence in things Australian by creating the Australian National University, and he created material wellbeing for all, being faithful to the Labor belief that creature comforts are the *sine qua non* of human wellbeing and happiness.'

Menzies, the man who took his place, was very different. He embodied the hope of Britain and the Anglo-Australians that the special relationship between Australia and the Mother Country was far from over, that the 'great betrayal' of the Second World War could be forgiven, and that Australia, despite its growing population of European migrants, could remain an outpost of England and English values in the South Pacific.

Like Chifley, like all great leaders, Menzies, too, had a vision for his country. One may not agree with his vision, but it was there and it drove him. He believed in free enterprise, personal liberty, the rule of law and contract and, above all, being 'dyed-in-the-wool British'. He loved the British governing class, would have liked to have been one of them – in 1941, with Australia's most dangerous hour approaching, he considered a political career in Britain and even flirted with the thought that he might replace Churchill. But above all he loved the Royal Family and the pomp and ceremony surrounding it.

Menzies was to rule from 1949 to 1966, the 'Menzies era'. Some saw it as a period of elitist, conceited, royalist policies, with Australia condemned to national immaturity, insecurity, insularity, censorship and 'cultural cringe'. Others saw economic stability, growth, general prosperity, low unemployment, university expansion, a suppression of Communist agitation, and a strengthening of ties with Britain. Lord Carrington, who was British High Commissioner in Australia from 1956 to 1959, recalled in 1999: 'Menzies was such an Anglophile that relations between the two countries were very tranquil indeed. I had hardly anything with which to concern myself.'

Menzies' reward from the Mother Country was a shower of honours and gifts, culminating in the autumn of his career with his investiture as a Knight of the Thistle and his succession to Churchill as Warden of the Cinque Ports, which allowed him to dress up in a wonderful, heavily embroidered uniform complete with cocked hat and a modest dress sword.

As if to confirm that the Australia-Britain link was once again safe, the new Queen Elizabeth made a grand tour of Australia in the summer of 1954, the first by a reigning monarch. She was greeted with proper deference by Menzies and wild enthusiasm by the ordinary people of Australia. True, there were dissenters. The Queen's staff had gone to great lengths to emphasize to the Australian media that the Queen of England was also Queen of Australia. (There is now some doubt that they were telling the truth: otherwise why did the Queen formally become Queen of Australia in 1973?) As a journalist on the Sydney *Daily Mirror* I was sent to cover the arrival of the royal couple in Sydney from New Zealand. I began my report: 'Elizabeth the Second, Queen of Australia, sailed into Sydney Harbour this morning for the start of her tour of this country.'

When the first editions of the newspaper hit the streets, 'Queen of Australia' had been deleted and 'Queen of England' substituted. I learned later what had happened. The owner of the *Mirror*, Ezra Norton, came from an Irish family that was fiercely anti-British. He had seen the page proof of my report and had changed it – 'There's going to be no fucking Queen of Australia in my fucking paper.'

It was unfortunate that Menzies took Australia back into the Mother Country's arms at a moment when a real sense of national identity was beginning to develop. Hazel de Berg, a pioneer in the field of recording oral history in Australia – 'oral history is the heartbeat of a nation' – had decided that since Australia was such a young country, if she could get people to record their memories 'we can touch the beginnings'. She began finding and recording anyone who had an interesting story to tell.

In 1958 she recorded Dorothea Mackellar, who wrote what is perhaps Australia's best-known poem, one that contrasts the delights of the English countryside with the harsh demands of wilful, lavish Australia. In 1996 I listened to the recording. A cultured voice begins, 'This is Dorothea Mackellar reading "My Country", a poem I wrote with sincerity.' The poet begins confidently: 'The love of field and coppice, of green and shaded lanes/ Of ordered woods and gardens, is running in your veins . . .'

But by the time she has reached, 'I love a sunburnt country, a land of sweeping plains . . .' the voice is shaky, and at the end, 'Wherever I may die/ I know to what brown country, my homing thoughts will fly,' Dorothea Mackellar is clearly weeping. The reason was not hard to find. Mackellar had written the poem in 1908 and many Australians thought she was already dead. But Hazel de Berg had tracked her down to a hospital where she had been for some time, quite ill. As de Berg listened to Mackellar reciting, she began to cry and by the time Mackellar had finished, both women had tears streaming down their faces.

De Berg then took the recording around Australian schools and played it for children who were learning Mackellar's work. Many of the children began writing to her in hospital to say how much the poem meant to them as Australians, and suddenly, in the last years of her life (she died in 1968) Dorothea Mackellar found fame anew and was flooded with letters of appreciation.

The list of people recorded by Hazel de Berg grew rapidly – her collection in the National Library, Canberra, now numbers 1,300 interviews – men and women from all walks of life who were born, or who had lived in, Australia. 'This is Sidney Nolan, and I just want to say a few words about how one begins to paint.' Or, 'This is George Johnston. I wrote *Clean Straw for Nothing*.' Or, 'Oh, dear. I forgot to tell you that it's Mary Gilmore speaking.' And Mary Gilmore then goes on to say that she could remember when there used to be kangaroos at a creek at the bottom of Kings Cross. And then to talk about her friendship with Banjo Paterson.

In her private life, too, Hazel de Berg was involved in a ground-breaking Australian initiative that can be seen in retrospect as a first shaky step towards a new attitude to Asia and Asians. While the government's postwar immigration policy made it clear that Asians were not welcome in Australia it nevertheless joined with Ceylon at a meeting of Commonwealth foreign ministers in Colombo in 1950 in setting up the Colombo Plan. The idea was to promote co-operative economic development in south-east Asia, with the richer nations offering money, education and trade opportunities to poorer ones in the region.

As well as contributing some $300 million by 1970, Australia oversaw special projects, offered expertise, and arranged for the education of 10,000 Asian students in Australia. It was a two-way street. Australians posted to other Colombo Plan countries mastered Asian languages. (It is a myth that Australians have no aptitude for languages. Australian-born Chinese still tell of the night a former Australian diplomat brought the staff of a Chinese restaurant in Sydney to tears by singing in Mandarin the tenor's role in a modern opera, *Sailing the Seas Depends on the Helmsman*.) Colombo Plan students being educated in Australia also experienced at first hand Australian generosity and the Australian way of life.

Hazel de Berg's daughter, Diana Rich, remembered in 1998 the day in 1956 when her mother received a telephone call from a friend in Canberra to say that some Colombo Plan students had arrived from Indonesia and he did not know where to find them accommodation. 'Mum said he had better bring them over to our place in Sydney. Mum threw herself into this like she threw herself into everything and in no time the place was full of Indonesians, with Mum busy learning Indonesian so that she could speak to them in their own language.

'She kept in touch with them all after they went back. I can remember her delight when one particularly bright student called Sumardi came to the house to see her many years later. When he was here as a student he used to get around on a little old motor-scooter. When he came back he proudly showed Mum his card. He was now the Indonesian Government Director-General of Information. Then he said, "I haven't got a scooter any more, Mummy," and he took her outside and there was this big, black, chauffeur-driven, Commonwealth government limousine.'

Australians remained as hedonistic as ever, taking their pleasures very seriously indeed. The final of the *Sunday Telegraph* beach girl competition could attract an extra 10,000 people to Bondi. The autumn racing carnival could bring nearly 100,000 punters to Randwick, all wearing hats – the standard wide brim felt hat for men, and exotic feathered hats for women, some with plumes so big they would not fit into a tram, car or bus. 'I hope this practice of feathered hats may finish itself quickly,' wrote the fashion editor of the *Sydney Morning Herald*. 'Many women looked as though their heads had been dipped in a feathered pillow.'

For some working women, however, life in a male-dominated society remained tough. It would be wrong to suggest that a court case in Western Australia was typical but the fact that it had happened at all makes it worth recording. Amelia Lilley, hotel keeper, was found dead in her hotel. She had been kicked to death and her husband was charged with her murder. A prosecution witness said he had seen the husband kicking Mrs Lilley but had not interfered. When asked why, he replied, 'It's normal to see a woman kicked in her own home. I wouldn't interfere. I'm married. It's a matrimonial right.'

Overseas travel was once again possible. Qantas and BOAC had an airline route to London, and Qantas was running Lockheed Super Constellations to the United States – if you lived in south Sydney, as I did, you could see the sky lit up by flames belching from their 18-cylinder engines as they took off from Mascot airport for San Francisco. But air fares were expensive; a better deal was to travel by ship. Those bringing migrants to Australia went back with Australians keen to see the outside world.

It is another myth that they all went seeking a cultural and intellectual life that Australia could not provide. Russell Drysdale was painting some of his best work, distilling important truths about his country. There was increasing recognition that in Norman Lindsay Australia had an outstanding painter. Lin Bloomfield, now a Lindsay authority, remembers, 'I was barely eighteen, a naive New South Wales country girl. You have to remember how very straight times were then. But when the art lecturer at Armidale Teachers' College in our first lesson held up a Lindsay watercolour and said, "I want you to appreciate this", we all went "Oooohhh!!" I'd seen pictures by Rubens before but I couldn't believe that here was an Australian actually doing this kind of work. It was so sensual and accomplished, and it inspired me to become interested in art, especially in Norman.'

Charles Chauvel, with *Sons of Matthew*, a pioneering epic about three generations of an Irish family who settled in Queensland, came as close to any indigenous film to portraying an Australian 'manifest destiny'. Frank Hardy had written a pioneering political novel, *Power Without Glory*, that not only shook the Melbourne establishment with its thinly disguised story of corruption and racketeering, but resulted in his being charged with criminal libel. (He was acquitted.) Eugene Goossens was running the Sydney Symphony, which he had transformed from a mediocre, parochial orchestra into one of world stature. He had also become a dynamic force behind a project for what eventually became the Sydney Opera House. (Unfortunately, in 1956, Australian Customs received a tip-off that Goossens, returning to Sydney after five months abroad, might have pornography in his baggage. He did indeed and amid widespread publicity was fined £100. He fled the country soon afterwards, never to return.)

Harry Seidler, an Austrian immigrant architect, was designing distinctive, modernist houses and clear, bold office buildings. Clive James, Germaine Greer, Barry Humphries and Robert Hughes, internationally the most famous Australians of the 1980s and 1990s, were gearing up for their assaults on the outside world. This nation of hungry readers (as early as 1870 Australia was importing one-third of all the books printed each year in Britain) was doing what it could to encourage local writing talent. The Commonwealth Literary Fund had been offering since 1939 a wide range of fellowships and grants to Australian authors and literary publications. But it was true that, unless, like Patrick White, an author had independent means and a British or American publisher, it was not easy to make a living from writing.

The Summer of the Seventeenth Doll, a play by Ray Lawler about two cane-cutters coming to terms with growing old, their relationship with their women and each other, was hailed as 'the best play ever written about Australia . . . the coming of age of Australian drama'. It was later to move to London's West End, where it was well received.

Perhaps, as the academic and writer Andrew Riemer has pointed out, what Australia lacked in the 1940s and 50s was not an intellectual life but *public* intellectuals. 'A public intellectual was an exotic species. Its natural habitat was one of those fuggy Paris cafés filled with smoke, loud talk and opinions on every subject. The temper of Australian intellectual life, such as it was,

was entirely different, inclining towards the much more isolated habits of the scholar and the academic intent on preserving and disseminating knowledge in universities. Admittedly one or two heirs of Central European intellectual traditions had found their way to Australia, most notably George Munster and Frank Knopfelmacher. But they were exceptional – too clever by half and profoundly un-Australian, according to their detractors.'

So why did so many talented Australians abandon their country during this period and make their way overseas, mostly to Britain, but some to the United States – apart from the gang of four mentioned above, there were Peter Finch, Shirley Abicair, Dick Bentley, Bill Kerr, Keith Michell, Leo McKern, Joan Sutherland, June Bronhill, Sidney Nolan, Arthur Boyd, Jack Brabham, Frank Ifield, Rolf Harris, Murray Sayle – and slip into a new life so smoothly that after a while hardly anyone remembered that they were Australians? (Who would ever have said 'the Australian, Robert Helpmann'?)

Some were undoubtedly escaping authoritarian Australia, a country where, although the Prime Minister lauded individual freedom, the authorities persecuted difference – the owner of the Moulin Rouge Café in Sydney's Kings Cross was fined £5 on an obscenity charge for exhibiting on the cafe's wall a copy of Toulouse-Lautrec's *Woman Adjusting her Stocking*. Some no doubt found Menzies' Australia politically stultifying. Some had strong personal reasons. Some had been unsettled by the war. Murray Sayle wrote in 1999: 'The thunderous noises off during our adolescence made everything Australian seem small-scale and parochial. Then there was our education which had given us foreign languages, and a grip of world history, against which our local historical figures, our lists of rivers and explorers looked insignificant.'

But the majority left because they were ambitious and wanted to make their mark in a wider world. If they were critical of Australia on their infrequent return visits it was because of the difficulty in admitting that perhaps they had not needed to go in the first place. 'Those who live in Australia and strive for our version of an intellectual and cultural life,' wrote critic Don Anderson, 'are accustomed to being lectured at by sophisticated English exquisites, or briefly-returned expatriates.' But, he implied, there was no reason for Australians to bother to appear grateful.

Others were grateful to get back. Hazel de Berg's work turned her into a lifelong Australian patriot. She went abroad only once – to London and New York to record Sidney Nolan and some others. She said later that she had enjoyed herself but that when she got back and the plane was approaching Sydney, she looked out of the window and saw the Southern Cross and said to herself, 'Well, that's it. I've been overseas. I've done it now and I don't have to go again.'

* * *

Like many expatriates I often pondered the question: what if I had stayed? On visits to Australia I looked for people of my age group who had considered leaving Australia at about the time I had, but had stayed on. I wanted to discover how their lives had worked out. Outside Hobart, Tasmania, in 1997 I met an historian, Peter MacFie, who brought up the topic himself.

'The wife and I were all set to go overseas in the fifties,' he said. 'We'd saved up and were going to tour Europe and then live in London. Then this terrific old stone house came on the market and in the end we put the money down on the house instead and didn't go.' I said, 'So you've never been to London?' He laughed. 'Mate,' he said. 'I've seldom been to Melbourne.'

His personal life was obviously very happy. He and his wife had clever and devoted children. They followed an active social round with friends and neighbours. He had devoted his talent as a historian to examining the minutiae of the convict settlement of Tasmania. He knew *everything* about the convicts and their jailers. And when interest in this – not only in Australia but in Britain – revived in the 1980s, his renown as an historian soared. And this was not just how many yards of roadway a convict gang could build in a day, but gripping social material.

Another Australian historian, Russell Ward, had advanced the theory that the Australian concept of mateship had its origins in convict chain gangs. Chained to a fellow convict, you became involved in his life and he in yours. You became 'mates' or 'cobbers'. MacFie contested this. He produced convincing evidence that the British jailers had a system which enabled a small number of them to control large numbers of convicts. It involved offering handsome inducements – extra food, tobacco, time off, money – to informers among the convicts. By maintaining a core of informers – 'dobbers' in Australian slang – the jailers had advance warning of escapes, uprisings, any impending trouble. His paper, *Dobbers and Cobbers: Informers and Mateship Among Convicts, Officials and Settlers on the Grass Tree Hill Road, Tasmania 1830–1850*, threw new light on Australia's early days and on the systems that Britain used to control her empire.

MacFie was able to live, work and win acclaim without moving out of the area in which he had been born. He would have gained nothing by having gone to Britain – a worrying conclusion for those of us who did.

Perhaps we were escaping a political climate we disliked, rather than a cultural one. The Chinese writer Zhang Xianliang says, 'Every thinking person has the choice of three different relations with the politics of their society: to participate, to flee, or to transcend.' What was it about Australian politics in the 1950s that made so many decide to flee?

CHAPTER 16

REDS

The fight against Communism was seen as a struggle of light against darkness, a titanic battle to save Australia from destruction.

The sun shone brightly on Australia in the early postwar years. Yes, it was a quiet backwater tucked away down there at the end of the world, as far from Europe as you could go without finding yourself on the way back again. But many thought that was a good thing. Australia was solid and safe with full employment, low inflation and one of the highest rates of home ownership in the world. So why were not all Australians happy?

The internationally acclaimed writer David Malouf put his finger on it. There was an anxiety at the centre of people's lives in Australia in the 1940s and 1950s, a free-floating fear that no one put into words but which took the edge off their natural optimism and enjoyment of their country. What was it? 'Communism, of course,' says Malouf. 'Reds, both outside and, more insidiously, within. The infection of Europe: all those recent horrors that we imagined might be brought in, like germs, with those who were fleeing from them.'

As we have seen, this was not the first time that this anxiety had surfaced – this worry that anti-Australian elements within the community were about to ruin everything that the country stood for (see Chapter 6). But the hatred and fear of Communism in Australia in the 1950s was rivalled only by the McCarthyite period in the United States.

It was fuelled first by the Korean War and later by experiences and stories of migrants. Korea, divided after the Second World War into a Communist North, supported by the Soviet Union, and an anti-Communist South, had long been a flash point in the Cold War. When fighting broke out in June 1950, Prime Minister Menzies, worried that Communism would march south, within a week had committed Australian military support to the United Nations force. The war was not an uplifting experience for most Australians. The brutal South Korean regime under Syngman Rhee appalled many of the correspondents covering the war, including Alan Dower of the *Herald*, Melbourne.

When he saw a column of women and children kneeling alongside a deep, freshly dug grave facing two machine guns, Dower said to a fellow correspondent, 'Hell, this is a bloody fine set-up to lose good Australian lives over. I'm going to do something about this.' Dower stopped the executions by threatening to shoot a South Korean officer and, by swearing that he would send a story that would 'rock the world', forced United Nations officials to intervene to end women-and-children death marches.

The war finished in a stalemate in July 1953 but it had repercussions in Australia for years. One Australian correspondent, Wilfred Burchett, had covered the war from the North Korean side for *Ce Soir*, a Paris left-wing newspaper. He wrote a story for it alleging that the Americans had been experimenting with germ warfare by dropping disease-infected insects over wide areas of North Korea. This was never proved and Burchett was accused of writing Communist propaganda.

He was also later accused of visiting Australian prisoners of war in Korea and helping interrogate them, a charge he strongly denied. In 1955, travelling in south-east Asia, Burchett's passport disappeared. The Australian authorities then punished him for his activities in Korea by using the 'Catch 22' ploy – it officially informed him he could be issued with a replacement passport if he returned to Australia, but refused to allow him to return to Australia without a valid passport. Gordon Barton, a wealthy anti-Vietnam War activist, solved this in 1971 by flying Burchett into Australia from Noumea, New Caledonia, in a light aircraft. This so outraged the anti-Burchett activists that Barton then had to arrange protection for Burchett in case someone tried to kill him.

The following year a Labor government under Gough Whitlam restored Burchett's passport, ending what had been a *cause célèbre* that had divided Australia on Cold War political lines – the anti-Communists who would have liked to see Burchett tried for treason, and left-wingers and civil libertarians who supported him.

Stories of Communist suppression brought to Australia by migrants were amply confirmed by the Soviet invasion of Hungary in 1956 and the arrival in Australia of 3,000 Hungarian refugees. That same year during the Olympic Games in Melbourne, Australians saw for themselves the hatred between the Hungarians and the Russians when fighting broke out during the water polo event.

Most Australians believed at this time that the country's strikes and industrial troubles were the fault of Communist trade union leaders. There were exaggerated rumours of the size and support that the Communist Party enjoyed and confusion over which organizations were Communist fronts, how many fellow-travellers there were, and who could be trusted. Former Labor Senator Jim McClelland remembered in 1994 that when he entered the Senate in early 1972, the Attorney-General, Ivor Greenwood, 'sincerely believed that a Communist grew on every rose bush and most Labor parliamentarians were either secret members or fellow travellers of the Communist Party'.

In many quarters, the fight against Communism was seen as a struggle of light against darkness, a titanic battle to save Australia from destruction. This had a profound and lasting effect on Australian life. It changed the political landscape. It sent the Labor Party into the wilderness for nearly a quarter of a century, ruined promising careers and came close to turning Australia into a mirror image of the system it was fighting against – at the height of the anti-Communist hysteria, the government was planning gulags in outback Australia for political dissidents. The two world wars are accepted as an integral part of the story of Australia. But we also have to accept that the Cold War, too, had an enormous influence on the country's history.

Keeping on eye on radicals of all hues had been the job of a military counter-espionage bureau set up in 1917, and then an Investigation Branch of the Attorney-General's Department which had been created in 1919. But in the postwar years, when the threat appeared to have expanded, something more effective was needed, a full-scale security service along the lines of America's FBI or Britain's MI5. A Labor government was instinctively against the idea – a secret police investigating people's political beliefs smacked of a Gestapo.

But Prime Minister Chifley came under increasing pressure from the United States. Washington cut back sharing secret intelligence information with Australia, saying that it did not believe that Australian security could be trusted. (The Americans were probably right. They had been breaking Soviet codes, and had intercepted and decrypted messages from the Soviet Embassy in Canberra to Moscow. These revealed that a Soviet espionage ring had been operating in Australia. There was no reason to believe that it was not continuing.)

The American ban threatened the British-Australian project to test missiles at the Woomera rocket range, which depended on a flow of classified technical information from the US to Britain, which then shared it with Australia. On these grounds, Chifley caved in and gave the go-ahead in March 1949 for the setting up of the Australian Security Intelligence Organization (ASIO) which, under its director general from 1950 to 1970, Brigadier Sir Charles Spry, was to play a controversial role in Australian politics.

Labor believed that ASIO considered that there was little difference between socialism and Communism, that from a security point of view, Labor would always be suspect and that many Labor politicians, while not Communists themselves, were sympathetic fellow-travellers. Labor saw ASIO as a pliant tool of the Menzies government, dedicated to keeping Labor out of office. Jim McClelland remembers, 'There was an ASIO file on me, I was assured by an ASIO officer. It became almost a badge of honour for Labor people to have an ASIO file.' (Not everyone thought so. Communist Party members in the Returned Servicemen's League, identified by ASIO, were purged from the League.)

Menzies wanted to hit Communism at its very heart. He had replaced Chifley with a platform that promised to deal with the Communist menace by legislation, and early in 1950 he moved to do so. Introducing the Communist Party Dissolution Bill he said that there were only about 12,000 active members of the Communist Party in Australia, but they occupied key positions in key industries. Most strikes were Communist-inspired. The choice was to ban the party and make it unlawful for anyone to be a member under penalty of imprisonment for five years, or 'we can do nothing and let a traitorous minority destroy us'.

But the law had been in force only five months when the High Court ruled by a majority of six to one that the government did not have the power to pass such legislation. Undeterred, Menzies took the matter to the people of Australia. In a bitterly fought referendum he asked them to vote 'Yes' to a proposal that the government be given the power to ban the Communist Party. 'The whole danger to peace in the world today springs from the policies, plans, underground activities and promoted local wars of the Communists. You can play your part in smashing this alien fifth column.' The new Labor leader, Dr H. V. Evatt, a distinguished jurist who had been the third president of the United Nations General Assembly, protested that the Act violated the basic UN principles of justice and human freedom.

Australians followed the 'Yes' and 'No' campaign avidly. They were concerned about Communist power, particularly in the trade unions. They felt that the Communist Party secretary, Lawrence Sharkey, was wrong to have told the Sydney *Daily Telegraph* that if 'Soviet forces in pursuit of aggressors entered Australia, Australian workers would welcome them'. But three years' jail for sedition for saying that seemed to many to be excessive – even if, as was the case, the prosecution had been launched by the Chifley Labor government. And now the Act banning the Communist Party provided for six years in jail for anyone convicted of being a member. That was a long time behind bars just for belonging to a political party. The Act put the onus of proof on the individual – once accused, a person would have to prove that he or she was *not* a member of the party. That seemed unfair. They weighed the arguments and to general surprise voted 'No' – the government would not be allowed to ban the Communist Party or any political party. Australians, put to the test, had allowed their libertarian streak to shine through.

It was not until 1994–5 that anyone properly appreciated how wise the Australian people had been. Les Louis, of Canberra, a retired schoolteacher who had become an independent researcher working in the government archives, came across a secret contingency plan to set up internment camps in the Australian countryside in the 1950–3 period that would hold up to 10,000 members of the Communist Party *and their sympathizers* (author's emphasis). 'ASIO was going to provide the names, the state police were going to make the arrests and the army was going to build and run the camps,' Louis told me in 1999. 'And it was not just Communist Party members who were going to be interned. ASIO was going to include members of any organization they regarded as a Communist front – all those new theatres, new housewives' associations and new literary groups. Nor was ASIO going to wait until war with the Soviet Union broke out – arrests and internment would take place as a preventive measure during an "emergency stage" if war looked likely.'

Louis also came across another secret anti-Communist plan in this period, 'Operation Alien'. This called for the armed forces to take over the running of essential services if 'Communist-inspired strikes' threatened the nation's security. Unlike the internment camps proposal, 'Operation Alien' was triggered several times.

It is unlikely that the Australian working movement would have tolerated the internment of fellow Australians, particularly trade unionists, or prolonged military intervention in the work place, so by voting 'No' to ban the Communist Party, the country was spared what could have been a period of industrial unrest and possible violent confrontation.

Labor's role in the 'No' campaign confirmed ASIO's view – Labor was a security risk. For its part, Labor remained convinced that ASIO was in the pocket of the Menzies government. Both prejudices were soon confirmed in full by the notorious Petrov Affair. Vladimir Petrov was listed as third secretary of the Soviet embassy in Canberra. He was actually a Soviet intelligence officer, as was his wife, Evdokia. ASIO had identified Petrov as such and had placed him under surveillance. His open enjoyment of the Australian lifestyle, including heavy drinking, led ASIO to conclude that Petrov was a candidate for defection.

ASIO joined forces with a Sydney doctor, Michael Bialoguski, an immigrant from Poland, to befriend Petrov, show him a good time, and encourage him to defect. There remains confusion over whether ASIO engaged Bialoguski for this role or whether, as Bialoguski told me years later, on his deathbed in

Britain, he had made the first approach to ASIO with the news that Petrov was ripe for a 'honey-trap' operation. And, like many intelligence operations, there is doubt over who recruited whom – Bialoguski always said he recruited Petrov for ASIO but Petrov was convinced that he had recruited Bialoguski for Soviet intelligence.

Either way, it was ASIO which came off best. Petrov, who was on probation in Australia after a poor showing in his previous Soviet intelligence posting, decided that he would have a better future in Australia than in the Soviet Union – especially when ASIO offered him, apart from all other sorts of inducements, a cash sum of £5,000 to defect. He met senior ASIO officers in Sydney on 3 April 1954 and handed them what documents he had been able to smuggle out of the Soviet embassy. ASIO whipped him off to a secret location and took a statement from him: 'I wish to become an Australian citizen as soon as possible. I no longer believe in Communism since I have seen the Australian way of living.'

Menzies announced the defection in Parliament ten days later. He deliberately waited for an occasion when he knew Evatt would be absent – he was guest of honour at his old school's annual reunion in Sydney – and made it at a time when the audience for the broadcast of Parliament was at its peak. He had told his ministers what he was about to do, swearing them to secrecy, but some of them had been unable to resist tipping off their favourite journalists. The Treasurer, Arthur Fadden, warning a reporter to make certain he was in the House at 8 p.m., added, 'I can't tell you more than that, but the big white bastard will be making an announcement and it's a winner.' Another was told to expect the biggest story of the century, one that would 'destroy Evatt'.

Moscow denounced Petrov as a thief and said he had been kidnapped by the Australian security authorities. It also sent a team of Soviet security officers to Australia to collect Mrs Petrov and bring her back to the Soviet Union. Having set the affair in motion, Menzies no longer needed to do anything to gain political capital from it. Migrants from Soviet-occupied countries turned out in force at Sydney airport and tried to stop Mrs Petrov from boarding her plane, claiming that she was being abducted. The newspapers and newsreels were dominated by scenes of a young, blonde woman, weeping and with one shoe lost in the scuffle, being half-carried across the tarmac by 'scowling Slavic gorillas'.

On the flight to Darwin, the first stop, the pilot radioed that Mrs Petrov seemed afraid. At Darwin, there were more exciting scenes as the heroes (Australian policemen in shorts) disarmed the villains (the Soviet security toughs in cheap suits) by force. The ending was scripted by Hollywood – Mrs Petrov spoke to her husband on the telephone and then announced that she wished to remain in Australia with him.

Menzies now decided that there should be a Royal Commission to investigate Soviet espionage activities in Australia. Petrov would be the principal witness and would give information about code names, contacts and agents employed by Soviet intelligence. The hearings, which dragged on throughout 1954, split the country and ruined many a career, among them that of Dr Evatt himself. Even before the Royal Commission began sitting Evatt had become convinced that he had been the victim of a conspiracy to prevent the Labor Party from winning the general election due that year.

Labor's chances had been steadily improving – even the conservative *Sydney Morning Herald* had concluded that Menzies would need to pull a rabbit out of the hat to win the election, set for 29 May. Evatt had every reason to believe that it was his destiny to be Australia's next Prime Minister. Then, just a fortnight before the nation went to the polls, Menzies had pulled out of the hat not a rabbit but the Petrov defection, raising all the old doubts about Communism and its links to the Labor Party. Evatt thought that the timing was suspicious, to say the least, and when the Royal Commission began its hearings and Petrov's evidence implicated members of Evatt's staff in Soviet espionage activities, he was absolutely convinced that there had been a conspiracy.

Had there? Although the Menzies government was returned to power, Labor did make significant gains, polling nearly 250,000 more popular votes than the Government, which suggested that it might indeed have won the election if it had not been for the Petrov affair. But the answer to the crunch question – Did Menzies deliberately time the announcement of the defection to gain maximum electoral advantage? – has to be no. Menzies, who died in 1978, always said he knew nothing about the planned defection when he set the election date. Political academic Robert Manne, who has examined the ASIO files for the period, says that they confirm Menzies' statement. And the official historian of the Labor Party, Ross McMullin, says: 'Even if one allows for the unlikely possibility that Menzies was given an unofficial inkling [of Petrov] before 10 February 1954 [the date of his first briefing by ASIO as shown in its records], the long and fluctuating process of cultivating the erratic, hard-drinking Soviet intelligence agent was hardly compatible with forward planning for a precisely timed election.'

This clears Menzies but not necessarily ASIO, no matter what files it produces to support its version of events. To get a Soviet intelligence officer into a 'honey trap' was a great coup for the Australian service. Way down there in the antipodes, a junior security service had snared a defector, proof that it, too, could strike a blow for the Free World. To get the most out of this, ASIO would have followed normal security service practice – keep the defector in place as long as possible, make him stay within his own organization but working for you, gathering documents, acquiring knowledge, answering your questions, so that when he does eventually 'come over' he brings with him as much valuable material as possible.

Thus the actual date of the defection was ASIO's to decide. It did not have to consult Menzies about this – it was an operational decision. Given what we know about Brigadier Spry's views on the Labor Party, his concern that it might win the election, and his unwavering support for Menzies, is it not highly probable that after Menzies had announced the election date, Spry himself decided the date of Petrov's defection so as to cause the biggest impact on the election campaign?

To try to settle this once and for all, I took the opportunity of a chance meeting over dinner in Germany in 1994 with Leonid Shebarshin, one-time head of the KGB, to ask him why my repeated requests to the KGB to see its file on Petrov had come to nothing. (At one stage, trying the back-door approach, I had an offer out in Moscow of $5,000 for a copy of this file.) Shebarshin said, 'There could be two reasons. The first is that there *is* no file. Petrov should never have been posted to Australia and whoever ordered him there might well have destroyed the file to protect himself from the repercussions when Petrov defected. Next, there could be something in the file about someone in Australia who is still alive. The Cold War is over but we still protect our friends.' So, conspiracy? Or coincidence? There is no conclusive answer.

Evatt, who had no access to ASIO's files, became so obsessed with the conspiracy theory that his behaviour at the Royal Commission was deemed intemperate and eventually he was banned from appearing before it. In October 1955, when Parliament debated the Royal Commission findings, Evatt pointed out that after all the publicity and all the expenditure, not a single spy had been uncovered or a single prosecution recommended. If he had stopped there, he would have made a telling point. But he then went on to reveal that he had written to the Soviet Foreign Minister, Vyacheslav Molotov, and had received a reply which supported his claims that key documents in the Petrov affair were ASIO forgeries. As Evatt should have expected, this caused laughter and ridicule in Parliament and Menzies, a clever politician, called an election the next day, nineteen months before it was due. He won decisively. Evatt never really recovered – some say the loss of the 1954 election had unbalanced his mind – and he died in 1965 after a long illness. Menzies, who hated him – the feeling was mutual – was one of the pallbearers at his funeral.

There were reasons other than the Petrov affair for Labor's loss in the 1954 election. One of them was a Melbourne lawyer of Italian descent called Bartholomew Augustine ('Bob') Santamaria, a major figure behind the scenes of Australian politics right up until his death in 1998.

Santamaria's parents had run a fruit shop in the Melbourne suburb of

Brunswick. He was educated at a Christian Brothers school and although the Depression sharpened his political consciousness, he rejected Marxism after he saw what happened to Catholics at the hands of Communists during the Spanish Civil War. He turned up at a students' meeting at Melbourne University, called to support the Republicans, and made a brilliant speech that ended with his pro-Franco battle cry: 'Long live Christ, the King.'

He was a pre-war member of Catholic Action, a lay apostolate to the working class. He developed his own distinctive version of Australian political Catholicism, which Robert Manne has described as upholding 'the virtues of Christian piety, duty, family, work; rural community and small business against the vices of Communism, big business, secularism and soulless modernity'.

In 1941 he was approached by a Labor Party leader in Victoria, Herbert Cremean, who asked him if he could organize a struggle against the influence of Communists in the trade unions. Santamaria thought he could. The crucial question was how best to mobilize the Catholic laity? His answer was to copy the Communists. Small, well-organized bands of Communists had been successful in winning control of trade unions. The Catholics could do the same. Santamaria set out to match the Communists 'cadre against cadre, cell against cell, faction against faction'.

It was best that this was done in secret. One recruit remembers his priest saying to him, 'I want your solemn word of honour that whether you join us or not, you won't reveal to anyone anything I say to you tonight.' It worked. Santamaria's creation, the Catholic Social Studies Movement, generally known simply as 'the Movement' (he was president from 1943 to 1957) was the animating force behind the industrial groups which took militant anti-Communism into the Australian workplace. Priests took on the task of persuading their working-class parishioners to turn up at union meetings and outvote the Communists and their fellow travellers. In five years, from 1945 to 1950, they broke the Communist Party grip on most of the strategic unions in the Australian Council of Trade Unions.

Now new, irresistible opportunities presented themselves. Historian Alan D. Gilbert points out, in *Australians from 1939*, that evicting Communists from trade unions had meant taking their places. 'This had created the possibility of winning great influence in Labor Party conferences where policy was decided. The Movement could now aspire to join the fight against international Communism by influencing the decisions of Australian Labor governments on defence and foreign policy.' Santamaria wrote confidently to his archbishop that for the first time in any Anglo-Saxon society since the Reformation, a government would soon be in a position to implement Catholic social policies.

All this had happened in a news blackout. Historian Donald Horne, recalling it in 1999, wrote: 'Santamaria's secret organisation, "the Movement", had been taken out of the church and given its secular name, the National Civic Council – all that was left of the ten-year conspiracy in which, supported by a network of parish branches, it had established a party within the Labor Party and was ready to unload, on demand, voting fodder to support anti-Communist union leaders, without one word in the newspapers. Meetings of up to 2,000 zealots, with Santamaria invigorating them with the force of Aristotelian logic, and not a word in the newspapers.'

Evatt, a Protestant, saw the danger, publicly blamed the Movement for contributing to Labor's defeat in the 1954 election and kicked its followers out of the party. Santamaria thus found himself the intellectual leader of a new party which eventually called itself the Democratic Labour Party (DLP) and which became a significant influence in Australian politics for 20 years – not because it ever became the government itself or was able to control the policies of the traditional Labor Party – but because it was in a position to keep the traditional Labor Party *out* of power.

Australia operates a preferential voting system and the DLP, by persuading a body of traditional Labor voters to give their second preferences to non-Labor parties, helped to keep the anti-Labor, Liberal-led coalition in power until 1972. It also gave the Liberals a constituency of single-issue voters – the anti-Communists – motivated by religion, and therefore not amenable to critical discussion. All of this had repercussions beyond Australian domestic politics and, it could be argued, indirectly drew Australia into the war in Vietnam, the most divisive war in Australia's history.

Murray Sayle, the well-known Australian war correspondent, recalled in 1998 how he was constantly asked in London why Australia had sent combat troops to Vietnam? After all, where were all those other allies who had fought with the Americans in Korea, the other big anti-Communist war? Where were Turkey, Canada, France and, most importantly for American public opinion, Britain? (The United States had on more than one occasion begged Britain to send at least a token British contingent to Vietnam.) The answer was obvious: the government of Australia depended on anti-Communist votes for its existence. So for internal political reasons as well as for foreign policy ones, Australia felt it had to send troops to Vietnam, especially as Menzies had always warned about the downward thrust of Asian Communism.

So anxious was the Menzies government to join the United States in Vietnam that it 'invented' an invitation to do so. On the night of 29 April 1965, Menzies stood in Parliament and announced that his government was committing frontline troops. Three years earlier Australia had sent 30 military training instructors. By the end of 1964 this had become 100. Now

Menzies said, 'The Australian government is in receipt of a request from the government of Vietnam for further military assistance. We have decided after close consultation with the government of the United States, to provide an infantry battalion . . .'

It took ten years for the truth to emerge. There had been no such request from Vietnam for military assistance. The Department of Foreign Affairs examined the official cables and memoranda of the day and reported that the initiative for an Australian battalion had not come from the government of Vietnam but from the United States. The report stated, 'The United States did not need the military aid, but it did desire the military presence of its friends and allies in order to show the world that the United States was not alone in its efforts against Communism in south-east Asia.'

But despite all the rhetoric, then and later, this Australian commitment was not really, as Prime Minister Harold Holt put it, 'All the way with LBJ' (one of the more disastrous slogans in political history, implying as it did that Australia was subservient to the United States and its President.) Throughout the war Australia never had more than 8,000 troops at one time in Vietnam. America regarded this as a useful contribution, but more suitable for propaganda purposes than for military, and certainly not equal to the importance of the cause. Sayle has pointed out (*Heraclitus 70*, a privately circulated magazine) that Australia's participation in the war was unknown, or as near as made no difference, to anyone outside Australia, even in Vietnam – as it still is.

Of course, it did not seem like that in Australia at the time. The government had introduced conscription for overseas service without a referendum as in the First World War. The issue of whether 20 year-old Australians should be compelled to risk death (508 Australians, including 7 civilians, died in Vietnam, of whom 200 were conscripts) to support America's anti-Communist cause in Vietnam, tore Australia apart.

From this one issue others grew – the right to civil disobedience, the right of students to protest, the role of the churches, the generation gap, the growth of a 'counter culture'. Australians who had never been much interested in foreign policy – or anything foreign at all – suddenly found themselves arguing about it in the press, on television, in the workplace, on the farms, in bars, in schools and universities.

The Vietnam War brought home to them more than anything else where Australia was located in the world and where her future would be decided. Some of her leaders understood this better than others. Menzies, who dominated foreign policy, both as his own Minister for External Affairs and as Prime Minister, was committed to an unrelenting fight against Communism and to relying on 'great and powerful friends' in London and Washington to protect Australia from Communist expansion. Others were more pragmatic.

John (later Sir John) McEwen, leader of the Country Party and Deputy Prime Minister (1958–71), realized that whatever Menzies and the Department of Foreign Affairs might think about Communism, Australia was a primary-producing nation and had to sell where she could in order to live. McEwen dominated Australian policy on trade, and turned the Australian Trade Commissioner Service into his own diplomatic and intelligence service – trade commissioners bypassed ambassadors and reported direct to McEwen's department. Then he went ahead and did deals with any country that wanted to buy Australian, irrespective of its politics and often in defiance of Australian official foreign policy.

'So while Australia was fighting Communism in Vietnam and asserting that China was the greatest threat to world peace,' says P. G. Edwards, official historian at the Australian War Memorial, 'McEwen was presiding over profitable wheat sales to China and developing new trade agreements with the Soviet bloc.' And while the Japanese remained the hated enemy who had butchered Australians in prisoner-of-war camps, McEwen's department negotiated the Australian-Japanese trade agreement, which was to turn Japan, at one stage, into Australia's biggest trading partner.

Did McEwen see the writing on the wall, scrawled in the familiar hand of Great Britain? In 1962, Britain made its first application to join the European Economic Community. France vetoed it, but it was clear that sooner or later Britain would be admitted, and that she saw her destiny in European terms, not in Commonwealth terms, not in terms of what had been her Empire east of Suez. Once in the Common Market, not only would all the Empire trade arrangements be over, ones on which Australian primary producers had relied since the colonies began to export wool, meat and wheat in the nineteenth century, but there would be the risk of Common Market surplus dumping in other markets. The special relationship would then be well and truly over. As the Secretary of the Australian Department of Trade, Sir Alan Westerman, aptly put it, 'We're a member of this far-flung Empire, but some say we've been flung further than anyone else.'

McEwen was right to look to Asia for new markets for Australia, but Menzies had a case for his foreign policy, too. If the Mother Country was retiring to her homeland and nothing east of Suez was going to concern her any more (a retirement that was completed with the handing over of Hong Kong in July 1997) then Australia would need to look to the United States not only for her security but for investment.

But American foreign policy since the Korean War had been to stay out of any land conflict in Asia, so how could Australia be certain of support from the United States if any of the crises in south-east Asia should threaten Australian security? One way was to throw open Australia for American military installations which the United States would have to defend against attack – the naval communications station at North West Cape (1963), the joint defence space research facility at Pine Gap (1966) and the joint defence space communications station at Nurrungar (1969).

Although the word 'joint' appeared in the title of most of these facilities, no one really believed that Australia had any control over their operations. Many thought that Australia was becoming a base in the South Pacific for the United States, that it was tying itself too closely to American defence policy and was losing its independence. This became an election issue in 1963, the Labor Party once again torn between principle and electoral expediency – the bases were contrary to its policy but the voters seemed to approve of them. The division contributed to Labor's loss of yet another election, Menzies winning easily.

But allowing the Americans to establish bases in Australia – which meant that Australia automatically became a target for Soviet nuclear missiles in the event of war with the United States – was not enough. The Australian government felt that if it did not show military support for the United States in south-east Asia then America might withdraw from the region altogether and Australia would be left on its own.

So in 1962 it sent a squadron of RAAF fighters to Ubon in Thailand and the 30 military advisers to help the South Vietnamese army. It was the beginning of an accelerating involvement that was to last ten years. The first Australian troops had been professional soldiers, but the government felt that since Australia had hardly any unemployment, it would not be able to recruit enough volunteers for Vietnam and would have to conscript them.

Remembering how divisive the issue of conscription had been during the First World War, this time the government did not go to the people in a referendum – it simply went ahead and passed legislation for compulsory two-year military service for young men turning 20. They were selected by 'call-up ballot', and could be sent into action anywhere in the world. This action was justified, Menzies told Parliament, by the threat of Communist military insurgency throughout south-east Asia.

On the eve of the departure of the first Australian troops, the Labor leader, Arthur Calwell, made some startlingly accurate predictions. He said that sending Australian troops to help the Americans was not in anyone's interests – eventually the United States would be humiliated by its involvement in the war in Vietnam, intervention would prolong and deepen the suffering of the Vietnamese people; and for Australia to 'exhaust our resources in the bottomless pit of jungle warfare . . . is the very height of folly and the very depths of despair'. Few listened to him.

In retrospect, Australia should have seen how its involvement in Vietnam would split the nation. It was not just the decision to 'add another flag' to the

Stars and Stripes so as to show that American policy in the region had the support of many democratic nations. So many other factors intruded. Conscription for service overseas without a referendum was itself a controversial decision, but the birthday ballot scheme was a public relations disaster. Marbles bearing every date of the year were placed in a barrel and drawn out just like a lottery. All young men turning twenty on the birth dates drawn were then called up for military service. When enough dates had been drawn to satisfy the armed services' requirements, the lottery stopped and no further demands were made on other young men who turned twenty that year.

It was the arbitrary nature of the system, its association with gambling and all the lotteries that were so popular all over Australia, but especially in New South Wales, that appalled many. When protests against the war reached their peak, one type of attack that the new Prime Minister Harold Holt had to endure was being bombarded with handfuls of lottery marbles by mothers from the 'Save Our Sons' anti-war movement.

The protests themselves became an issue. How far was it legitimate to go in order to make known your attitude to the war? After an anti-war meeting at Mosman Town Hall, Sydney, Calwell, who had spoken at the meeting against the war and against conscription, was shot in the face by a youth with a sawn-off shotgun. Calwell was not badly injured and Prime Minister Holt deplored the shooting, saying that, despite disagreements, Australian public life had been singularly free from violence.

But not any more. In October 1966 President Johnson arrived in Australia to show support for Holt and Australia's help in Vietnam. In Sydney his motorcade was met by thousands of anti-war demonstrators, some of whom lay down in the roadway to prevent the President's car from continuing. When a police officer in an escort car told this to the Premier of New South Wales, Robert Askin, who supported the war and who was travelling with Johnson, Askin told the police officer, 'Run over the bastards.' Fortunately, the police officer ignored the order.

Police in other cities used horses like cavalry to charge and break up demonstrations. At one protest in Melbourne, anti-war demonstrators tried to set fire to the United States consulate, tore down and burned the American flag and tried to hoist the flag of the Vietnamese National Liberation Front. They smashed windows of police cars, threw firecrackers at police horses, and then tried to storm police headquarters where those arrested during the demonstration were being held.

In some cities, demonstrations took place to support the right to demonstrate. After the Queensland government refused to allow street meetings to take place without an official permit, 4,000 students and lecturers from Queensland University marched to protest what they called this infringement of their democratic rights. Intended to be a peaceful demonstration – marchers were told by their leaders not to resist arrest – it soon turned violent. Every police officer in Brisbane had been called to duty. They confronted the marchers and began arresting them, dragging women to the police vans by their hair and punching and kicking men. Onlookers said that the police had been inflamed by chants of 'Nazis . . . police state . . . we hate cops.'

The Seamen's Union made an important protest, more symbolic than practical. A freighter, the *Boonaroo*, had returned to Melbourne in May 1966 after its first trip to Vietnam carrying munitions. A union official, Roger Wilson, recalled what happened: 'I received a phone call from a shipboard delegate. He said, "We've just been advised that when the ship finishes discharging its cargo, we are to move down to the powder grounds at Point Wilson and load bombs for Vietnam. We've had a meeting and decided not to do it." That was the beginning of a long campaign and people accused the union of being unpatriotic and against the soldiers. The union was not against the soldiers. They were opposed to an unjust war and were refusing to ship weapons which would contribute to its immorality.'

Curiously, anti-war protests in Australia were more widely reported in the Australian media than the Australian role in the war itself. The Americans had assigned the Australians to the coastal province of Phuoc Tuy, east of Saigon and of minor military significance. Although 508 Australians were to die there – 17 in one encounter with the Viet Cong in August 1966 – no Australian medium kept a regular, full-time correspondent in Saigon. Sayle remembered in 1998: 'One by one, bored young men from the Australian Associated Press arrived to write "Home-towners" ("Private 'Curly' Jones from Curl Curl, New South Wales, loves to surf") did their time and left little the wiser.' It was the reaction to the war back home and the attempts to justify Australian participation in it that was newsworthy, rather than how the war itself was being fought and how the Australians were acquitting themselves. Vietnam was no Gallipoli.

When the war dragged on and the light at the end of the tunnel grew dimmer, support for it began to evaporate. As Sayle recalled in 1988, 'Theological debates pitting democracy against nationalism soured many a Sydney drink-in, but they did not alter the fact that one side had a tested plan for eventual victory and the other side, while seemingly the more powerful, did not. "This," I reported in 1968, "is a war that cannot be won." (But it can be lost, I should have added.)'

Protest against the war reached its peak in Australia in the Moratorium rallies of May and September 1970 when at least 200,000 people took to the streets in state capital cities, probably the biggest demonstrations in the nation's history, bringing traffic and business to a standstill. Speakers from a wide cross-section of the community said they were at a loss to understand why Australian youths were being conscripted to fight and die in another country's war.

Despite predictions from the government and the media that there would be widespread violence and extremism, the protests were peaceful and impressive. The Melbourne *Age*, a conservative newspaper, caught the mood. The scale of the protests had given a new meaning to the notion of peaceful public dissent. It said in an editorial: 'A legitimate expression of opinion by a substantial section of the population that included eminent religious leaders could not be written off by the government as "the antics of Communist-inspired fools".'

Actually, the churches had been deeply divided over the war. Nine Anglican bishops and four retired bishops had written to Menzies in March 1965, asking the government to seek a peaceful settlement. But an equal number of bishops had declined to sign the letter. A Methodist spokesman, Reverend Alan Walker, formed the Canberra Vigil Committee to co-ordinate ecumenical protest against the war. A group of young priests and lay Catholics established Catholics for Peace and published a journal, *Non Violent Power*, which advocated a combination of Christian and Gandhian methods of peaceful protest against the war. But the Roman Catholic hierarchy discouraged anti-war protest and at least one Sydney priest was threatened with suspension from his religious duties because of his anti-war activities.

Was the failure of the churches, particularly the Catholic Church, to take a stronger moral stand on the Vietnam War and conscription issue responsible for their subsequent loss of influence on Australian life? A survey quoted by Gilbert shows that in 1966 the number of Australians who had not been to church during the previous year was about 23 per cent. Ten years later the figure was over 50 per cent. In one decade a million and a quarter Australians ceased to be churchgoers. Many even ceased to believe in God. In the past, virtually every Australian at least professed to believe, but by the late 1970s one in five said they did not.

The Vietnam War was certainly a factor in this change, in that it called into question militant, unthinking anti-Communism. Australians saw where a hawkish anti-Communist foreign policy could lead and they did not like it. There were other factors. Rising standards of living moved the poorer Irish-Australian Catholics into the affluent classes and the church ceased to be the main focus of their lives. Catholics among the immigrant communities – except, perhaps, among the Asian Catholics, who were very devout – had a more relaxed view of their church and their religious duties. And the great cultural revolution of the sixties and seventies changed nearly everyone's views on life and its meaning.

Bob Santamaria and the DLP were both victims. The DLP just collapsed in 1974. Santamaria soldiered on, living through what he saw as the modern fall of the Roman Empire. Robert Manne, who knew him, described his mood in his last days: 'The West had, he thought, moved from its Christian period to the Enlightenment, and from the Enlightenment to the Kingdom of Nothing.' After the fall of the Soviet Empire and the collapse of Communism, Santamaria, this old Cold War warrior, decided that democracy and Australian economic sovereignty had been more threatened by capitalism than Communism. He began attacking the naked economic self-interest of the capitalist class with as much venom as his old left-wing enemies. Fittingly, this led to a reconciliation with some of them before his death in 1998.

The increasing disillusion with the war in Vietnam was reflected in the fortunes of Australia's political parties. In 1966, Harold Holt, campaigning on the Liberal government's Vietnam record, won a huge victory, better than anything achieved by Menzies. But then on Sunday, 17 December 1967, Holt went for a swim at Cheviot Beach, Portsea, near Melbourne, and vanished.

No trace of him has ever been found and his disappearance has sparked almost as many conspiracy theories as the death of President John F. Kennedy. (The former Reuters' correspondent in China, Anthony Grey, wrote a book, *The Prime Minister was a Spy*, in which he said that Holt was a long-time spy for China and had been taken off Cheviot Beach by a Chinese submarine.)

Holt's successor, John Gorton, won the 1969 election by only a narrow margin and in 1972, the Labor Party, led by Gough Whitlam, was swept into power for the first time in 23 years. Whitlam immediately announced an end to conscription and the withdrawal of the last Australian troops still in Vietnam. (Most had come home the previous year, following the decision by the United States to withdraw the bulk of its own troops.) The Whitlam era, an exciting time of new directions, fresh attitudes, hope – and then despair – was about to begin.

CHAPTER 17

THE WHITLAM ERA

These were exciting times. Everything seemed to be happening at once. The government's transformation of the political scene was accompanied by basic changes in the traditional Australian way of life. Women were on the march. Whitlam had shown sympathy for women's issues and many women believed that real equality was just around the corner.

When Labor MP Gough Whitlam made his maiden speech to the House of Representatives in Canberra in March 1953, a Labor colleague, Clyde Cameron, came up to him afterwards to congratulate him. Cameron admitted later that he had perhaps been too effusive – he had told Whitlam that one day he would be Prime Minister. The interesting thing was that Whitlam did not seem the least bit surprised at this prediction.

True, Edward Gough Whitlam was cut to fit the Prime Ministerial mould. The amazing thing is that he was a Labor man, not a conservative. Born in Melbourne in 1916, brought up in Canberra, where his father became Commonwealth Solicitor-General, he served with the RAAF in the war, completed a law degree at Sydney University and was admitted to the Bar in 1947.

Tall, confident, erudite and pedantic, he was fussy about his appearance – he wore white tie and tails at a Parliamentary reception for the Queen in 1954 when some of his Labor colleagues were dressed in their best lounge suits. He spoke well, was photogenic on television, had ability, energy and drive. But he looked like a younger version of Robert Menzies and, if background, education and social standing had any influence on political beliefs, Whitlam should have been a natural Tory, in Australia a member of the Liberal Party.

There is further evidence for this. An English friend of mine has a theory, more often right than wrong, that men from the Australian ruling classes look like koalas and men from the Australian working classes look like kangaroos. On this test, Whitlam is definitely ruling class and there is something of a mystery behind his devotion to Labor. Some say his father imbued him with a moral sense of public service, an almost patrician view that those favoured by fortune had a duty to work for the greater good of society. Others that he was a natural radical, particularly in matters of health, education and culture, and considered that the best chances of achieving reform in these areas was through the Labor Party. Others point to a specific instance that led to his conversion – the cynical behaviour of the anti-Labor parties during a referendum in 1944 which might have transformed Australian postwar politics. The Commonwealth government under Curtin wanted the states to transfer power to the Commonwealth for the duration of the war and five years afterwards so as to deal on a nationwide basis with a whole range of issues from demobilization to unemployment, national works, health, a uniform railway gauge, air transport, family allowances and Aboriginals.

A Yes vote would not only have brought Australia in line with parliaments in countries like Britain, Canada and South Africa, but might well have led to greater national cohesion and less rivalry between the states. Whitlam, who campaigned within the RAAF for a Yes vote, was appalled by the 'hysterical negativism' and 'deliberate misrepresentation' by Labor's enemies. A vision for their country's future appeared to mean less to them than immediate political advantage. Whitlam reacted by joining the Labor Party the following year.

Not everyone in the party was happy with their future Prime Minister. Whitlam was a modernist who was determined to bring Labor into the television age by reforming the party's organizational structure, encouraging more academics, intellectuals and businessmen to join, developing more up-to-date policies, embracing equality for women, distancing the party from union power, appealing to a broader band of voters, making it a modern, electable party – a Down Under Tony Blair-cum-Peter Mandelson twenty years before the two British politicians transformed the British Labour Party.

Whitlam found a party with a strong traditionalist core, wedded to the 'purity of the socialist objective', where many members were still fighting feuds that had started dozens of years earlier. These often concerned Labor MPs who had changed sides – 'Labor rats' – such as the First World War politician William Hughes. Eddie Ward, a tough, outspoken Labor MP from East Sydney, once refused to attend a party celebrating Hughes's fifty years as a Parliamentarian, citing dietary reasons – 'I don't eat cheese.'

Whitlam became leader of the Labor Party in 1967 and immediately set about overhauling its structure. He believed that the following two or three years would determine whether the party 'continued to survive as a truly effective parliamentary force, capable of governing and actually governing'. Some of his advisers put it even more bluntly. If Labor could not win office soon, then it would cease to exist at all.

By the time the 1972 election came around Labor was more than ready. In 1965 the party had finally got rid of its historic attachment to the White Australia Policy. Calwell and other die-hards opposed change to the end, arguing that they were not racists, just traditional Australians worried that too many coloured people in Australia would generate racial strife.

America's second thoughts about the war in Vietnam had showed that Labor's opposition to it had turned out to be right. Whitlam himself had opposed the war but had been careful not to associate himself too closely with demonstrations against it, probably in the belief that it would harm the party if he were seen to be taking part in a march that turned violent.

But the one foreign policy initiative that did more to establish Whitlam's stature both in Australia and internationally was his visit to China on 6 July 1971. He brought this about by sending the Chinese leader, Chou En-lai, a telegram that must rank as one of the most important diplomatic telegrams in Australia's history. It read:

> Australian Labor Party anxious to send delegation to People's Republic of China to discuss terms on which your country is interested in having diplomatic and trade relations with Australia. Would appreciate your advice whether your government would be able to receive delegation.

There are two points to make about this initiative. It was the first step towards ending more than 150 years of Australian fear, mistrust and racial hatred of the Chinese, and it was taken by a party not yet in power that had a vision of what Australia's relations with China could be. And it was *before* President Nixon's historic 'opening to China'. When China replied to Whitlam's telegram, saying that an Australian Labor Party delegation would be welcome, Whitlam went, and while he and Premier Chou En-lai held wide-ranging discussions in Peking, Henry Kissinger was waiting in the wings to arrange Nixon's visit. But thanks to Whitlam, Australia was there first.

The election campaign was a highly professional one, lavishly financed by a whole new class of donor, including one Rupert Murdoch, an up-and-coming media baron following in the footsteps of his father, Sir Keith Murdoch, who may well have stirred in his grave at the news that his son was supporting a party he had bitterly opposed. Rupert went further than providing campaign funds. He swung his newspapers behind Labor – including Australia's only national newspaper, the *Australian* – met Labor leaders and offered tactical and public relations suggestions. In all its history, Labor had never had such wholehearted support from a major newspaper group.

Show business personalities joined the Labor campaign and helped promote its slogan 'It's time' – 'borrowed' from New Zealand Labour leader Norman Kirk's campaign in 1969 – and the jingle that went with it. Although party organizers had produced a series of slick television advertisements, Whitlam decided to make as many public appearances and speeches as possible. He went to Sydney's western suburbs, a rapidly expanding working-class area, to make his policy speech. He silenced a tumultuous reception at the Blacktown Civic Centre and linked himself to John Curtin by using Curtin's wartime greeting: 'Men and women of Australia.'

Then in simple, sincere words he set out his and the Labor Party's vision for a new Australia. 'There are moments in history when the whole fate and future of nations can be decided by a single decision. For Australia this is such a time. It's time for a new team, a new programme, a new drive for equality of opportunities; it's time to create new opportunities for Australians, time for a new vision of what we can achieve in this generation for our nation and the region in which we live. It's time for a new government, a Labor government.'

Whitlam's speech writer, Graham Freudenberg, had ushered Whitlam from the wings on to the stage and had heard his last words before he faced the crowd – 'It's been a long road, comrade, but we're there.' Now, listening to the roar of the crowd, watching a weeping old lady, a lifelong Labor supporter, grab Whitlam's hand and kiss it, Freudenberg decided that what he had witnessed was 'not so much a public meeting as an act of communion and a celebration of hope and love'.

As the campaign progressed, expectation of victory grew. In Victoria there were still sceptics – something would happen to thwart Labor at the last moment. This was understandable because the party's history there had been so disastrous that, as Labor's historian Ross McMullin points out, no Labor voters under forty had ever voted in any election won by their party. The Victorian state secretary of the party, Jean Melzer, later recalled that in the last week of the campaign 'it was like a great big party throughout Melbourne'.

The campaign climax was a meeting at St Kilda Town Hall with all the razzmatazz now part of new Labor – a band, dancers, singers, television celebrities, streamers and another inspiring speech by Whitlam. This time he associated new Labor with wartime Prime Minister Ben Chifley, the one-time engine driver and working man's hero. In 1949, just two years before his death, Chifley summed up his political credo: 'I try to think of the Labor movement, not as putting an extra sixpence in somebody's pocket, or making somebody Prime Minister or Premier, but as a movement bringing something better to the people, better standards of living, greater happiness to the mass of the people. We have a great objective – the light on the hill –

which we aim to reach by working for the betterment of mankind, not only here but anywhere we may give a helping hand.' Whitlam said that the light on the hill had almost gone out. 'Let us set it aflame again,' he said. Covered in streamers, he had to make several curtain calls. McMullin comments, 'It was surely the most ecstatic political meeting in Australian history.'

Victory was almost an anticlimax. It was not a landslide but it did produce a comfortable majority. It was also a triumph for young Australia, in that voters still in their twenties saw in Whitlam a chance for their country to break from its predictable path and find a new role in a new world. One campaign worker, a young lawyer called Gareth Evans, a future Labor minister, recalled 2 December 1972 as 'probably the most exuberant and totally joyous night' of his life. Many other Australians, young and old, felt the same. My father, a Labor supporter all his life, wrote to me in London suggesting I should hurry home – 'Great things are going to happen.'

Many Australians did hurry back. In 1998 Gordon Bilney, a South Australian who in 1972 had been working with the Australian contingent at the United Nations, recalled: 'Shortly before the election I had been offered a job on Whitlam's staff. I had been non-committal. But by the time Whitlam had been in power a week, all I wanted was to get back to Australia as soon as I could. There were thousands like me, accustomed to cringing culturally – an Australian abroad was thought of either as an Austrian or as a variety of South African – but who quickly found reason to take pride in what the new government was doing.'

Whitlam's policy speech had made nearly 150 specific promises to voters, probably the most comprehensive blueprint for reform that any Australian political party had ever presented to an electorate. Determined to maintain in office the excitement the election campaign had produced, Whitlam broke with precedent. Usually a new government waited to choose its ministers until the results of marginal seats had been declared which, with the Australian preferential voting system, could sometimes take weeks. Then, unlike Britain, where the Prime Minister appoints his ministers, all government MPs vote to decide who the ministers will be.

Impatient to get to work, Whitlam formed an interim ministry of two men – himself and his deputy Lance Barnard – and called it by an ancient Roman title, the Duumvirate, originally meaning two equal magistrates, but later applied to any partnership of equals. Whitlam described it as the smallest ministry with jurisdiction over Australia since a temporary British administration under the Duke of Wellington in 1844. From these two Duumvirs came a stream of decisions and announcements that made headlines day after day, not only in Australia but in many other countries. Suddenly, Australia mattered.

Conscription was abolished, draft resisters released from prison and the remaining Australian soldiers in Vietnam recalled. The imperial honours system was scrapped and Whitlam announced that he would not accept appointment to the Privy Council, an honour traditionally accorded to all Australian prime ministers and Opposition leaders. And, another blow to British prestige, Whitlam said he would not be using the Prime Minister's British Bentley but the Ford Galaxy he had when Opposition leader.

The case for equal pay for women, which had stalled, was re-opened before the Commonwealth Arbitration Commission and fourteen days after the election the Commission ruled that women doing the same job as men would be paid the same wage. The granting by the government of mining leases on Aboriginal reserves was halted, and a new ministry, the Federal Ministry of Aboriginal Affairs, under Gordon Bryant, established. Australia formally recognized the People's Republic of China and diplomatic relations began at ambassadorial level. Explaining this, Whitlam said, 'While it has been long recognized that Australia's geographical position gives it special interests in the Asian region, up until now we have not come to terms with one of the central facts of that region, the People's Republic of China. This serious distortion in our foreign policy has now been corrected.' The country woke up each day wondering, 'What's Gough going to do today?'

Nothing appeared too minor to have escaped the Duumvirate's attention. Excise duty on wine was removed and sales tax on contraceptives abolished. Racially selected sporting teams were banned from competing in Australia, from visiting it, and even from transiting it – to the fury of the South African Rugby Board, at whom the ban was principally aimed. Wheat exports to Rhodesia were banned and the government ordered the Rhodesian Information Centre in Sydney to close. The Australian contingent at the United Nations was instructed to support any issue that would give real effect to Australia's new non-racial stance, and to pledge support to any measures that would help forge new relationships for Australia in south-east Asia. A start was made on a new look at Australia's relations with its colony, Papua New Guinea, one that was eventually to lead to its independence.

This cracking pace continued once the ministerial elections were over. Ministers, gathered to celebrate, surprised themselves by singing popular Australian songs like 'Waltzing Matilda' and 'The Road to Gundagai', one minister leading the revelry dressed in a purple suit and sunglasses – a scene more removed from Westminster tradition would be hard to imagine. But in its first year this unlikely Cabinet made well over a thousand decisions and established more than a hundred expert commissions to report on every aspect of national life. In 1972, the last year before Whitlam came to power, Parliament had passed 700 pages of legislation. In Whitlam's first year the figure was 2,200.

If the workload was heavy, the scope was breathtakingly broad. Whitlam took on the most conservative of all professions, the doctors, and after nearly three years of struggle, succeeded in setting up Australia's first universal health insurance scheme. Academic Robert Manne says, 'This was by far Whitlam's most popular and most enduring legacy.' He shook up the Australian education system, abolishing fees for university students, increasing government spending on technical colleges, and directing special funding for the education of children in isolated areas, Aboriginals, and those with handicaps. McCullin says, 'In no other sphere did the government take such giant strides towards Whitlam's fundamental objective, equality of opportunity for all Australians.'

Following the abolition of the imperial honours system, Whitlam cut some other ties to Britain. 'God Save the Queen' ceased to be Australia's national anthem and was replaced with 'Advance Australia Fair' – although Whitlam said 'God Save the Queen' could still be played on occasions when it was important to emphasize Australia's links with Britain. This caused some confusion when Whitlam's successor, Malcolm Fraser, not only reinstated 'God Save the Queen' but said that apart from royal and vice-regal occasions when 'God Save the Queen' would be mandatory, Australians could decide themselves whether the national anthem should be 'God Save the Queen', 'Advance Australia Fair', 'The Song of Australia' or 'Waltzing Matilda' – making Australia the only country in the world with four national anthems.

The Whitlam government was not only distancing Australia from Britain but also from the United States. The era of Holt's 'All the way with LBJ' was well and truly over, and verbally violent attacks on an erstwhile ally over President Nixon's December 1972 decision to resume bombing North Vietnam shocked Washington. Whitlam sent a private message to Nixon, but three of his ministers protested publicly. Jim Cairns said the renewed bombing was 'the most brutal, indiscriminate slaughter of defenceless men, women and children in living memory'. Clyde Cameron said that maniacs seemed to be in charge of American policy, and Tom Uren said that Washington was dominated by 'a mentality of thuggery'.

These were exciting times. Everything seemed to be happening at once. The government's transformation of the political scene was accompanied by basic changes in the traditional Australian way of life. Women were on the march. Germaine Greer's seminal work, *The Female Eunuch*, had been published in 1970 and she had returned to Australia to speak at rallies and lead marches on issues like abortion law reform. Elizabeth Evatt had become the first woman to be appointed to the bench of the Commonwealth Conciliation and Arbitration Commission. Whitlam had shown sympathy for women's issues and many women believed that real equality was just around the corner.

The old morality was weakening. The Victorian vice squad could still seize and destroy 414 copies of Philip Roth's novel, *Portnoy's Complaint*, but Australia's first sex shop opened in Sydney's Kings Cross in 1971, a new magazine for women, *Cleo*, with a nude male centrefold, was launched in 1972, and two bathing beaches in Sydney were opened for legal nude bathing in 1976. They were opposed by local residents who were worried that they would attract 'undesirables and peeping Toms'. Homosexuals, however, still had a long way to go. At a gay rights march in Sydney in 1978 police arrested 93 people for taking part in an unlawful procession.

The cultural cringe was dying. Patrick White won the Nobel Prize for literature, the first Australian to receive the highest honour in the literary world. The Australian government was in the process of becoming a significant patron of the arts, with funding available for opera, ballet, theatre, orchestras and arts education bodies. And in October 1973, amid the most exuberant celebrations ever seen in the nation's oldest and largest city, the Sydney Opera House was opened by the Queen herself.

The 15,000 people at the Opera House were outnumbered by the 750,000 on and around Sydney Harbour who wanted to prove that Sydney's reputation for throwing a party was justified. Hundreds of white racing pigeons and 30,000 'champagne bubble' helium balloons soared into the sky, as tugs and fire floats – sirens hooting – filled the air with jets of water, and a formation of F-111 jets shot through the sky. An Aboriginal actor, Ben Blakeney, in the role of Bennelong, the first Sydney Aboriginal to learn English and whose name was given to the point on which the Opera House is built, declaimed lines intended to remind everyone about the original owners of the land. 'Two hundred years ago fires burned on this point, the fires of my people; and into the light of the flames from the shadows all about, our warriors danced.'

The Queen, not usually one to mention the negative side of anything, acknowledged that the Opera House story was not one of unqualified success. 'I understand that its construction has not been totally without problems, but every great imaginative venture has had to be tended by the fire of controversy.'

That was certainly true. The Sydney Opera House may be one of the seven wonders of the modern world, a breathtakingly beautiful landmark whose outline instantly says 'Sydney' to all who see it, an original creation the likes of which the world had never seen before, but its history is a shameful one, a catalogue of all the meanest traits in the Australian character of that period. The idea of an Opera House in Sydney came from Eugene Goossens, the conductor of the Sydney Symphony Orchestra, as early as 1947. It took another ten years to get around to holding a design competition. The Danish architect Joern Utzon won, but was lucky to have done so. He was a foreigner and the government of New South Wales had hoped that a local, or at least an Australian, would get the prize. Next, Utzon's design was nearly discarded in an early pruning and was rescued only when the final shortlist proved so unimpressive that some of the judges, including the brilliant Finnish architect Eero Saarinen (the designer of the TWA terminal at J. F. Kennedy airport, New York) suggested that the selection committee went through the pile again.

The Opera House was budgeted to cost $7 million, but Utzon was a visionary and many features he incorporated had never been built before. 'So what?' said his supporters. 'It's not the architect's job to worry about how to realize his design. Utzon is pushing forward the frontiers of the construction industry. It's a challenge and if we're real Australians we should be prepared to meet it.' This was correct. Many building techniques took a giant leap into the future, because some small Australian company came forward, after others insisted it could not be done, and said, 'We'll give it a go.'

But all this cost money. The budget exploded, the completion date was put off and then off again and the Opera House became a political football. Cheap points could be scored by attacking whichever government was in power when a new budget was announced, a new completion date fixed. No Australian leader had the vision to say, 'Hang on a minute. What does it matter what it costs? We're funding it from the Opera House Lotteries. No work of art is ever really finished, but one day we'll have an Opera House that'll put Sydney on the map, that'll reshape Australia itself, that'll shame all the bastards who knocked it, and make the rest of us proud for the rest of our lives.'

It did get finished – at a cost of $102 million – but neither Goossens nor Utzon were there for the opening. Australia had run Goossens out of town after Sydney Customs found some pornographic material in his baggage (see Chapter 15) and he had died in 1958. Utzon was also run out of town – although the state government manipulated matters to make it look as if he had resigned. It failed to pay him all he was due, and made so many changes to his original design that Utzon later said he did not feel the building was any longer really his and therefore had no desire to return to Sydney to see it. He may yet relent.

Australia redeemed itself just a little. The first performance in the Opera House was well before its official opening. On 17 December 1972 the Sydney Symphony Orchestra gave a baton salute to Goossens and put on a programme that began with the overture from the *Merry Wives of Windsor* and included pieces by Ravel, Mozart and Beethoven.

Yes, Sydney society was there – the Governor of New South Wales, the State Premier and a few other nobs. But the Opera House was filled with the 2,000 ordinary construction workers who had built it – the Australian concrete labourers, the Italian tilers, the Yugoslav carpenters – all dressed up in their best suits, with their wives, their mothers and their girlfriends. And as that great writer on cities, Geoffrey Moorhouse, says, 'It is probably true that such a thing could have happened only in Australia.'

The sky brightened for Aboriginals under the Whitlam government. The first land rights campaign started in 1963 when a collection of Aboriginal clans known as the Yolngu, from north-eastern Arnhem Land, sent a petition written on bark to the government in Canberra. It objected to a big Swiss-Australian company called Nabalco mining bauxite on their land without their permission. The Yolngu lost, but the very fact that they had protested spread quickly throughout Aboriginal Australia.

A new generation of Aboriginal workers on cattle stations became aware of their rights and their power. This coincided with a more sympathetic attitude among young white Australians. But, ironically, the most important factor in what now occurred had its origins in the attitude of the station owners themselves. By denying their Aboriginal workers access to living conditions enjoyed by white Australians and by keeping them outside the cash economy (they were paid largely in rations), the station owners had unwittingly encouraged Aboriginals to maintain their traditional lifestyle.

So, although slow to anger and not prone to carry resentment for long, when they nevertheless felt by the mid-1960s that they had had enough, they were sufficiently independent of the station owners to take the dramatic step of going on strike. At cattle stations all over the Northern Territory, Aboriginal stockmen simply walked off the job. Most of these strikes received little or no publicity but one became national news.

On 23 August 1966, Vincent Lingiari led 200 members of his Gurindji tribe off the cattle station at Wave Hill owned by the British Vestey family. Lingiari died in 1984 but one of the striking stockmen, Bill Bunter Jampijinpa, recalled later: 'We were fed up. We were treated just like dogs. We were lucky to get paid the fifty quid a month we were due, and we lived in tin humpies you had to crawl in and out of on your knees. There was no running water. The food was bad – just flour, tea, sugar and bits of beef like the head or feet of a bullock. The Vesteys were hard men. They didn't care about blackfellas.'

This assessment was confirmed by journalist Christopher Forsyth, who wrote in the *Australian* at the time, 'The accommodation provided by the Vesteys for the Aboriginals was rusted iron shells, four walls leaning towards each other, pieces of hessian covering the worst holes . . . They were about five feet wide and eight feet long. There appeared to be no water and certainly no sanitation. I have never seen a more desolate place, nor such a disgusting sight in a country which prides itself on giving people a fair go.'

Jampijinpa remembered how the strike began. 'Old Vincent came back from hospital in Darwin and said he had decided that he would pull us out.

He pulled everyone out that Tuesday and we walked with the kids and our swags to the Victoria River and set up camp there. After we'd been there a couple of days, the Vestey mob came and said they would get two killers [slaughtered beasts] for us and raise our wages if we came back. Old Vincent said, "No, we're stopping here." So we stayed there until the New Year and then we walked to our new promised land – we call it Daguragu, back to our sacred places and our country, our new homeland.'

This apparently minor incident in a remote part of the Northern Territory is now recognized as the beginning of the Aboriginal land rights movement, probably the most significant event in Australian history since Federation. The point was that Daguragu was known to white Australians as Wattie Creek and it was within the boundaries of Wave Hill station. The Gurindji, Vestey employees, had staked out a claim to land that within the Australian legal system belonged to their bosses, the Vesteys. Under Australian law, the Aboriginals were squatters and the Vesteys would have been within their rights to have ordered the Aboriginals to leave, and if they had refused, to have forced them to do so.

But by now the Australian metropolitan media had discovered the story and trade union, student, church and writers' organizations moved in to lend the Aboriginals their support. Union and political activists who had never seen an Aboriginal in their lives before trekked into the outback to show solidarity with their new-found striking comrades. They were filmed and photographed and written about. Suddenly, Aboriginals, a non-people for 180 years, were big news. (They hesitated about appearing in a British documentary until Melbourne writer Frank Hardy persuaded them to do so. He acted out for them an imaginary scene in which Lord Vestey would be sitting in front of his television set in his English castle one night. Suddenly the Wave Hill Aboriginals would appear on the screen and Lord Vestey would spill his whisky and shout, 'That bloody Gurindji mob again. Can't get rid of the bastards.')

The emotional appeal of the Aboriginals' case was strong, as an incident described by Hardy illustrates. Peter Morris, the manager of the Vestey station, went up to Vincent Lingiari at Wattie Creek and demanded, 'What are you doing on Vestey land?' Vincent thought for a moment and then replied, 'I dunno. This belong to my grandfather.'

So the Vesteys did not force the issue. They pointed out, correctly, that they only *leased* the land from the Federal government, so it was not for them to decide whether the Aboriginals' claim to it as their tribal land was right. They were prepared to go along with any solution the government proposed.

Encouraged by this and the support they were getting from metropolitan Australia, the Aboriginals submitted to the government a request for 500 square miles of Wave Hill's 6,158 square miles of land, saying that it was tribal territory. At first their prospects looked good. The government of the day, the Liberal-Country Party coalition, was on record as saying that it favoured policies to help the Aboriginals to help themselves.

But by the time it came before Cabinet, ministers had realized the dangers. Many Country Party voters were landowners themselves and sympathized with the Vesteys, not the Aboriginals. Next, Australia was on the brink of a minerals boom – there appeared to be no limit to the extent and nature of minerals discovery. What if the government recognized Aboriginal land rights and then an enormous mineral strike was made on tribal land – what would happen?

The government funked the challenge and announced that it was not in favour of granting land rights to the country's original owners. The finality of this decision seemed to be confirmed when another group of Aboriginals lost an action against a mining company, Mr Justice Blackburn, judge of the Northern Territory Supreme Court, reaching the curious conclusion that 'the Aboriginals belong to the land more than the land belongs to the Aboriginals.'

The Aboriginals did not give up hope. Even before its victory in December 1972, the Labor Party had announced in general terms its thinking on the Aboriginal question, and had promised to reverse the history of 'despoliation, injustice and discrimination which have seriously damaged and demoralised the once proud Aboriginal people.' When the head of the Vestey family, Sam, Lord Vestey, was asked by a Melbourne journalist about Labor's promise, he began by blasting city-dwelling Australians who had supported the Gurindjis at Wave Hill. 'The trouble with Melbourne people is that they have never been to the Northern Territory,' he said. 'You killed all your Abos [in Victoria]. Now you have a guilt complex and you meddle in things you don't understand. All the trouble about the Northern Territory Abos is stirred up by Communists and the Melbourne papers. People down here wouldn't know a Gurindji from a Wailbri.'

They were about to learn. Two months after Whitlam's election win, his government commissioned Mr Justice Woodward to decide the best means of recognizing and establishing the traditional land rights of Aboriginals – note, not *whether* land rights should be recognized and established, but the best means of doing so. Woodward's report of April 1974 recommended that Aboriginal land rights legislation should be introduced into the Australian Parliament, that it should be immune from any interference by ordinances of the Northern Territory, and that an Aboriginal Land Commission should be established to assist Aboriginals in laying claim to their traditional land.

Since the Gurindjis had already made a claim, in 1975 the Labor government gave them 2,000 square kilometres of tribal land from the Vestey leases. In an inspired move, Whitlam went to Wattie Creek himself to hand over the titles deeds to Vincent Lingiari. As Whitlam later remembered it: 'We gathered near Wave Hill station, west of the Victoria River, some two hundred miles north-west of Tennant Creek. All of us there that day had a keen sense of history in the making, of something crucial to our national spirit and identity, of a process whose origins stretch back to the very beginnings of human settlement on this continent.'

Whitlam had arranged a little surprise. The founder of Melbourne, John Batman, the son of a New South Wales convict, had entered into some sort of sale treaty with local tribal elders, and when in 1834 he took possession of the site of what became Melbourne, he wanted to impress on the Aboriginals that the land was now his and no longer theirs. So he got a tribal elder to take a handful of earth and pour it into the palm of Batman's hand. At Wattie Creek, Whitlam reversed the process. He bent down, picked up some of the dry soil, took Vicent Lingiari's hand and poured it into his palm.

The idea had come from Herbert 'Nugget' Coombs, Australia's most prominent postwar economist, who, before he died in 1997, embarked on a second distinguished career as an 'activist and interferer' on behalf of Aboriginal rights. When Whitlam enthusiastically adopted it, Coombs wrote the speech to go with the gesture. Recalling it in 1999, the political columnist Alan Ramsey wrote, 'It was one of those simple speeches than genuinely lift the spirits and make you think that professional politics isn't such a festering pus-ball after all.'

Whitlam began, 'On this great day, I, Prime Minister of Australia, speak to you on behalf of all Australians who honour and love this land we live in. For them, I want: first to congratulate you and those who have shared your struggle on the victory you have won in that fight for justice begun nine years ago when, in protest, you walked off Wave Hill station. Second to acknowledge that we have still much to do to redress the injustice and oppression that has for so long been the lot of black Australians; third to promise you that this act of restitution we perform today will not stand alone.'

And then, as the dry, dusty soil trickled through Vincent Lingiari's fingers, Whitlam concluded, 'I want to give back to you formally, in Aboriginal and Australian law, ownership of this land of your fathers. And I put in your hands this piece of the earth itself as a sign that we restore them to you and your children forever.' Vincent Lingiari responded by saying, 'We are all right now. We are all friendly. We are all mates.'

CHAPTER 18

THE CROWN TRIUMPHS

Just as every American of a certain age can remember where he or she was when they heard that President Kennedy had been assassinated, so every Australian of a certain age can remember where they were and what they were doing and how they felt when they heard that Gough Whitlam, the democratically elected leader of the nation, had been dismissed by the Queen's representative, the Governor-General.

Some Australians argued that it was not only Aboriginals who had lost their birthright – white Australians had surrendered control of Australian resources to outsiders. Too many foreign investors had deprived Australians of the right to exploit their own country's riches. It was time to 'buy back the farm'. No one expounded this view in stronger terms than the Minister for Minerals and Energy, Rex Connor.

Connor was a staunch Australian nationalist, proud of the fact that he had been born on Australia Day, 26 January. A big, burly man who always wore braces and a hat, and known to his enemies as 'the Strangler', Connor had a vision of great and profitable enterprises, Australian-owned and operated, selling gas, oil and minerals to an energy-hungry world.

Connor's dream coincided with the 1973 oil crisis and the increase in the price of oil. The oil-producing nations of the Middle East and, in many cases, their rulers suddenly became enormously rich. The world was awash with 'petro-dollars' and since there was a limit to what their owners could spend them on, these dollars could be borrowed at low rates of interest. Japan, Britain and France had jumped in and negotiated large petro-dollar loans on favourable terms. Why could Australia not do the same? A big loan would help the Government out of its current financial difficulties and the rest could be used to pay for Connor's grand schemes.

Just before Christmas 1974, the Whitlam government authorized Connor to borrow up to $4 billion. Since the whole idea of a petro-dollar loan was breaking new ground, Connor made enquiries about approaching the lenders direct instead of through the traditional banking channels in London or New York. He discovered that, yes, it was possible to bypass the usual loan underwriters, but that the Middle East had its own financial customs and these would have to be observed. Oil sheikhs preferred to deal through a middleman, usually someone of the Islamic faith, and to keep details of the loan secret from rival sheikhs. In no time an Adelaide businessman, an acquaintance of Connor's, had come up with just such a middleman, Tirath Khemlani, a Pakistani commodities dealer working out of London. Connor authorized him to go ahead and see if he could raise the $4 billion loan.

In retrospect it is difficult to believe that no one checked out Khemlani's background in greater detail. By one of those strange coincidences of which life is sometimes full, my first wife, Eva Hajek, was at that time working as a teleprinter operator in Khemlani's London office. There were two other employees – an Indian accountant and an Australian disc jockey who was hoping to break into the British radio world. Khemlani's main business seemed to be buying cargoes of sugar while they were on the high seas and then diverting the ships to markets where an unexpected shortage of sugar had forced up prices. This involved lots of telex messages – hence my ex-wife's job.

But as the loan negotiations progressed, the nature of her work changed. It now involved travel. Her duty was no longer simply to punch out telex messages on the teleprinter machine in the London office, but to travel to various destinations in Europe and the Middle East and send telex messages under Mr Khemlani's name from there to Australia. The purpose was never explained to her, but she guessed that the recipients of these telexes would assume that Mr Khemlani himself had sent them from the city where they were lodged and believe that he was on the trail of the loan.

Perhaps he was. Perhaps he did have contacts in the Middle East who knew sheikhs with $4 billion to lend to a reputable government at a reasonable interest rate, and it would be racist and wrong to write off Khemlani's chances of pulling off such a deal because he was a) a Pakistani, b) a small-time commodities dealer and c) not a regular banker. But the fact is that in the end he did not deliver.

The young MP Paul Keating, later to become Prime Minister, remembered finding Connor morning after morning in March and April 1975, asleep in his Canberra office, waiting for the rattle of the telex machine that would tell him that Khemlani had succeeded. Keating pleaded with him, 'For God's sake, Rex, this is no way for a Minister to behave.' But Connor was under increasing attack from powerful enemies. Opposition MPs had decided that an unrelenting attack on the government's morality would pay political dividends. They were helped by public servants – Treasury officials who were appalled by the government's unorthodox borrowing procedures – who leaked documents to a press that had become uniformly anti-Whitlam, especially the newspapers owned by Rupert Murdoch.

Why this change of heart? Murdoch had been enthusiastic in his support of Whitlam and Labor in 1972. Labor historian Ross McMullin offers a simple if disturbing explanation. 'The government had denied Rupert Murdoch financial enrichment in the Alwest project [a mining venture in Western Australia] as well as rejecting his request to be High Commissioner in London.' McMullin also points out that investigative journalism was all the rage in Australia at that time, following the success of Bob Woodward and Carl Bernstein, of the *Washington Post*, in bringing down President Nixon.

With the government under siege from the media, Whitlam made a determined effort to put the loans affair behind him. He revoked Connor's borrowing authority on 20 May, and in July recalled Parliament for a special debate on the affair. Masses of documents were assembled, including a box full of telex messages between Connor and Khemlani. Connor, a shadow of his former self, made a strong speech: 'I am an honest man. I deal with honest people. I have stood in the path of those who would have grabbed the mineral resources of Australia. I have no apologies whatever to make for what I have done.'

Connor *was* an honest man but he had made a mistake. Melbourne *Herald* reporters ploughing through the loan documents found a telex from Connor to Khemlani dated 23 May reading: 'I await specific communication from your principals for consideration.' This proved that three days after his borrowing authority had been revoked, Connor was still trying to raise the petro-dollar loan. Whitlam insisted that there was nothing wrong with maintaining contact with Khemlani, but felt obliged to ask for Connor's resignation, another blow to a government already reeling from inflation and soaring unemployment – the worst since the Depression.

Whitlam's enemies now moved in for the kill. The Government did not have a majority in the upper house, the Senate, so there was a political tactic available to the Opposition to force an election. It could block supply in the Senate – withhold approval for the money supply the Government needed to run the country. True, this had never been done in Australia before, but it was legal and there is a first time for everything.

Like the last scene in a Western, the two sides now faced each other for the showdown. Whose nerve would break first? Whitlam was not about to call a general election, because he knew he would lose. The Opposition knew that if it continued to refuse supply, by the end of November the government would run out of money. Whitlam tried to lift the issue above party politics.

He said it was in the line of 'great constitutional struggles of the past – of 1640, 1688, 1832 and 1910'. It was Australia's misfortune to have adopted its inflexible constitution at Federation, a mere decade before the British crisis of 1910, when the House of Lords was stripped of its power to block supply. He pointed out that in 1975 Australia was the only parliamentary democracy in the world with an upper house able to take such an action.

Australians responded to Whitlam's appeal. They not only followed the confrontation with passion, they understood the issues. Opinion polling showed that 70 per cent of voters disagreed with the blocking of supply and that the government's approval rating had climbed back to 47 per cent. Newspapers began to change their attitude and Whitlam told colleagues that victory against the Senate was near – some senators were starting to wilt.

But Whitlam's optimism was misplaced. One of his Cabinet colleagues, Jim McClelland, met an Opposition front-bencher, R. J. Ellicott, QC, and, as lawyer to lawyer, took the opportunity to warn him that blocking supply was not going to work. 'You won't get away with this,' McClelland said. 'Oh yes, we will,' Ellicott said. 'In the end this will all depend on Old Silver and he'll do the right thing.'

'Old Silver' was the Governor-General, Sir John Kerr, the Queen's representative in Australia. That same afternoon McClelland realized exactly what Ellicott meant, when Ellicott went public and said that his considered constitutional view was that if the Prime Minister could not obtain supply and refused to call an election, then the Governor-General should sack him. Whitlam and his colleagues dismissed this as a joke. They were adamant that the Governor-General was constitutionally obliged to follow the advice of the Prime Minister only, and since Whitlam was not about to advise Kerr to sack him there was no way it could happen. In fact, the reverse applied – Whitlam could sack Kerr simply by advising the Queen to do so. Anway, Kerr was on Labor's side. He was a 'mate' who had gone out of his way to assure McClelland and others of his support. Talking to Whitlam about Ellicott's opinion that the Governor-General should sack the Prime Minister, Kerr said, 'It's all bullshit, isn't it?'

Yet, for anyone prepared to look around a little more closely, there were disturbing indications that all was not well. Kerr was receiving a lot of visitors at Yarralumla, his official residence. Kep Enderby, the Attorney-General, called to point out to Kerr that no government had been dismissed by a British monarch or a vice-regal representative since 1783 and to assure him that the crisis would be over within a week. Bill Hayden, a former policeman who had switched to politics, came to tell the Governor-General of Labor's contingency plans for alternative financial arrangements. Hayden came away so worried that he hurried back to Parliament House to warn Whitlam. Kerr, he said, had shown little interest in what he had to say. He was preoccupied and evasive. Hayden told Whitlam that his 'copper's instinct' made him certain that Kerr was up to something.

This was the point at which the Labor leaders should have reminded themselves about Kerr's background and personality and have asked themselves whether Hayden's 'copper's instinct' could be right. Kerr was the son of a boilermaker who had fought his way up through the ranks of lawyers in Sydney to become the Chief Justice of New South Wales. He had a very high opinion of himself and expected everyone else to recognize and respect his genius.

When Whitlam offered to put him forward for the job of Governor-General, Kerr spent six months negotiating the salary, pension and benefits he felt appropriate for 'the most important public figure in Australia'. Many Labor leaders hated his airs and graces and referred to him behind his back as 'the Liberace of the Law' (after the camp Hollywood entertainer), or 'Goldilocks' because of his mane of silver hair. Kerr sensed this antagonism and blamed Whitlam for not treating him with the deference and respect he felt his office demanded. But no one in the Labor Party noticed that Kerr was simmering away and might do something drastic. Everyone, including Whitlam, the man who made him, took Sir John Kerr for granted.

Five days later, when Whitlam learnt that polls showed Labor's approval rating continuing to rise, he decided to make one last effort to break the deadlock. He met the leader of the Opposition, Malcolm Fraser, and the leader of the Country Party, Doug Anthony, and offered them a deal. If the anti-Labor senators continued to block supply then Whitlam would call a half-Senate election before Christmas. If they agreed to allow supply, then he would postpone such an election until the middle of the following year, 1976. Fraser and Anthony turned him down flat.

So at lunchtime on that historic day, Armistice or Remembrance Day, 11 November 1975, Whitlam went to the Governor-General's residence, and was shown into Kerr's study. He had just begun to tell Kerr of his election plans when Kerr cut him short, handed him a formal notice of his dismissal, and told him that the Chief Justice, Garfield Barwick, had been consulted and had agreed it was the right thing to do. 'We shall all have to live with this,' Kerr said. Whitlam replied, 'You certainly will.' Whitlam was barely out of the front door when Fraser emerged from a waiting room and was sworn in as the new Prime Minister. A political assassination, the consequences of which still divide Australia 25 years later, had taken barely a few minutes to execute.

Whitlam went back to the Prime Minister's lodge, ordered a steak for lunch, and had his staff call his senior ministers to a meeting. As one by one they arrived he greeted them with 'The bastard's sacked us.' There then followed a series of Parliamentary manoeuvres that bordered on farce. In the Senate just after lunch, the supply bills came up for debate again, and to the amazement of the Labor Senators, who had heard nothing about their leader's dismissal, the bills were passed. In the Representatives, Fraser waited until he heard what had happened in the Senate, then stood up and announced that he had been commissioned by the Governor-General to form an interim government pending a general election for both Houses of Parliament.

Amid uproar, Labor moved a motion of no confidence in Fraser and a request to the Governor-General to reappoint Whitlam as Prime Minister. When this motion was passed, the Speaker was dispatched to the Governor-General's residence to tell Kerr. Kerr countered this move by simply refusing to see him. In the meantime the Governor-General's secretary had arrived at the front entrance to Parliament House to read a proclamation dissolving Parliament for the double election.

By now the nation had heard news flashes of what was happening in Canberra and a crowd of furious Labor supporters and journalists had gathered at Parliament House. The lasting image of the moment was captured by photographer Maurice Wilmott. It shows a young man, immaculately dressed in a black suit with a silver tie, reading Kerr's proclamation while microphones hover around his face. Standing immediately behind him and towering over him is Gough Whitlam, staring grimly into the future.

Then the still image comes alive, and as the young secretary finishes reading the proclamation with the words 'God Save the Queen', Whitlam steps forward in front of the microphones and says, 'Well may we say "God save the Queen", because nothing will save the Governor-General. The proclamation you have just heard . . . was countersigned "Malcolm Fraser", who will undoubtedly go down in Australian history from Remembrance Day 1975 as Kerr's cur.' Then he added, 'Maintain your rage . . . until polling day.'

Just as every American of a certain age can remember where he or she was when they heard that President Kennedy had been assassinated, so every Australian of a certain age can remember where they were and what they were doing and how they felt when they heard that Gough Whitlam, the democratically elected leader of the nation, had been dismissed by the Queen's representative, the Governor-General.

Let us begin with the photographer on the spot, Maurice Wilmott. 'I was the staff photographer in the Canberra press gallery for the *Australian* and the Sydney *Daily Mirror*. I'd been down in Sydney on a visit and I was having a drink in the pub near the newspaper office. When I came out I noticed a lot of official cars, including the Governor-General's, parked near the office entrance. I thought to myself, "Oh, oh. Something's going on."

'I'd barely got back to Canberra when I got a phone call from Brian Hogben, one of the paper's executives. He said, "Maurie, I want you on the top step of Parliament House tomorrow morning from 7am onwards. Don't ask why. Just be there and don't move." So I was. And I waited and waited and then Gough and Smith [the Governor-General's secretary] arrived and I banged on a twenty millimetre wide-angle lens and started belting away. I must've shot a whole roll.

'How did I feel? Well, Gough was not one of my favourite people and I'd had a couple of clashes with him over photographs I'd taken of Labor politicians at play. I've no sympathy for any politician so I didn't feel much at all. I knew it was an important picture and I was glad I was in the right place at the right time.

'In fact that was my uppermost thought. How did Hogben know it was going to happen? What were all those official cars doing at the newspaper office? Had someone tipped off Murdoch?'

Jim McClelland was eating a sandwich for lunch while he worked on some ministerial files when his telephone rang. It was Lionel Murphy ringing from the High Court to tell him his daughter had called to say that Whitlam had been dismissed. McClelland remembered, 'I told him not to believe it. There were all sorts of silly rumours flying around. But a few minutes later a member of my staff put his head around my door and said, "No need to worry about those files, boss. We're all out of work." I went to my drinks cabinet and poured myself a triple whisky.

'An hour later I got a phone call from Howard Nathan, then a barrister and an expert in industrial law and whom I used as a troubleshooter to put out industrial brush fires. He was doing just that out in the Pilbara [Western Australia]. "I'm not doing any good," he said. "Our writ doesn't run any more." The news had reached the farthest corners of the continent.

'That night there was the famous wake at Charlie's Restaurant, that went on until 5 a.m. Next morning I repaired, a little less chirpy than usual, to the Lodge for a discussion on our election strategy. All the heavies were there, looking like stunned mullets. Gough was walking around like a zombie. He had not slept at all. We all put on as brave a face as possible, but most of us suspected that it was the end of the Whitlam era.'

Gil Appleton was working at the International Women's Year Secretariat in Canberra when she and her colleagues heard the news. 'Our first reaction of utter disbelief soon gave way to rage, rage that a government which had swept away years of stuffy, backward-looking conservatism and had done so many good things for Australian society should be dismissed after less than three years in office. We stopped work and set about making protest banners and signs, listening all the while to nonstop coverage of the crisis on the ABC radio.

'Then we all went down to Parliament House, where a huge crowd was gathering. At one point Ivor Greenwood, who had been a very unpopular Liberal minister during the Vietnam War, appeared on one of the upper balconies of Parliament House and the crowd saw him and started shouting: "Jump! Jump! Jump!" When Gough emerged to hear the proclamation and then to utter his famous line that nothing would save the Governor-General, people wept openly. There were rumours of civil strife, that the army was to be called out. In the atmosphere that day almost anything seemed possible.'

Wayne Goss, then a young lawyer working for the Aboriginal Legal Service in Brisbane, was politicised by the dismissal and the events that followed that day. He joined a crowd outside the Liberal Party headquarters. 'We could see that inside the building beaming Liberals with slicked-down hair were drinking champagne out of long-stemmed glasses.' He joined the Labor Party the next day and rose to be Premier of Queensland.

Margaret Simpson was heading back from tennis to her house in Castle Hill, a Sydney garden suburb in the heart of Liberal land, listening to the radio. 'I was one of the many women who voted for Whitlam because I thought it was time for a change and I liked his policies on women's issues. I drove into the garage but I couldn't get out of the car because I was frightened something would happen before I could get into the house and and switch on the radio there. Then I heard Gough speak and accept his dismissal and then I still couldn't get out of the car because I couldn't see through my tears.'

Throughout Australia people took to the streets to express their anger. A Labor Party rally in Melbourne turned violent and windows of the Liberal Party headquarters were smashed. But it never got beyond that. There was no national strike. The workers of Sydney and Melbourne did not march on Canberra. In 1968 French students had brought their country to the brink of revolution over issues of much less importance. Why did Australians take the dismissal of their Prime Minister with such relative calm?

First, because Whitlam himself accepted it. When Kerr handed him his notice of dismissal, Whitlam could have laughed in his face – Whitlam's wife, Margaret, has said she would have torn up the document on the spot. But as McMullin points out, 'Whitlam's instinctive response was to contest the dismissal in the arena where he was pre-eminent, the House of Representatives.' But that did not work and within hours Whitlam had accepted that he was out of office and was giving his followers a pep talk about the election campaign that lay ahead.

Next, Bob Hawke, then a trade union leader, along with other union officials, urged their members to remain calm and non-violent. Widespread unrest and demonstrations that could turn violent would work against Labor at the forthcoming election. But at the back of both Hawke's and Whitlam's minds – and everyone else old enough to remember – was Governor Sir Philip Game's dismissal of New South Wales Premier Jack Lang in 1932 (see Chapter 9). If Lang had not gone quietly, secret anti-Labor armies were ready to mobilize and march on Sydney. They would have been resisted by armies of Labor supporters. Civil war could have been around the corner. Whitlam was certainly thinking about this risk on the day Kerr dismissed him, because he said to one MP: 'Kerr's done a Game on me.' So was Hawke. He made an emotional appeal for calm, saying that he had an apprehension that Australia could be 'on the edge of something terrible'.

Could it? Strictly speaking, Kerr, as Governor-General, was commander-in-chief of Australia's defence forces. It was more a ceremonial position than one of real power, but if Whitlam had refused to go and Australians had taken to the streets to support him, then Kerr could have ordered in the army to maintain order. Whether he would have done so or not we will never know, but the thought that he might have must have been a factor in the way Whitlam and his colleagues bowed to Kerr's will. As well, there were still secret armies and paramilitary organizations around in the mid-1970s, including the Civil Defence League in Sydney and the Society for Law and Order in Melbourne, and Whitlam could well have been worried that resistance on his part would provoke bloody clashes on the streets in both cities.

It is possible in all great historical events to see the makings of a conspiracy. Kerr's action was out of character – as Whitlam told Hayden when Hayden warned him that Kerr could be contemplating drastic action, 'Comrade, he wouldn't have the guts.' Was it possible, therefore, that there was someone behind Kerr, directing and guiding him and working to a bigger agenda? The usual suspect, America's Central Intelligence Agency, sprang immediately into the frame.

The behaviour of the Whitlam government in the months after its election in 1972 appalled Washington. Not only did its foreign policy appear to undermine everything that the United States had worked for in the region, but its attitude to security and intelligence matters could only be described as alarming. The CIA and the FBI might have many enemies within the American administration but they were treated with respect. So the 'raid' by Australia's Attorney-General, Lionel Murphy, on the Melbourne headquarters of ASIO was shocking and incomprehensible to Washington.

Of course, the 'raid' was not quite as it was presented as in the media at the time. Murphy, concerned about the security arrangements for the impending visit of the Prime Minister of Yugoslavia, had asked to see ASIO's file on right-wing Yugoslav extremists in Australia. ASIO said that unfortunately he could not see it because it was not held in Canberra but in Melbourne. Murphy, suspecting that ASIO was trying to stall him, said that was not a problem – he would fly to Melbourne on the first plane and read the file there.

On the flight to Melbourne, Murphy learned (he never said how) that there was another passenger, a Canberra-based ASIO officer, carrying in his briefcase the very file that Murphy had been told was in Melbourne. The Commonwealth Police – who detested ASIO – had been briefed about Murphy's mission. They not only arranged for Melbourne colleagues to be present at ASIO's office, but tipped off the media. The Commonwealth Police escorted Murphy into ASIO's office, he read the file on Yugoslav extremists, still hot from the ASIO officer's briefcase, enquired about his own file but did not press to see it, and left.

But melodramatic photographs in the press the next day, along with stories emphasizing the historical antipathy between ASIO and the Labor Party, made it appear to Washington that the 'raid' was the first shot in a war Murphy was going to wage on his own security service. He would have been justified in doing so. ASIO was obsessed with Communists, their fellow travellers, left-wing sympathizers, 'pinkos' and 'ratbags', particularly in the media. It kept files on journalists, writers, radio announcers, actors, poets and artists. Comments in their files ranged from 'overheard to make derogatory remarks about royalty' to 'has a particularly biting tongue and some early trace of Communist sympathy' and 'has worked with security risks in the entertainment field'.

Even before Murphy's run-in with ASIO, Whitlam had antagonized the CIA by ordering an end to participation by ASIS (Australian Secret Intelligence Service) officers in the CIA operation to overthrow the Allende government in Chile. Now the American intelligence community began to worry about their Australian colleagues. If the Australians were saddled with a radical, left-wing government – one which was considering not renewing agreements for American bases in Australia – perhaps they could do with some help?

The CIA's tactic was to let ASIO and ASIS know how concerned it was about developments in Australia and then to leave it to the two Australian services to apply political pressure for change. That way the CIA was not seen to be interfering in the internal affairs of a friendly nation. This was so effective that the Australian Defence Department's assessment was that Australia was facing the most serious risk to its security in its entire history – Uncle Sam might leave Australia to fend for itself. But Whitlam was unmoved and, to the CIA's surprise and dismay, he sacked the heads of ASIO and ASIS in quick succession.

Where this would have all ended we shall never know because the dismissal solved everything. When the Liberal-Country Party coalition under Fraser went on to win the December 1975 election, Whitlam and Labor were out and an America-friendly government was back in power. The military bases agreements were renewed. Any idea of buying back the farm was abandoned. Australian and American foreign policy once again went hand in hand. ASIO and ASIS resumed their close relations with the CIA and the FBI and with MI5 and SIS, and all was well again in the spy world.

So was there a conspiracy? Of course there was. The principal conspirators were Sir John Kerr and Malcolm Fraser. Their aim was to get rid of the Whitlam government. But questions about the role of the CIA are not so easily answered, much less proved. I have learnt during thirty years of writing about the international intelligence community that proof about anything in the secret world is hard to come by. That is why, unless the accused has confessed, it is almost impossible to get a conviction in most spy cases.

After all, the whole point of a covert intelligence operation is to leave no trace that it ever took place, much less who organized it. So there is no paper in the CIA archives setting out how the Whitlam government could be destabilized. There was no need for one. Every CIA station chief around the world knew that Washington regarded the Australian government as unfriendly and understood that it was his duty to do anything he could to change that. In deciding whether or not the CIA had a role in the conspiracy to get rid of the Whitlam government, all we can do is examine the circumstances at the time and draw some probable conclusions.

The Whitlam government was not as pro-American as previous Australian governments, to put it mildly. It had withdrawn its support for the Vietnam War. Its ministers had been virulent in their criticism of American actions and American leadership. It had gone from 'All the way with LBJ' to 'Nixon is an arrogant, double-dealing hypocrite'. In recognizing China, signing a big trade deal with the Soviet Union, and discussing with India the chances of forming a south-east Asian organization free from big power politics, it had shown that it was prepared to follow its own, independent foreign policy.

Its 'buy back the farm' idea endangered American investments and commercial interests in Australia – a hundred years earlier such a move alone would have brought in the US marines. Its promise to look again at the granting of leases for American military bases in Australia could threaten American security – these were said to be essential to American defence if war broke out with the Soviet Union. And finally, the Whitlam government was not paying the intelligence community proper respect. Whitlam had shown that he was not only prepared to emasculate his own intelligence and security services, but put the CIA presence in Australia at risk by identifying and naming in Parliament CIA officers working in the country. The CIA director, William Colby, later wrote that the threat posed by the Whitlam government was one of the three world crises in his career.

Given all this it is surely reasonable to draw the conclusion that Washington desired the downfall of the Whitlam government as soon as possible. But Australia was not Chile. There were no dissident military officers to back, no Australian General Pinochet to lead a coup. There were, however, lots of plotters who could be encouraged, advised and supported. For instance, I do not think that the CIA steered Khemlani to Connor, but once the contact had been made and the CIA knew about it, I have no doubt that the CIA took the opportunity to encourage the relationship because of its potential to embarrass the Whitlam government.

I see the CIA's role in the conspiracy as an accumulation of minor interventions – a rumour planted here, a discreet but threatening letter there, a confidential briefing, an expression of disquiet, a use of influence, a calling-in of favours, a promise of reward, a story leaked to the press, a promise of sensational revelations that never eventuated, much of it done through dupes and agents, carpetbaggers and opportunists, so that none of it could ever be attributed to the agency. All this created an atmosphere bordering on hysteria in which any damaging rumour would be believed, and it encouraged an impression of a government out of control, a danger not only to itself but to the free world.

Finally, a few of the undisputed facts. In mid-November 1975, the CIA in Langley, Virginia, sent a cable to ASIO in Canberra saying that recent behaviour of the Whitlam government was causing acute concern about the implications for Australia's future relationship with the United States. The cable was not intended to be shown to Whitlam but the sentiments of the cable did reach Kerr, either directly from ASIO or via the British intelligence liaison officer in Canberra – because Kerr was the Queen's representative. During the war Kerr had been in a specialist research unit that had some connection with the American Office of Strategic Services, the forerunner of the CIA. On 11 November Kerr sacked Whitlam. He may have sacked him anyway, but to believe that the CIA's role in the preceding months and its cable to ASIO had not the slightest influence on his decision defies logic.

In the general election held on 13 December 1975, the Labor Party suffered the biggest landslide in the House of Representatives since Federation. Bob Hawke said, 'We've had the guts ripped out of us.' Whitlam had believed that voters would be so outraged by his dismissal that they would vent their feelings via the ballot box. But they did not. What had gone wrong?

The almost universal hostility of the media towards Labor was one factor. The Murdoch press was particularly virulent and some journalists working for Rupert Murdoch's national daily the *Australian* went on strike at its flagrant anti-Labor tone. I asked Murdoch about this at a meeting in London in 1981.

He said, 'I'd backed Whitlam in the 1972 election and the journalists thought I'd do the same in 1975. Couldn't they see how things had changed?'

Another factor was 'too much too soon' – the speed of the government's reforms, although McMullin says rapid reform, of itself, was not the main problem, and that what cost the government dearly was its failure to inform the voters sufficiently about the benefits they received from its activities.

Inexperienced and inefficient ministers were a problem, as usually occurs when a party that has been long out of power finds itself in government. Members who have stuck loyally to the party over its years in the wilderness have to be rewarded, even though they may, by now, be too old for the job – and quite likely not the best people for it anyway. Luck was a factor. The economic downturn overseas that hit the Whitlam government's reform programme and then threatened its very existence would have come at a different moment in the government's history, if only it had won the closely contested 1969 election.

But we have to look into the hearts of Australians to find the real reason they kicked out Labor in 1975. Gil Appleton remembered in 1999: 'I went along to all the big "maintain the rage" rallies during the campaign and I became convinced that Whitlam couldn't lose. But there was this great, silent mass of people who were never enraged and just wanted a quiet life. We fanatic Whitlamites just did not know they were there. The election result was a terrible shock for us.'

It certainly was. Sandra Darroch recalled in 1999: 'I got back to Australia from Britain just after the dismissal and went to the beach the first Sunday with my sister. She was absolutely distraught about it, burst into tears, and said "How will we all go on after this?" I was very surprised because I was unaware of how deep the feeling ran. I said, "I think we'll all manage fine. Things will go on as usual." I think that not having gone through the Whitlam era had made me unable to be a True Believer like so many of my friends and colleagues.'

The True Believers suffered twice. There was the loss of what might have been. They had voted Whitlam into power because he promised new ideas in government, new initiatives, new directions and because he had a vision of the future. They thought most Australians felt like they did. Imagine then their grief to discover that many of their fellow compatriots, the swinging voters, were in their heart of hearts conservatives with middle-class values. They were prepared to vote for social reform but not to pay for it. The moment the dream of a good and just society soured under the realities of a worldwide economic downturn, the very people who had put Whitlam into power chucked him out again. As their window stickers put it: 'He has had his chance and he has stuffed it.'

★ ★ ★

The remaining years of Sir John Kerr's life were miserable ones. He was subjected to relentless harrassment whenever he appeared in public and, hoping to end this, he resigned in 1977. When it continued he moved to London, where he could be seen most days, usually the worse for drink, at one or other gentlemen's club. He died in 1991, professing to the last to be bewildered about the antagonism his dismissal of Whitlam had caused.

Whitlam continues in Australian public life and usually draws a round of applause when recognized. As Sandra Darroch predicted, life went on, even for the True Believers, but over the next quarter of a century, when circumstances brought even a few of them together, they would talk about that night at Blacktown in 1972, or the St Kilda Town Hall rally, or Whitlam's meeting with Vincent Lingiari, and all the other high points of that brief but memorable Labor government. They know there will never be another one like it, but they rejoice that at least they were there.

CHAPTER 19

COMING OUT

What turned Sydney in 20 years from the bigoted, murderously homophobic place it was into the gay capital of the world, where the 'Sisters of Perpetual Indulgence' (men in nun's habits) are protected from assault by a girls' motorcycle gang called 'Dykes on Bikes', and a visiting English journalist reported that he felt terribly out of place, 'not because I was an effete Pom, but because I was married'?

In a society that originally had only two types of citizens – criminals and their jailers – it is perhaps understandable that a culture of police and public service corruption developed in Australia. Most Australians knew about corruption but did not greatly care. It seldom impinged on their lives, involved mostly criminals and police, and usually came about because of repressive laws relating to the pursuit of pleasure – the sale of alcohol, drinking hours, nightclubs, prostitution, gambling, homosexuality and abortion.

Every now and then a reforming government would set up a royal commission into police corruption. For a while the public would be scandalized by what these commissions uncovered. There would be a damning report and promises of action. But nothing would happen and the subject would be forgotten until the next time. It is no exaggeration to say that there have been so many Royal Commissions of different types into police corruption in Australia this century that no one really knows the total. All that can be stated with certainty is that the number and their scope grew rapidly in the 1970s when the drugs scene in Australia suddenly exploded.

Until the Vietnam War years Australians had a fairly relaxed attitude to recreational drugs. Aboriginals knew of various plants and herbs that altered mind and mood and traded them between tribes. The best known was probably pituri, a psychoactive drug made from a bush called *Duboisia hopwoodii*. White Australians preferred rum. After the nation was born in one big rum and sex party, the first white settlers drank so much of it that it held the status of a currency.

Opium was freely available in the late-nineteenth and early-twentieth centuries and even after opium dens in big cities disappeared, patent medicines with a potent opium content could be bought across the counter at most chemists and remained very popular, especially among suburban housewives, until the 1950s. Cocaine was a widespread recreational drug among the 'smart set' in Sydney in the 1930s and 1940s. Good quality cannabis was on sale for years at tobacconists in the form of 'Cigares de Joy'.

None of these drugs was seen as causing the problems that alcohol did, and the full weight of the churches and the Salvation Army and the 'wowsers' and the 'do-gooders' was directed at drinkers and the places in which they drank – pubs, nightclubs, restaurants and 'sly grog joints'. Governments passed laws to regulate their trading hours, the type of alcohol they could serve and to whom they could serve it. Australians broke these laws all the time and attempts to enforce them led only to an enormous waste of energy, manpower and resources and widespread police corruption.

I travelled in outback Australia in the 1950s, when illegal after-hours drinking involved a ritual almost as rigid as at the Masonic Lodge. At closing time the publican would usher everyone out the front doors, which he would then ostentatiously lock. Everyone would then make their way around to the back of the pub and enter a bar with blacked-out windows where they would continue drinking, often until the early hours. At some stage during the night the local police would call by and join the party.

Jim Ratcliffe, who as a young man lived at several properties in outback Queensland, recalled in 1998: 'The Station provided the drovers with food and lodging. It was pretty basic – an iron cot in a tin-roofed shed, mutton chops for breakfast, a flap sandwich for lunch and a joint for dinner and they were happy. But then there was the drink. They helped themselves from a two-and-a-half gallon plastic container of rum – anything from half a bottle to a bottle a night. They'd mix it with cola and drink themselves senseless most nights. You could tell how long they'd been on the job by the state of their teeth – the sugar in the rum and the cola rotted them. So black stumps meant an old hand.'

As a generalization it would be reasonable to say that alcohol was always a social problem in Australia but hard drugs were not. That changed after the Vietnam War and many place the blame squarely on the United States of America.

The California gold rushes of the nineteenth century brought Chinese and opium to the US and sparked off a passionate – some would say irrational – fear of the drug. In the early 1900s, led by the churches, America launched a mission to protect the world, especially the white world, from the scourge of opium and later all recreational drugs. The Americans convinced Britain, once the world's biggest opium dealer, to join the mission, and turned it into an international effort with a conference in Shanghai in 1909. Australia signed up for total prohibition in Paris in 1919 and a few years later in Geneva without a second thought, because, 'It doesn't really concern us. We haven't got a drug problem.'

That remained true for 50 years. Growing up in Australia in the thirties, forties and fifties, drugs were something prescribed by your doctor. No parents warned the children of my generation about drug addiction. The idea that anyone would inject a drug into a vein for pleasure appeared so ludicrous and improbable that a warning was unnecessary. When I returned to Australia in the 1980s after a long absence, things had obviously changed. I saw drug-related scenes I had not witnessed anywhere else in the world – a middle-aged driver slumped unconscious over the steering wheel of his truck, parked in a busy Sydney street, one hand still on the wheel, and protruding from a vein in the back of that hand, a half-empty syringe; a line of debilitated young men and women queuing, anxious and trembling, at the counter of an inner-city chemist shop for their methadone dose; a well-dressed couple injecting each other as they sat on the grass in the sunny park at lunchtime, right below my Sydney window. Half my friends of my generation seemed to have children in rehabilitation programmes. What had gone wrong?

No one now denies that during the war Vietnam was awash with drugs. They were a currency that could buy guerrilla armies, bribe officials, smooth political paths. They were relatively cheap and easily available and many American GIs developed drug habits. When the war began to wind down, the Drug Enforcement Administration (DEA) in Washington became worried that American soldiers would bring back their drug habit with them and, since their suppliers would not want to lose them, heroin would follow the troops home.

So the DEA flooded Vietnam with its agents and they successfully prevented this from happening. In his book, *Drug Traffic, Narcotics and Organised Crime in Australia*, academic Alfred McCoy describes how the DEA agents did it. '[They] compelled the syndicates to sell heroin originally produced for American addicts in alternative markets. In short, the DEA simply diverted south-east Asian heroin from the US into European and Australian markets, evidence for what we have called the iron law of the international drug trade.' (The iron law is that as fast as one market is closed by anti-drug authorities, another opens up.)

Heroin poured into Australia. The treaties Australia had signed proved worthless. They did nothing to stop the supply and instead inhibited the

Australian authorities from taking radical action to avert a disaster. In less than ten years Australia had at least 20,000 heroin addicts, with the rate of addiction rising every year. And, as has always been the case, in the wake of heroin came addiction to other, less dangerous drugs, until by the late 1990s Australia, like most other western countries, was a nation of drug-takers and – again like most other countries – drug-taking had become an established practice in its culture.

Australia has always prided itself on its innovation, and since the 1970s there have been many original and thought-provoking ideas for tackling the drug problem. But none of them got very far, because this independent parliamentary democracy found that when it came to drugs it might as well have been the 51st state of the United States of America. Because of its treaty obligations Australia had to follow America's lead, and American policy was absolute prohibition of all recreational drugs except cigarettes and liquor. (Many American anti-drug campaigners would not even concede that Prohibition failed.) Concerned Australians were appalled that even though the war on drugs had failed, Australia continued to sign up to more treaties that stopped it from trying new approaches.

The former Chief Judge of the Australian Capital Territory, Russell Fox QC, begged the government in 1988 not to sign the Vienna Convention. 'Australia should not at this time reaffirm (and strengthen) unsuccessful treaties of twenty and thirty years ago which tie our policy on most drug use to complete and unqualified prohibition. The inevitable result is a most dangerous illegal market which is by definition uncontrolled. There can be no quarrel with the making of international arrangements or with an attempt to eliminate illegal traffickers. The real question is how this can best be achieved.' The government ignored him and ratified the treaty.

There was no chance even of any Australian state within its own borders trying out policies that might reduce the cycle of addiction, corruption, crime and death. The Commonwealth of Australia had signed the treaties, the government in Canberra was responsible for honouring them and, if any state stepped out of line, it was Canberra's duty to pull them back again. Therefore it consistently vetoed state plans to run heroin trials or to experiment with decriminalizing marijuana.

Australia was not prepared to defy the United States, because it had shown that it would employ diplomatic and commercial pressure and the threat to cut off US aid to force even friendly countries to stick to the total prohibition policy. It mostly did this through the International Narcotics Control Board (INCB) in Vienna. The staff of this United Nations agency were career anti-drugs campaigners, and they had the power to cut off supplies of life-saving pharmaceuticals to countries which appeared to be backing away from total prohibition.

The INCB and the United States had another way of deterring governments from considering new approaches to the problem. The state of Tasmania proved an ideal place to grow opium poppies and by the early 1990s had become one of the safest and most efficient producers of crude morphine and morphine-based drugs in the world. This industry was an important source of income for the poorest state in Australia, but its continued success depended on the INCB and the United States. They could offer the carrot: an increased quota for more opium poppies – or wave the stick: you will lose your licence altogether. As a result Tasmania became the most vociferous of the Australian states in opposing heroin trials, decriminalization of marijuana or any other experiments to reduce Australia's drug problem.

So before the drugs era, police corruption in all states – but particularly New South Wales and Queensland – was widespread but petty. Illegal gambling joints, illegal bookmakers, after-hours drinking clubs, abortionists, brothels and petty criminals paid comparatively small sums of money for the police to look the other way, or for protection or favours in court. The huge sums of money involved in the drugs business changed all that.

In 1991 Sydney police stopped a car they suspected was being used in the drugs trade and found nearly a million dollars in cash in the boot. As one senior officer said, 'You get a group of young officers concerned about interest rates and mortgages like the rest of us. If they were to stuff their pockets full of cash before handing the money in, the criminals would be in no position to say, "Hey! They knocked off twenty thousand dollars."'

Or as former judge, Jim McClelland, put it to me in 1988: 'There's a level of corruption in New South Wales that people are prepared to tolerate. The basic trouble is that bribery is such a flash crime, one where the briber is as guilty as the person he bribes. You'll never get a briber to give evidence because he will incriminate himself in the process. That's why most crooked policemen get away with it, as do most corrupt politicians.'

One of the bodies tackling this problem, the Independent Commission Against Corruption (ICAC), a sort of Australian version of the 'Untouchables', thought it had overcome this attitude when it revealed a corruption case that really touched the public nerve like no other. Bear in mind that at the time the ICAC announced its coup, elsewhere in Australia a coroner was trying to find out who had gunned down the nation's number two police officer in his car outside his own home, and a detective sergeant was awaiting sentence for running a drugs ring.

The ICAC revelation was that as many as one in three driving licence examiners had accepted money to pass learner drivers. The amounts ranged from $5 left on the seat of the car, to the case of one examiner who accepted

$20,000 in a lump sum plus $1,000 a week from a driving school to pass all its students. The ICAC explained why this was an important case. 'This reaches everyone. Everyone is a driver or a passenger. Until now when people have read about other corruption cases, they've said to themselves, "I'm not a land developer. I'm not a politician. I'm not involved with drugs." But now they realize that there's every likelihood that the driver coming down the road towards them got his licence by slipping the examiner fifty dollars. Now at last they understand what corruption is all about.'

In 1995, the Wood Royal Commission of Inquiry into Police Corruption 'turned' a corrupt officer, Detective Sergeant Trevor Haken. Using the miniature video cameras which media and sports magnate Kerry Packer had developed for cricket, Haken filmed corrupt officers accepting bribes, pocketing drugs and ordering child pornographic videos. One miniature camera installed in the dashboard of a car showed Detective Inspector Graham Fowler, known to his colleagues as 'Chook' Fowler, taking from Haken his share of a bribe from a drugs dealer and complaining that it should be more.

Wood reported: 'Corruption in New South Wales embraces receipt of bribes, green-lighting [where the police give a criminal or a gang the green light to go ahead with, say, a robbery, on the understanding the police will receive a share of the proceeds], franchising, protection or running interference for organized crime, releasing confidential information and warning of pending police activity, gutting or pulling police prosecutions, providing favours in respect of bail or sentencing, extortion, contract killings, stealing, supplying drugs, and other forms of direct participation in serious criminal activity. That is without being exhaustive.'

Police reporters on Sydney newspapers were fond of telling visiting colleagues that corruption in the New South Wales police had reached its final stage, that where the various police squads had taken over the activities of the criminals they were supposed to be investigating. That is, the robbery squad had become involved in carrying out robberies, the fraud squad in planning fraud, and the murder squad in committing murder. Journalists may have stretched the facts but some of the murders and robberies in Sydney in the 1970s and 1980s had strong hints of police participation. (*Blue Murder*, a television drama series based on these cases and suggesting that a highly commended detective sergeant was involved in them, was widely viewed in every other state in Australia, but could not be shown, or the video bought, in New South Wales because of the libel risk.)

One of these cases, the disappearance of newspaper editor Juanita Nielson, remains one of the great unsolved mysteries of Australian crime. Nielson was heiress to a Sydney department store fortune and had put some of her money into a local newspaper in the inner suburb of Kings Cross. Annoyed at the way developers were tearing down nineteenth-century Sydney to build apartment blocks she used her newspaper to campaign vigorously against them, rallying the building unions to her side.

On the morning of 4 July 1975, she visited the Carousel Cabaret, otherwise known as Les Girls, a transvestite club run by Abe Saffron, an old-time Sydney nightclub owner. She was noticed leaving at 11.30 a.m. and was never seen again. Environmentalists were convinced she had been abducted and murdered by a 'hit man' hired by one of the developers whose project had been endangered by her campaigns. At her inquest, which lasted more than three months, a friend said he had received an anonymous telephone call telling him that Nielsen had been abducted to scare her off and that her death had been an accident. But there was also a suggestion at the inquest that a well-known Sydney detective had killed her. The jury found that Juanita Nielson was dead, but was unable to say when, where, why or at whose hands she had died.

Finally, in 1996 the New South Wales government decided that all efforts to tackle police corruption in the state would come to nothing if the head of the force, the commissioner, continued to be appointed from within the service. It would have to find an outsider, someone not inured to the culture of corruption going back more than 200 years.

It hired a firm of headhunters and after a worldwide search settled on Peter Ryan, who had served with the Greater Manchester police force, then as a chief superintendent at Chelsea, London, dealing with terrorist attacks in the early 1980s, and then as chief constable of Norfolk.

In 1997 Ryan recalled: 'When I was first approached by the headhunters, I was struck by the challenge. I went out to Sydney to get an idea of just how big a challenge it was, so I had a point of reference. The sheer size of the task, of dealing with a state four times the size of the United Kingdom, the tremendous operational demands of tackling Australian organized crime, Asian crime, the Mafia, drugs, European-influenced crime – let alone preparation for the Olympic Games – must make it the most demanding job in the English-speaking world. I thought I'd like to try to do my best.'

At first it must have seemed as if nearly all New South Wales was against him. He was greeted by Police Association members chanting 'Pom go home.' Ryan was not intimidated and quickly dismissed more than 20 officers and suspended 150 more. But the police are an integral part of politics in New South Wales, and commissioners, powerful though they are, have always had to tread a delicate line, especially since the division of responsibility between the Commissioner and the Police Minister is ill-defined. In March 1997, the Police Minister, Paul Whelan, wound up the New South Wales Special Branch on the day its head admitted to the Royal Commission that it had kept illegal files on barristers who were guilty of nothing more than defending criminals, that it had for years maintained an office bar which opened in the mid-afternoon, and that its officers regularly returned to work drunk after long lunches.

Whether Whelan consulted Ryan about closing down the Special Branch is disputed, but a month later an advertisement appeared in the *Police Service Weekly*, calling for recruits for a new department to be called the Protective Security Response Group. Whelan accused Ryan of trying to reconstitute the Special Branch under another name. Ryan said that the Group was needed to protect diplomats, politicians and visiting dignitaries and gather intelligence about terrorism, and that anyway the move was an operational matter and therefore a decision for him as Police Commissioner.

Behind the dispute was the story that, in its eagerness to end police corruption once and for all, the New South Wales government might have given Ryan too free a hand and created a new power base in state politics. Ryan, for his part, said that there were 'forces of evil' within the police which were trying to damage his reputation. He claimed that he had a lot of support from ordinary Australians. 'Letters came in their thousands to wish me luck and I'm stopped on and off duty by people shaking my hand and introducing me to their children . . . I think everyone knows that I can do the job. That is if the government doesn't go weak at the knees.'

It would be pointless to deny that some crime in Australia is linked to migrant communities. A factor here is that many migrants come from countries where the government trusts none of its citizens to tell the truth and demands proof for everything. The Australian system, where the authorities generally assume that a citizen is telling the truth, but provides penalties if they are then caught lying, tempts some migrants into illegal acts, particularly fraud, as the following case shows.

Basic third-party car insurance in Australia is included in the cost of the road tax – thus ensuring every motorist is insured – and is administered by the Government Insurance Office (GIO). When someone is injured in a car accident, he or she has their injury assessed by a doctor and then, making his or her selection from a panel of insurers, goes to a tribunal to make the claim. But what had once worked as an efficient accident compensation scheme turned into a nightmare, as ethnic gangs began faking accidents and making false claims.

The GIO's anti-corruption task force discovered witnesses who say they saw two old cars collide head-on in a quiet, uncongested street. One witness described what happened next. 'From nowhere about six people came running up, got *into* the wrecked cars, and started screaming in apparent agony. I couldn't believe what I was seeing. It was like a scene from a Mack Sennett comedy.'

All eight people, in this case of Turkish origin, then went to a co-operative doctor who diagnosed whiplash back injuries – medically hard to disprove – and sent them privately for expensive x-rays, pathological tests and physiotherapy before they made claims on the GIO. But the GIO refused to pay, and when the cases came to court the judge dismissed them and 22 similar cases worth a total of more than $2.2 million, noting that some of the claimants were facing criminal charges of conspiracy to cheat and defraud and perverting the course of justice. The GIO initiated 60 arrests concerning 200 fraudulent claims with a potential total payout of $20 million.

Yet, despite all the foregoing, corruption in Australia is insignificant compared with many – often much-admired – countries in the rest of the world. In Mexico, 90 per cent of income tax remains uncollected. In Ecuador, the Taxation Department costs more to operate than it collects. Russian elites have looted their own country of several trillion dollars, causing the greatest impoverishment of a nation since the Allies exacted reparations from Germany after the First World War.

In the United States there was a seminal moment in November 1999, when the amount of money placed offshore reached a total greater than the amount held in the United States itself. American law students are now taught that if a lawyer provides a service or a support structure for a company or organization that engages in criminal activity, he or she is not really part of a criminal conspiracy. The Center for Public Integrity in Washington claimed in a book it published in January 2000, *The Buying of the President 2000*, that not a single presidential candidate was financially above reproach.

In Britain in 1995, after Gordon Foxley, a senior civil servant in the Ministry of Defence, was convicted of having received £1.5 million in kickbacks from German and Italian defence contractors, there were allegations that the Ministry was reckoned to be 'among the most corrupt in Europe'. And in January 2000, a British parliamentary inquiry into housing benefit fraud revealed that 382 local government councillors and officials had fiddled the system by making fraudulent benefit claims, and that many councils had not bothered to prosecute them.

The cutting of Australia's ties with Britain continued in the 1980s. On 16 August 1985 the Australian government announced that it was ending all appeals to the Privy Council and making the High Court of Australia the highest court in the land, from which there would be no further possible appeal. This ended the powers of the United Kingdom parliament over Australia and its legal system that had existed for nearly 200 years.

Although the announcement followed long consultation involving the Australian and British governments, Australian state governments and the

Queen herself, the credit for ending this relic of colonialism should go to an Australian judge, Sir Owen Dixon, who, 22 years earlier, had the courage, quite rare in lawyers, to admit that he had been wrong and change his mind.

As Chief Justice of the High Court of Australia, Dixon was the author of a key passage in a judgment in 1963 that was described in the *Criminal Law Review* at the time as 'one of the most striking statements of judicial policy, and one of the sharpest rebuffs by one superior court to another, of which there is record.' Many thought that this Australian son of British migrants from Yorkshire was the greatest judicial lawyer in the English-speaking world. Dean Acheson, U.S. Secretary of State and advisor to four American presidents, said Dixon would adorn the Supreme Court of the United States if only it were possible to appoint him to it and English colleagues urged him to take chambers in London and practise at the English Bar.

But this austere, witty, learned (he knew several languages) judge was wedded to Australia and its people. He opposed the High Court having a permanent seat in Canberra or anywhere else, arguing that the court should not be removed from the people, the judges or the legal profession. It should be 'an all-Australian court, going to the people, rather than requiring the people to come to it'. He did not believe that retired judges should take part in public life and declined to allow his name to go forward for the office of Governor-General. He would have made an ideal first President of Australia, but he died in 1972.

The case in which his judgment changed the course of Australian legal history was *Parker v The Queen*, an appeal against a conviction for murder involving a love triangle. In brief, on the day of the killing, Mrs Parker informed her husband that she was leaving him for one Daniel Kelly. Kelly arrived on a bicycle to pick up Mrs Parker and they rode off together, Mrs Parker sitting sideways on the bar, Kelly pedalling, an arrangement known in Australia as 'a doubler'. Parker followed them in his car, caught up with them, and then deliberately ran them down. When he got out of the car he found his wife lying in the roadway drain. Thinking he had killed her, Parker flew into a rage and stabbed Kelly in the throat with a knife, killing him.

Parker was convicted of murder, but appealed all the way to the High Court. The High Court needed to determine whether or not to follow the principles established by the House of Lords in *The Director of Public Prosecutions v Smith*. These had to do with the intention or the mental element in murder. Put simply, the House of Lords held that where a man did something that caused another's death, it did not matter whether that man had actually considered whether his action would cause that death. The point was that he *should* have considered it and have taken the probable consequences of his act into account, and should therefore be legally liable. In deciding this, the House of Lords considered what test should be used to determine what the probable consequences of an act would be. It believed the only test would be what the ordinary, reasonable person, in these circumstances, would have thought the natural and probable result would be.

However, this House of Lords decision was directly inconsistent with an earlier decision of the High Court of Australia, which held that 'the introduction of the maxim or statement that a man is presumed to intend the reasonable consequences of his act is seldom helpful and always dangerous'. Parker's defence followed these lines, in that he claimed his attack on Kelly was unpremeditated and an instantaneous reaction on his part to an emotional crisis of overwhelming intensity.

The key passage in Dixon's judgment, the passage that not only changed Australian legal history, but was another step towards full independence from Britain, reads as follows:

'Hitherto I have thought that we ought to follow the decisions of the House of Lords at the expense of our own opinions and cases decided here, but having carefully studied [*DPP v Smith*], I think that we cannot adhere to that view or policy. There are propositions laid down in the judgment which I believe to be misconceived and wrong. They are fundamental and they are propositions which I could never bring myself to accept . . . I wish there to be no misunderstanding on the subject. I shall not depart from the law on the matter as we had long since laid it down in this Court and I think that Smith's case should not be used as an authority in Australia at all.'

This was a complete about-turn by Dixon, hence his phrase 'Hitherto I have thought . . .' In 1942 he had held that 'where a general proposition is involved the court should be careful to avoid introducing into Australian law a principle inconsistent with that which has been accepted in England'. This had meant that the High Court of Australia would even overrule itself, if necessary, to follow a decision of the House of Lords.

But from the moment Dixon changed his mind, decisions of the House of Lords were no longer binding on Australian Courts because the House of Lords was no longer part of the Australian judicial hierarchy.

Their Lordships eventually accepted this with good grace. Four years later, the Privy Council in *Australian Consolidated Press Ltd v Uren*, held: 'There are doubtless advantages if, within those parts of the Commonwealth (or indeed the English-speaking world) where the law is built upon a common foundation, development proceeds along similar lines. But development may gain its impetus from any one and not only from those parts. The law may be influenced from any one direction.' Their Lordships went on to say that it was up to the High Court of Australia itself to decide whether a decision of the House of Lords compelled a change to what was a well-settled judicial approach to the law of libel in Australia.

As a young police rounds reporter in Sydney in the 1950s I would occasionally come across a legend in the New South Wales Police Force, Detective Sergeant 'Bumper' Farrell, of the Vice Squad. Farrell was a legend for two reasons. The Rugby League Football authorities had charged him with biting off the ear of an opponent during a scrum, but Farrell had beaten the accusation by proving that he had left his false teeth in the dressing room. The other reason was his unrelenting persecution of homosexuals, whom he claimed to be able to recognize instantly – 'They always wear suede shoes.'

In those days homosexuality was a criminal offence punishable by 14 years in jail. Some psychiatrists considered it an illness and felt that with new advances in psychiatric medicine they would soon be able to 'cure' it. So as late as 1968, clinics in Australia were using electric aversion therapy machines. The self-confessed homosexual sat in front of a small screen and pressed a button to project various words and images. According to the manufacturer, 'If the words refer to pleasures he is not meant to enjoy, he receives a small electric shock.'

But by the late 1970s pressure for some sort of reform of the laws relating to homosexuality was increasing. It probably began in 1975 in Adelaide, the most conservative city in Australia, after a gay man was found drowned in the Torrens River. The local gay community claimed that the area was notorious as a spot where local police took gays to beat them up for sport, and that the man had been murdered by the police. (Two former vice squad officers were acquitted over the death 13 years later.) The Adelaide tragedy coincided with constant police harrassment of gays in Melbourne and police beatings of homosexuals at various meeting places in Sydney.

But none of this deterred the organizers of Sydney's first Gay Mardi Gras from going ahead with their protest march-cum-celebration party in the streets of Kings Cross, the city's bohemian quarter, on 24 June 1978. The idea was to mark the anniversary of a brawl in New York between police and drag queens, 'the Stonewall Bar incident', that had radicalized the US gay rights movement. It was a boisterous but non-violent event until about midnight, when the New South Wales police decided that they had had enough and began arresting people.

The writer David Marr remembered in 1998: 'The worst of the anti-Vietnam War rallies were not as nasty as this. One policeman dragged an unconscious woman by her hair through the police station door. Four or five beat a young man's head against the station's iron gates. A man already in the cells was beaten so badly that he was taken after some hours to St Vincent's Hospital. A scrappy, ugly demonstration continued all next morning. I saw a woman tossed down the steps of the court and kicked in mid-air by a senior policeman.'

We scroll forward to 1998. Sydney is celebrating the 20th anniversary of that first Mardi Gras. Today's Mardi Gras is now the biggest gay party in the world, an event of international renown, and a crowd of a quarter of a million people turn out to party and dance until dawn, after cheering the 270-float parade as it makes its way through the main streets of the city. And the loudest cheer of all is reserved for . . . the marchers from the New South Wales Police Service 'in their smart blue uniforms, faces glowing under their peaked caps'. These are not police officers on duty to maintain order. These are police officers *taking part in* the Gay and Lesbian Mardi Gras Parade, as it is now called. 'Bumper' Farrell, long since dead, would be entitled to ask: what happened? What turned Sydney in those short 20 years from the bigoted, murderously-homophobic place it was into the gay capital of the world, where the 'Sisters of Perpetual Indulgence' (men in nun's habits) are protected from assault by a girls' motorcycle gang called 'Dykes on Bikes', and a visiting English journalist reported that he felt terribly out of place, 'not because I was an effete Pom but because I was married'? (True, he was in Adelaide and he was exaggerating, as English journalists visiting Australia for the first time tend to do.)

What happened was the arrival on the political scene of Neville Wran, a Sydney lawyer, a Labor civil libertarian, who was on record as saying that the antiquated laws on homosexuality would have to be reformed. He was not the first lawyer to feel this way. Another, John Dowd, later a Supreme Court judge, had seen the misery that repressive laws on homosexuality had caused, and when he became a Liberal Party MP announced his support for reform. But he was ahead of his time and his action cost him the leadership of his party, something he says he does not regret: 'This is human life we're talking about. If my political career was worth more than human life, then I'm a piece of shit.'

It was one thing for Wran to support reform, another to get it through the State Parliament. He had support from many quarters – two lecturers at Sydney University, Lex Watson and Craig Johnston, organized a Gay Rights Lobby as an umbrella pressure group – but all the main Sydney churches were against him. Sydney's origins must have been a factor here, because in Adelaide and Melbourne both the Catholic Church and the Church of England quietly accepted reform. But if you are a religious leader in a city that was born in a frenzy of drunkenness and fornication, then it would be understandable to regard it as a city steeped in sin and any relaxation of the laws of God and man a dangerous move.

Wran was well aware of this. One of his political mentors had warned him that 'Labor governs New South Wales with the permission of the Catholic Church', and since Wran was the first non-Catholic leader of the Labor Party for 50 years, he was not about to do anything to alienate the church or his Labor colleagues. One, Ann Symonds, said in 1998, 'The women of the Labor Party are radical. That's where I come from. But coming into Parliament was like being thrust into the company of all my Irish uncles. Those Catholic boys! Half the poor blokes went to Christian Brothers and Marist Brothers schools. No wonder they're all so confused about sexuality at any level.'

So Wran's promised reforms went into abeyance. The years passed. The Mardi Gras parade got bigger and brasher. The beating up of gay men continued, as did police prosecutions. It needed something spectacular to break the deadlock. Then in April 1984, the Council for Civil Liberties (CCL) held its twenty-first annual dinner. Wran, as a founding member, was guest of honour and the tables were filled with the cream of Sydney's legal profession. Everyone expected Wran to make a witty but anodyne speech, probably about the highlights of the CCL's history.

Instead he launched into a vitriolic attack on the CCL and all its members. His launching pad was a newspaper story based on police tapes of telephone conversations which suggested that there might have been a criminal conspiracy involving the state's judicial system. One of the voices on the tapes was that of Lionel Murphy, a High Court judge and a close friend of Wran's. Wran's speech attacked the illegal way the police had recorded the conversations. This was a civil liberties issue of utmost importance, Wran said, and the CCL and its members had offered nothing more than 'a squeaky voice of protest'.

When there were jeers and catcalls and bread rolls began to fly, Wran shouted, 'This is a very abysmal night for the Council of Civil Liberties.' Stung, the guests began to attack Wran for his record on homosexual law reform. Who was he to talk? He had promised changes to the law years ago but had not kept those promises. When was he going to deliver?

Two days later, Wran left for a fortnight in Europe with his wife. On his return he did not wait until he had reached his office, but announced at Sydney Airport that he would immediately introduce into Parliament his own Bill for homosexual law reform. At last everything came together. The new leader of the Opposition, Nick Greiner, was trying to modernize the Liberal Party and make it more appealing to younger voters. The Catholic priest who had spearheaded the opposition to reform, Cardinal Sir James Freeman, had retired and his successor conceded that 'all sins are not necessarily crimes'.

There were two days of debate in Parliament in May. The homophobes had one parting shot. The Liberal member for Wagga Wagga, Joe Schipp, drew attention to the AIDS epidemic and said, 'It may be as well if they do not find a cure.' But he was in a dwindling minority and the Governor of New South Wales signed the Bill into law on the Queen's Birthday weekend. 'Wran's little joke,' said David Marr.

CHAPTER 20

ESTRANGEMENT

'Just as Great Britain some time ago sought to make her future secure in the European Community, so now Australia vigorously seeks partnerships with countries in our own region. Our outlook is necessarily independent.'

– Prime Minister Paul Keating at a reception for the Queen, Canberra, 1992

James McClelland, known to Australians as 'Diamond Jim' because of his stylish clothes and dandyish ways – he was once voted one of Australia's ten best-dressed men – was a flamboyant judge, politician and writer. He was not the sort of man one would think to appoint to the most difficult royal commission to involve Britain and Australia this century. Yet in August 1984, the Labor Prime Minister, Bob Hawke, gave McClelland the job of presiding over an inquiry into British secret nuclear tests in Australia between 1952 and 1957.

This was delicate ground. In its quest to maintain its role as a world superpower Britain had been determined to build an independent nuclear deterrent, and in those five years had exploded no fewer than twelve atomic weapons, most of them at Maralinga in South Australia. Then, from November 1958 to September 1961, there was a second wave of tests, this time of weapons components, that involved the burning, explosion and disbursement across the test range of plutonium and other radioactive materials. These tests, too, were held in great secrecy, because Britain was party to an international nuclear test moratorium at that time and she did not want to be accused of breaking it.

In fact, the way Britain kept the facts about the tests secret from Australia – even the Prime Minister at the time, Robert Menzies, was not fully informed – was a major scandal that Australia seemed reluctant to confront. It was only persistent claims in the Australian press, that Australian ex-servicemen had contracted cancer and other fatal illnesses from being exposed to radioactive fallout, that forced the Hawke government to appoint the Royal Commission.

Menzies had justified allowing the tests by saying in 1953, 'No conceivable injury to life, limb or property could emerge from the test that has been . . . conducted in the vast spaces in the centre of Australia . . . with all our natural advantages for this purpose.' Now it appeared that this was not so. But so little was known about the tests. In the years during which they took place a series of international incidents – the Suez crisis; the Soviet invasion of Hungary; the Olympic Games in Melbourne; the Soviet launch of Sputnik, the world's first space satellite – helped push Maralinga out of the news.

At first it appeared that the McClelland Royal Commission would be able to reveal all. Then Britain's attitude became all too apparent. The Conservative government of Margaret Thatcher flatly refused to send anyone to Australia to represent Britain. During the first four months of hearings in Sydney, the seat for the British representative was pointedly left vacant. Australian public interest in the hearings was intense and by the time the inquiry moved to Britain in December 1984, the British government could ignore it no longer.

McClelland had decided that he would go on the attack right from the beginning, that he would court publicity for Australia's cause, that he would not allow the British to 'duchess' him as had happened so often to his fellow countrymen in the past. As he put it later: 'Frankly, my attitude was "Fuck you bastards, we'll get what we can out of you, even if we have to shame you into it by going public and showing what you're about." And it worked.' In his opening address, McClelland accused the British government of 'dragging its feet' and failing to produce documents buried in vaults at the British Atomic Weapons Research Establishment at Aldermaston.

'Secrecy in the national interest has always been a convenient alibi for failure of disclosure,' he said. 'We're not here to poke our noses into British technical secrets, but there is a certain minimum of information to which, as host country to nuclear tests, we feel entitled to have access.' He won the support of *The Times*: 'This is not principally an issue of science or medicine, but of whether Britain can convince a friendly ally, to whom we owe a debt, that when we say we will help we mean what we say.' An avalanche of documents poured in and more than forty people appeared as witnesses, some to feel the sharp edge of McClelland's wit. 'I was born a scientist,' said nuclear physicist Sir Ernest Titterton. 'Lawyers are a little less fortunate,' McClelland shot back. 'They have to be trained.'

It was a bravura performance and at the end, three months later, it was left to the *Guardian* to sum up its significance. Paying tribute to 'the persistently inquisitive and refreshingly informal Australians' who had managed to prise 38 tons of documents from the death grip of the Official Secrets Act, the *Guardian* said, 'The whole sorry story amounts to another swingeing indictment of British official secretiveness, and it is to our shame that it was left to the Australians to expose it.'

The inquiry produced no conclusive proof that cancers in servicemen were caused directly by the tests, but it did highlight for the first time the plight of the Maralinga-Tjarutja Aboriginals, who were forced off their traditional lands for thirty years. McClelland took the inquiry back to Maralinga, where he and the lawyers sat in the red dust and listened to the Aboriginals' stories. The Aboriginals have since received millions of dollars in compensation and are gradually rebuilding their communities adjacent to the test sites.

As for the sites themselves, Britain refused for a long time to accept McClelland's call in the Commission's final report for Britain to bear the full cost of a proper clean-up. In the end it agreed to collaborate with Australia and contributed equipment and experts to a state-of-the-art nuclear rehabilitation of the area.

But the Royal Commission turned out to be more than just an exercise to get Britain to admit responsibility for the nuclear pollution of part of Australia. It revealed the willingness of the Menzies government in the 1950s to go along with whatever Britain felt was necessary to develop her nuclear defence system. Behind this willingness lay a consistent theme in Australian foreign policy – the search for great and powerful friends to guarantee Australia's defence.

Curtin had turned away from Britain in the dark days of 1942 and announced that henceforth Australia looked to the United States for its defence. Menzies was trying to reverse this decision. It did not work. Britain tested her bombs in Australia and then decided her future lay in Europe. The Maralinga story thus represented a final episode in Australia's traditional foreign policy subservience to the Mother Country.

In 1994 I wrote a long article called 'The End of Great Britain' for an Australian magazine. The emphasis was meant to be on the word 'Great'. I said that there came a time in the history of all nations that were turning points, and that whatever Britain might be as the end of the Millennium approached, great was not an option.

Her industrial might had gone and manufacturing – except of arms, where Britain ranked second in the world – had ended. Areas like Sheffield, that had once housed the densest concentration of heavy industry in the world, were now rubble-strewn wastelands. Most of Bradford's spinning mills had been closed or sold to India. The last big British car plant, Rover, had just been taken over by the Germans – who later sold it back for £10 – and the British merchant fleet that once dominated the high seas had practically vanished.

In 1994, the British people worked harder and longer for less money than just about everyone else in Europe. More than a third of the British workforce earned less than the minimum the Council of Europe considered necessary to exist decently. More children lived in poverty than in any other European country except Ireland and Portugal, and beggars and the homeless had reappeared on the streets of London like ghosts from a Dickensian past.

I complained that many aspects of British life seemed frozen and change difficult if not impossible. For example, Australia was in the middle of a programme of law reform of a nature and breadth that would astound the conservative British legal establishment, if it were to take the trouble to learn of it.

I ended with a quote from a book on cricket, *Anyone But England*: 'The imperialists who had always presided at Lord's, who once preached the gospel of cricket's civilising mission, now grit their teeth in rage as ex-colonial countries transform cricket into an instrument of national assertion. At the close of the twentieth century, English cricket, like England itself, is increasingly on the margins of the new world balance of power.'

Things in Britain have looked up since then – except for cricket, where they have become worse. But the reason for summarizing the 1994 article is not to admit that I was overly pessimistic, but to recount the reaction. Although the article was not published in Britain, I was bombarded with letters from people in Britain who had been sent it by relatives and friends in Australia, frequently accompanied, so they said, with a letter along the lines of – 'Why are you wasting the years over there when you can enjoy the good life here in Australia?'

Few of the letters to me queried the facts or my stance. Most were of the 'more in pain than anger' category and contained the questions: 'What have we ever done to you?' And 'Why do you hate us so?' The tone was that of a mother whose favourite son or daughter has left home, leaving her wondering what did she do wrong to deserve such treatment.

In reply I said, in effect: 'Hang on a minute. We didn't leave you. You kicked us out.' I ran through the three main events of the twentieth century that made Australia realize that the Mother Country had changed the locks – the betrayal of 1942; Britain's decision to join Europe; and the 1971 Immigration Act. (The 'Great Betrayal' is discussed in Chapter 11; the Common Market and the ending of traditional Empire trade arrangements in Chapter 16.) Some Australians made excuses for both. Britain was fighting for her survival in 1942 and there were sound military reasons for concentrating on defeating Germany first, even if it meant the loss of Australia to the Japanese. Britain's entry to Europe was good for Australia in that it forced her to find new markets. But no one could find any excuses for the 1971 Immigration Act, and many found it the most painful betrayal of all.

At a stroke it ended the right Australians had enjoyed since the founding of the country in 1788 to free entry into Britain and full equality there with their kith and kin. They no longer had the automatic right to enter the country at will, stay and work as long as they liked, and enjoy all the privileges of UK citizens. In 1994 the restrictions on Australian entry to Britain were tightened still further – to bring British immigration law into line with the rest of the European Community – making it even more difficult for Australians to enter and stay for any more than a few months. Even for young working holiday-makers, the new laws limited the type of work they were able to do to non-professional and non-vocational.

But the restrictions were nothing compared with the emotional blow that the 1971 Act struck at Australians – they could no longer enter the immigration channel reserved for UK citizens but had to go through one marked 'Others'.

I knew personally several elderly Australian ex-servicemen who doggedly refused to accept this and who argued with embarrassed immigration officers that they had fought for Britain from 1939 to 1945, their fathers from 1914 to 1918, that they had been born British subjects and that there was no way that they were going to queue with foreigners to enter what they still regarded as the Mother Country.

They found support from at least one Englishwoman. She wrote to the London *Daily Telegraph* to say that this was the sort of thing that was helping the Republican cause in Australia. 'I know that Australians feel peculiarly irate at having to line up at Heathrow with those who are "others" (aliens) while people from European countries, which throughout history have been regarded, and acted, as enemies of Britain, are allowed entry very freely.

'I believe this single factor has had a tremendous influence on the way Australians now view their connection with Britain. So many have wondered why they are treated as aliens, when they, or their ancestors, actively fought alongside British men and women in the past, and when they share the same head of state, at least for the present.'

A former editor of the *Telegraph*, William Deedes, agreed with her. 'She is right. She thinks historical links with Britain would be strengthened if Commonwealth citizens travelling to Britain could enter by a particular gate headed by the words "Commonwealth Peoples" or a similar term. We shall do nothing about it, of course, partly because our nationality arrangements are still in an inextricable tangle.

'But when we next hear an Australian politician talking about a republic, we must not accuse him of playing politics. We should acknowledge that for our growing estrangement from Australia we carry most of the blame.'

The two Australian Prime Ministers for whom this growing estrangement merited hardly a second thought were Bob Hawke and his successor Paul Keating. Hawke epitomized the model that many Australians believed to be the ideal national type – a mixture of muscle and mateship, pragmatism and polish, an intellectual with a saloon-bar style. Writer Michael Dean described Hawke as 'Nuggety and handsome, he looks like a middle-weight boxer who's been skilful enough to keep his nose straight.'

Hawke was a Rhodes Scholar and at Oxford won a prize for being able to drink a glass of beer faster than any other student, a record he still holds. As president of the Australian Council of Trade Unions he proved to be a brilliant official, winning favourable settlements for unions before Australia's arbitration commissions. When he became Prime Minister in 1983, his reputation with the unions enabled him to bring some long-overdue harmony to Australia's industrial relations and damp down the country's eternal struggle between bosses and workers. Hawke won four elections in a row, making him one of Australia's more successful politicians, but he resigned in 1991 after a worsening economy led to a greatly reduced parliamentary majority.

He was an Australian nationalist with little time for Britain and English ways. I interviewed him during the election campaign of 1984, when Britain was negotiating to sell some naval vessel or other to Australia, a deal which Hawke referred to 'the Poms trying to rip us off again'. The interview was informal – Hawke was lying on the lounge in his hotel room wearing only a pair of shorts and eating iced cherries from a bowl perched on his chest – so as I left I asked him if I could carry a message from him to the British Labour Party, which was also about to go into an election. 'Yeah,' said Hawke. 'Tell 'em good luck. They're gunna need it.'

Paul Keating, who had been Treasurer since 1983, succeeded him and was Prime Minister until 1996. During that time he became probably the best-known Australian politician abroad, especially in Britain, where the tabloid press christened him 'The Lizard of Oz', after he placed his arm across the Queen's back when shepherding her among various politicians at a reception in King's Hall, Canberra. 'Hands orf cobber!' the London tabloids said.

This obsession with protocol – 'Never touch the Royal family unless they touch you first,' advised Monty Python – caused the British press to miss the real story, one that encompassed in that single incident in 1992 the dying relationship between Britain and Australia, Keating's economic policies and his decision to work towards a federal republic by 2000. The headline in Britain should have been: 'Queen Herself Sets Australia on Path to Republic'.

Keating is an unusual Australian in that he is a man of passionate conviction in a country where to feel strongly about anything is to be considered a bit eccentric. That was a disadvantage in a brilliant political career that took him from a Catholic boy who left school at fifteen to Prime Minister of Australia at only 47. 'You see, in Australia you're not supposed to be convinced of your view or proud of it,' he told me. 'That's called arrogance. You're expected to dissemble, appear not too certain of what you're doing, and add a tinge of obsequiousness to it all.' This was not for Keating. He spoke of his vision for Australia with emotion and even referred to it as a Crusade. His fellow treasurers in the Western world thought he was the most effective treasurer Australia had ever had – removing exchange controls, floating the dollar, reforming the taxation system, ending protectionism, allowing the entry of foreign banks, trying to encourage a proper industrial base.

But by the time of the Queen's visit in February 1992, Keating was Prime Minister of a government under siege. There was a recession, nearly one million Australians were unemployed, the country was hurting and Keating was refusing to apologize for the 'recession we had to have', believing that it would sound insincere and be a gift to his political enemies. Instead he spent most of February travelling the country and consulting with businessmen, associations and state governments about a ten-year plan for economic recovery and development. Into this gruelling schedule he had to sandwich his duties to the Queen.

At none of their meetings did the Queen ask Keating about the state of the economy or his plans for it, presumably because she thought it was none of her business. (Or perhaps she had heard Gareth Evans's story of trying to make small talk with the Duke of Edinburgh, explaining how he had spent a lot of time when Attorney-General travelling to and from London putting everything in place for the Australian Act. The Duke had nodded dismissively and uttered only two words, 'Big deal.') So Keating told her what his government was trying to do – to modernize and broaden the base of the economy so as to arrest the alarming deterioration in the secular terms of trade that had resulted from Australia's reliance on commodities such as wool, wheat and mining. And the Queen with a grasp of politics few give her credit for, replied, 'Ah, yes. They always took us to the country.'

So when a few days later Keating spoke at the reception in King's Hall, congratulating the Queen on the 40th anniversary of her accession to the throne, he felt she would understand the sentiments behind his speech. 'These days we must both face the necessities of a global economy and global change of staggering speed and magnitude. We must also face regional realities. Just as Great Britain some time ago sought to make her future secure in the European Community, so now Australia vigorously seeks partnerships with countries in our own region. Our outlook is necessarily independent.'

As Don Watson, an historian and Keating's speech writer and adviser, the source for this account, said, 'Her Majesty did not blink at these unexceptional remarks.' But Opposition politicians quickly told the media that the Prime

Minister had insulted the Queen. The Royal media pack now had a story – Keating's wife Annita, who is Dutch-born, had not curtsied to the Queen at Sydney Airport (she had checked the protocol and found it was not necessary); Keating had put his arm around the Queen in King's Hall; Keating had insulted the Queen in his speech.

When news of the headlines in the British press reached Canberra the next day, the Queen advised Keating to ignore them. He might well have done so if the Opposition had not decided to make it a political issue. The Opposition leader, John Hewson, accused Keating of not understanding the meaning of respect. Watson says that this was a primal insult, 'full of Protestant condescension and toadying'. Hewson now stood side by side with the British tabloids and with those who had jeered Annita Keating. 'For Paul Keating, his opponent was not only on the wrong side in the post-colonial debate, he had stepped on to consecrated ground.'

Observers in Parliament that day heard only the savagery of Keating's attack on the Opposition, the belly roars of approval from the Labor backbenchers. But in retrospect we can see the coming together of several strands in his thinking about the way Australia had divided in the past and the way the Australia of the future was forming. 'I was told I did not learn respect at school,' Keating said. 'I learned one thing: I learned about self-respect and self-regard for Australia – not about some cultural cringe to a country which decided not to defend the Malayan peninsula, not to worry about Singapore and not to give us our troops back to keep ourselves free from Japanese domination. This was the country that you people wedded yourselves to, and even as it walked out on you and joined the Common Market, you were still looking for your MBEs and your knighthoods and all the rest of the regalia that comes with it.'

Watson says that from then on Keating saw 'the nice threads' of a good policy and he took to it as he had taken to economic reform a decade earlier. The last symbols of Australia's colonial past, and those unwilling to change them, were inimical to a clear-eyed understanding of Australia's reality. 'He wanted it to be universally understood that we were, as never before, on our own, and that the future depended on our having the courage to make radically new choices,' Watson says. 'He thought from this clearing of the national decks might come the energy that would make the whole Australian experiment exciting again.'

Keating took the message everywhere and kept adding new threads. After a meeting with Aboriginal leader Charles Perkins, he said to his aides, out of the blue, 'We'll never be any good until we set things right with them.' He took it abroad, especially to Asia, where, says Watson, 'He thought we'd do much better if we went neither cap in hand nor with the ghost of Empire about us, but unambivalent, sure of who we are and what we stand for.'

Many Australians, especially younger ones, went along wholeheartedly with Labor's new policy and older ones told pollsters that they were 'astonished by the turnaround in their own feelings about Keating'. At his policy launch for the 1993 election, Keating declared Labor's intention to create the Federal Republic of Australia by 2000. Watson says, 'The Queen had done her bit to give Keating a new context for his Prime Ministership – a new story and new policies. She had helped to set the republic rolling.' He might have added, 'With a little help from Fleet Street.'

Keating won the 1993 election but lost in 1996. The Australian electorate splits roughly into two, with a swinging vote of a few percentage points deciding elections. These uncommitted voters tend to think, 'Well, Labor's been in for a couple of terms so we'll give the other lot a go.' There is also the down side of egalitarianism – 'If I'm his equal then I can do his job as well as he does. So why has he got it and not me?' When the politician is a conservative, the questioner consoles himself with the thought that the politician came from a family destined to rule, someone like Menzies. But when the politician is a Labor man, the thought is that he must have been corrupt, or had influential mates.

Keating knew that sooner or later Australia would reject him. He told me in 1989, 'There's no thanks in this game. I don't expect any. There's good and bad in the Australian psyche. There's a healthy disregard for authority. That's the good quality. But there's the tall poppy syndrome as well. They want to cut achievers down to size. In the past few years this has grown to epidemic proportions and a politician bringing bad news in bad times becomes a lightning rod. That's the role I've had to play.

'But a good government provides leadership. If you think that something is in the nation's best interests, then you go ahead and do it, even if it's temporarily unpopular. You have to have the courage to say, "We're going here because it's for the good of Australia. You can either like it or lump it."'

Out of office, Keating retreated from public life, but on his rare speaking engagements Australians queued to hear him. By 1999, even the British press was missing him. They lauded his leadership qualities and agreed that he had left his mark on history. The *Guardian* described Keating's vision for Australia – a street-smart country maximizing its chances in a difficult world – and it quoted what could well be his political epitaph, 'When a small nation has inherited a Garden of Eden, your footwork has got to be exemplary.'

CHAPTER 21

THE DOORS OPEN

'Australia abolished the White Australia Policy by stealth, not by reason of debate and decision. There are tactical advantages in such methods but they leave unresolved tensions at the nation's heart.'

– Paul Kelly, political historian

The crumbling of the White Australia Policy began not in Australia itself but 15,000 miles away in Europe. The number of displaced persons coming to Australia had tapered off at the close of the 1950s and the Immigration Department, anxious to maintain the flow of migrants that had transformed the country and helped take its population from seven and a half million in 1947 to over ten million by 1959, had to bend its own rules to tap new sources.

True, every now and then some upheaval in the Old World would provide a short boost to immigration figures – the Hungarian Revolution in 1956, for example – but to keep the figures high, Australia had to abandon the idea that migrants from southern Europe were too swarthy for Australian tastes and encourage Sicilians, Greeks, then Turks, and eventually Lebanese to come to Australia.

Then a new factor interfered with Australia's plans – the success of the European Economic Community (EEC). As European countries which had provided Australia with its ideal migrant – blond, blue-eyed, skilled and hardworking – tasted the fruits of economic success, and began enjoying full employment, high wages, and social services as good as Australia, the idea of paying your own fare to travel halfway across the world to start a new life grew less and less appealing. EEC countries even began to compete with Australia to attract migrants from Turkey. The German government implemented a guestworker scheme which, because Germany was close and the wages high, thousands of Turks preferred to a life down under, such a long way from home.

Canberra historian A. W. Martin summed up Australia's problem. 'In the late 1960s, just as pressure was mounting in Australia for higher immigration targets, the old sources of settlers, except for the United Kingdom itself, were drying up. To attract the desired numbers the government had to adopt new policies and look for migrants in communities not tapped before.'

So Australia looked upwards – to Asia. It was wrong to believe, as many did, that Australians were ignorant about Asia. The war had made many well aware of who their near neighbours were and of their relationship with Australia, and authors of the calibre of Christopher Koch, Robert Drewe and Blanche d'Alpuget had written about Asia in their fiction long before Asia became central to Australian life. Now the government looked with fresh eyes at Indonesia, Malaya, Singapore, Thailand, Vietnam, Cambodia, the Philippines, India and Pakistan, where hardworking and often well-educated people would jump at the chance to migrate to Australia. But what about the White Australia Policy? What about nearly two centuries of racial prejudice against Asians? Leaders from both main political parties moved to change all that.

At the Labor Party's federal conference in 1965, reformers tried for the third time to remove from the party's platform its commitment to a white Australia. They had failed in 1961 and again in 1963, mainly due to the vehement opposition of former Immigration Minister Arthur Calwell. This time Don Dunstan, the colourful South Australian politician, made an impassioned speech to delegates. He said he had spent the early part of his life in Fiji, which had native, Indian, white and Chinese communities. He loathed racist overtones of any sort and for Labor to continue to promote a white Australia as party policy would cost it the support of many members, especially the young, who found it offensive. He was backed by Gough Whitlam, a future Prime Minister, but they were successful with their colleagues only by agreeing to a compromise – Asians would be welcome if they could show an ability to assimilate.

What now occurred in Australia's immigration policy became possible only because white Australia ceased to be a party political issue. The very next year the Liberal Minister for Immigration, Hubert Opperman, the former champion cyclist, allowed a substantial increase in the intake of part Asians. By 1970 the number of Asian and part-Asian migrants settling in Australia was running at nearly 10,000 a year.

Not that many knew it. As we have seen, postwar governments did everything possible to sell their European immigration programme to old Australians – 'Italians make good Australians', said a poster in 1951 – and a whole section of the Immigration Department devoted itself to finding and publishing migrant success stories. But there was no political capital to be had from drawing everyone's attention to the influx of Asians.

It was almost as if politicians hoped that no one would notice that there seemed to be more Asian faces on Australian streets than before, and that if they did, they would think that they were students or tourists. Political historian Paul Kelly says that Australia abolished the White Australia Policy by stealth, not by reason of debate and decision. 'There are tactical advantages in such methods but they leave unresolved tensions at the nation's heart.' Others go further. Katharine Betts, in her anti-immigration polemic, *The Great Divide*, argues that the dismantling of the White Australia Policy was not a victory for reason because 'the new policy was a political victory that left the racist parochial unconverted but outmanoeuvred'. The 'racist parochial' is defined by Betts as the ordinary 'ocker' Australian who enjoys materialism, tacky consumerism, commercial television and sport and who is an enthusiastic nationalist. She accuses Australian 'elites' of having conspired to engineer a large immigration programme over the heads of such people.

Yet there was a groundswell in the 1960s and 70s for change in Australia's immigration policies. People were looking at an Australia that was already very different from the one of their pre-war memories, which were of a white, British and largely Protestant country. Postwar governments had sold them on the idea of white European migration with the assurance that these new people would be quickly and easily assimilated. Many *had* been successfully assimilated, but it was never easy for the older ones and they often encountered irrational prejudice. They sought comfort among their own kind, and meeting places sprang up to provide it – Hungarian clubs, Italian clubs, Greek clubs, Russian clubs. The fact that migrants gathered there, ate their own food, spoke their own language and reminisced about the country that had given birth to them did not make them any less loyal to their adopted country – many of them if asked would speak glowingly of Australia and the chances it had given them.

'The best bloody country in the world, mate,' an Italian taxi driver said to me in Sydney in 1999. 'And I ought to know. I left Palermo when I was eighteen and went to Argentina. I got married and had kids and stayed for fourteen years. We left there in the end because it was too violent politically and I didn't feel it was a safe place to bring up a family. Then we came to Australia. It's a real democracy here, where you can live your own life. They leave you alone. Your kids get a good education, you can always count on a fair go, there's work if you want it and we feel safe. The climate and the beauty are a bonus. We're never going to move again. Australia's a paradise and Australians don't realize how lucky they are.'

It seems to have occurred to sufficient important people, at about the same time, that it might be a good idea to get away from assimilation and move on to something better. Professor Jerzy Zubrzycki, a Canberra-based sociologist, suggested in 1968 that instead of trying to force migrants to make English their main language and to conform to the sterotype Australian, Australia should welcome cultural diversity and encourage migrant communities to maintain their own languages. The Whitlam government's Minister for Immigration, Al Grasby, technically a migrant himself, wrote a paper in 1973, *A Multi-cultural Society for the Future*, which argued that migrants' cultures and customs were precious and should be respected and preserved, rather than washed away in the controversial process of 'assimilation'.

Grasby was born in Brisbane of an Irish mother and a Spanish father who had come to Australia via Chile. The family left Australia in 1932 and wandered the world for the next fifteen years, Grasby's education spanning thirteen different schools. They were trapped in Britain by the outbreak of war and Grasby's father was killed in an air raid. Grasby himself returned to Australia as a ten-pound English migrant in 1948 and eventually settled in Griffith in southern New South Wales, a farming area with a large southern-Italian population. He represented the area in the state and then in the Federal parliament.

Grasby's paper was a 'from-the-heart' plea for a new sort of Australia, a manifesto for the plural society of which he and others dreamed. 'Where is the Maltese process worker, the Finnish carpenter, the Italian concrete layer, the Yugoslav miner – or dare I say it – the Indian scientist? Where do these people belong, in all honesty, if not in today's composite Australian image?' Grasby's title appears to have been the first use of the word 'multi-cultural' in this context in Australia, although it had been used before in Canada to describe similar policies.

By 1977 the word was in common usage and the Australian Ethnic Affairs Council, an advisory body set up by the Fraser government, entitled one of its publications, *Australia as a Multi-cultural Society*. The Fraser government also asked a prominent Melbourne lawyer, Frank Galbally, to see whether migrants were being properly looked after. Galbally handed in his report the following year, recommending that the government spend $50 million over the following three years on migrant welfare services, and that ethnic cultures should be preserved, not repressed.

When Fraser tabled this report in Parliament he did so in ten languages, a symbolic acceptance of the fact that multi-culturalism was now Australia's official policy. Confirmation of this was the creation in 1980 of the Australian Institute of Multi-cultural Affairs, to develop multicultural policy as well as to encourage awareness and understanding of the diverse cultures in Australia. But, as we shall see, not everyone welcomed it.

After the fall of Saigon in 1975, the trickle of Asian migrants into Australia became a flood. It was like the end of the war in Europe in 1945, only this time in Asia, as nearly a million Indo-Chinese people sought resettlement abroad. The Whitlam government was reluctant to take too many, believing that as strong anti-Communists they would have conservative political beliefs, and because it was anxious about growing unemployment at home. The Fraser government agreed to a small intake from refugee camps in Thailand, but then in 1976 an old Australian nightmare came true – Asians began landing on Australia's sparsely-inhabited northern coastline.

They came in ramshackle boats across 4,000 miles of dangerous sea. Australia felt she could not turn them away. Where would they go? And if they refused to go, what could the Australian authorities do – shoot them out of the water? Within a year 5,000 Vietnamese refugees, almost 1,000 of whom were boat people, had been permitted to enter Australia. In 1978, the Vietnamese intake was 9,000 and by 1982, 65,000 had been admitted. From then on, no matter what government was in power, immigration from Asia was an accepted part of Australian policy.

This created problems that had not occurred during the European migration of the 1940s and 1950s. Asian migrants were highly visible and they preferred to settle in cities rather than country towns. It was inevitable and natural that they should want to live together and 'ghettos' sprang up – Richmond in Melbourne became known as 'Little Saigon' and Cabramatta in Sydney as 'Vietnamatta'. But even these 'ghettos' were seldom confined to one nationality. Auburn, a deprived working-class suburb in Sydney's west, has about 54 different nationalities in the area, with many Vietnamese but a lot of Turks and Lebanese as well.

One characteristic that distinguished Asian migrants was their determination and ambition. Their children did very well at school and soon Vietnamese, Indian, Chinese and Thai names began to appear in the newspapers as winners of all sorts of academic prizes, way ahead of their white Australian colleagues. But at the same time, the police and the press began to talk about 'Asian' gangs, 'Asian' drug dealing, the 'Asianization' of Australia and before long 'Asians Out' signs appeared on lavatory walls and railway line cuttings. It seemed that the underlying attitudes that had supported the walls of White Australia were still in good order.

Still, the subject could well have remained one for academic study had not the conservative historian Geoffrey Blainey, addressing a meeting of Rotarians in the country town of Warrnambool, Victoria – an audience more representative of old white Australia would be hard to find – said that the pace of Asian migration was ahead of public opinion (true, but are not governments meant to lead?), that this was dangerous at a time of rising unemployment and was bound to stir up racism. Interviewed later, he accused the Hawke Labor government of being biased against British migrants and treating Asians as a 'favoured minority'.

Asian migration was now way out in the open and the media was in the thick of the controversy. There were polls, surveys and interviews; letters, editorials and articles; literary contributions, lectures and debates. The academic world was split on the issue – most academics were against Blainey but a lot were too scared to speak out. Politicians and the public were also divided on the issue. From whether Australia should continue to admit Asian settlers at the same rate, the question moved on to whether it should have any new migrants at all, or no more than necessary to replace those who left. What should the optimum population of Australia be – 20 million, 30 million, 50 million? Had it already exceeded a figure that it could comfortably continue to support at the standard of living to which its citizens had grown accustomed?

There were complaints that Australia appeared to have no long-term immigration policy and seemed reluctant to develop one. No one seemed to know, for instance, whether the benefits of immigration outweighed the costs. There were arguments over what percentage of Australians were Asians and what this figure would be, say, 20 years hence. Some said Australia had always been a land of immigrants and should continue to be so – the more the better, no matter where they came from. Others said that all the frontiers had been occupied and the immigration era was over.

Dr Mary White, author of *Listen . . . Our Land is Crying*, said, 'If our population is not capped now, Australia will be unable to feed its own. It has no choice, as I see it [but] to abandon its role as a food exporter; that will inevitably be forced upon it by the land itself over the next couple of decades.'

While this debate went on, many Vietnamese simply got on with their lives. Some opened restaurants, and Sydney and Melbourne soon boasted a Vietnamese cuisine arguably better than in Vietnam itself. Others entered fields traditionally reserved for white Australians. Hung Le's family fled Vietnam in 1976 in a leaky prawn trawler and eventually settled in Melbourne. He went to school, played Australian Rules Football, was a successful fast bowler in the local cricket team, and then became a violin-playing street busker and finally a stand-up comic, taking the mickey out of Australians and Vietnamese-Australians with an act he called *Wog-A-Rama*. He also wrote a show for a tour of London, Edinburgh and New York called *Noodle Frontity* and an autobiography entitled *The Yellow Peril From Sin City*. All his work is politically incorrect but it makes people laugh. Le said in 1998, 'By bringing stuff out into the open, I can make people reconsider. I'm doing positive things. I'm trying to tell people that we don't all sell smack.'

Of course not. Vi Le left Vietnam in 1975 with her parents and four other children because they wanted a country that would offer stability and education. 'We had lived in Laos, Hong Kong, France and Bangkok, so settling in Australia was never a problem. I did a Bachelor of Commerce at Melbourne University, then a graduate diploma in corporate finance and then joined the ANZ bank.' She was with the bank for 14 years when Vietnam opened up again, and was offered the job of Australian Trade Commissioner in Hanoi. She said in 1998, 'It's been a tremendous experience, though as far as the Vietnamese were concerned, I was an Australian, apart from the fact that I spoke their language.' This was essential. Part of her first bank deal in Vietnam involved a state enterprise. A lot of the negotiation had been in English, but then the head of the state enterprise turned to Vi Le and said in Vietnamese, 'I'm happy to sign but can we clear up one thing first? Is this the sort of loan we have to repay or not repay?' She told him, 'The former, thanks very much.'

By accident rather than by design, but hand-in-hand with turning Australia into an Asian country with a substantial Asian population, another revolutionary change was underway – a reconciliation with Aboriginals. As we have seen, Aboriginals had been making progress over land rights, following the well-publicised strike by Aboriginal stockmen at the Vestey's Wave Hill station in 1966 and their success in persuading the Whitlam government to give them back their tribal lands. The Vesteys, with the acumen that had made them the richest family in Britain, saw the writing in the dust and started to move out.

They could not do so overnight, but in June 1992, in one of the biggest land sales in Australia's history, nine Vestey cattle stations, including Wave Hill, went to auction in Brisbane. Six were sold at the auction for $37.55 million and the remaining three by private treaty the following month. All the buyers except one, a Greek shipping magnate, were Australian individuals or companies. The Vesteys, absentee English landlords who had been in the country for more than 75 years and whose Australian cattle stations had once covered an area bigger than Europe, had gone for ever.

The crucial point about the Wave Hill strike was that it was not a strike at all in the usual sense. It had not been a withdrawal of labour in an effort to obtain better wages or improved conditions. The Aboriginals had withdrawn from the workforce altogether and returned to their roots – their land. In 1975, when the Whitlam government handed over to Aboriginals 2,000 square kilometres of land the Vesteys had held under lease, and Whitlam himself poured earth into the tribal leader's hands as a symbolic gesture, Aboriginal expectations were raised throughout Australia. They *could* get their land back. What was new was the way they went about it – not so much by political agitation but by using the white man's own institutions, in this case his courts.

Seven years later, 'Eddie' Koiki Mabo and some of his fellow tribesmen from the lovely but remote island of Mer, part of the Murray Islands group in the

Torres Strait, went before the Supreme Court of Queensland to claim that the island of Mer did not belong to the Crown, but to them. It took ten years for the case to make its way to the High Court, which delivered its decision on 3 June 1992. Known as the Mabo decision, or just 'Mabo', it ranks as one of the most important in Australia's short history.

It was the beginning of a brave, complicated, divisive but ultimately redeeming attempt to admit that the first white Australians had stolen the country from Aboriginals and that the time had come to hand it back to them. Mabo and its subsequent and associated case, Wik, allowed Aboriginals, if they so wished, to claim back nearly 80 per cent of the whole of Australia. It was as if the Supreme Court of the United States had ruled that North America had been stolen from the original Indians and that they could now legally claim large chunks of it back.

The seven judges in the Mabo case (one dissenting), white, middle-class, conservative jurists, altered the course that Australia had followed since its founding. They looked with fresh eyes at European imperialism, at the way Britain and other European powers regarded places like Australia in the eighteenth century. 'The great voyages of European discovery opened to European nations the prospect of occupying new and valuable territories that were already inhabited. As among themselves, the European nations parcelled out the territories newly discovered to the sovereigns of the respective discoverers.' But what about the people already occupying these territories? The legal view in Britain was that, provided the indigenous inhabitants were not organized into a society 'united permanently for political action', they were backward peoples and if they left their land uncultivated, then they were *terra nullius* (without land) and anyone could claim it. (To his credit, the great English jurist, Sir William Blackstone, expressed his doubt about this argument.)

But, according to this doctrine, the Australian Aboriginals (then about 750,000), were nomads who did not cultivate the land. Therefore, the British legal argument went, they had no claim to the land and it could be legally taken by the white settlers. The invaders' courts of law endorsed this opinion time and time again.

So all rights to land in Australia were formally vested in the Crown – in effect, the British government. Colonial officials then sold or granted the land to the settlers, usually on the basis of their having sufficient capital to develop it. This system was important to the early economic success of the Australian colonies. It boosted the Australian wool industry and its principal customer, the mills in Britain, against the older German industry because the Australian graziers had the advantage of cheap or free land. The only ones who missed out were the traditional owners of the land, the Aboriginals, because of *terra nullius.*

Yet the early settlers had learnt very quickly that Aboriginals actually had strong affinities with the land and would fight to the death to defend some areas. In fact, the Colonial Office in London became alarmed at the slaughter of Aboriginals in these clashes and put pressure on the state governments in Australia to recognize Aboriginal land rights, to guarantee them access to sacred areas for ceremonies, even when these were on settled land, and to compensate them for land which they had lost for ever. The states resented this pressure, pretended to comply, and then ignored the issue.

The Australian judges now ended once and for all the legal deceit that when the First Fleet arrived to settle Australia the country was empty. They saw in the First Settlement 'a conflagration of oppression and conflict which was, over the following century, to spread across the continent to dispossess, degrade and devastate the Aboriginal people'. They were left as 'intruders in their own homes'. But Australian law should not be 'frozen in an era of racial discrimination'. Eddie Mabo and his fellow islanders had shown a long association with Mer, of continued cultural and economic ties. Any idea that the island was *terra nullius* was indefensible. Aboriginals had common law rights to native title and the same principles applied on the mainland. A new era in Australian history was about to begin.

It took the Labor government of Paul Keating seventeen months to prepare legislation in response to the High Court decision, but the Native Title Bill finally made its way through Parliament just before Christmas 1993. Keating, realizing the history-making nature of the Bill – even if the rest of Australia as yet did not – made one of the best speeches of his career. He said that no self-respecting democracy could deny its history and to deny what had been done to Aboriginals would be 'to deny part of ourselves as Australians'. The many challenges of the High Court's decision had been comprehensively dealt with in the Bill, but he wanted to stress the wider significance.

'For today as a nation we take a major step towards a new and better relationship between Aboriginal and non-Aboriginal Australians. We give the indigenous people of Australia, at last, the standing they are owed as the original occupants of this continent. The standing they are owed as seminal contributors to our national life and culture: as workers, soldiers, explorers, artists, sportsmen and women – as the defining element in the character of this nation . . . We can make the Mabo decision an historic turning point: the basis of a new relationship between indigenous and other Australians.'

The Bill trod a delicate path through all the problems that telling Aboriginals, in effect, that Australia was suddenly theirs again, had created. The government's view was that under common law past valid freehold and leasehold grants extinguished native title. This placated Australian suburbia, which had been worried that Aboriginals would lay claim to their quarter-acre house plots.

There was no obstacle to renewal of pastoral leases in the future, whether validated or already valid. To get native title, Aboriginals would have to prove continuing connection with the land. This somewhat, but not entirely, calmed the farming community who held long-term government leases of land and who had feared that they might lose them overnight.

But mining leases did not extinguish native title rights, which could be exercised in full after the leases expired. The mining industry saw this as discrimination, but Keating pointed out that some state laws already provided that when a mining lease expired, the owner of the land could take possession forthwith. 'How can we offer native title holders any less?' asked Keating. Finally, the Bill provided for a system of specialized and accessible tribunal and court processes for deciding native title claims and negotiating and reconciling Aboriginal and white Australian interests.

The one thing neither the courts nor the government could do was to define who or what was an Aboriginal. Despite the best efforts of government officials and religious leaders to keep the two races apart, two hundred years of living in the same country meant that many Aboriginals had white blood and vice versa. (There were other inter-racial marriages. In some places like Darwin and Broome, for example, there were Australians with mixtures of white, Aboriginal, Chinese, Malay and Japanese blood.) Put baldly, how much Aboriginal blood did someone need to be an Aboriginal?

Cultural differences were of some help but not a lot. Many urban Aboriginals behaved in the same manner as their white urban counterparts. And on visits to Australia over the years I noticed that many white Australians were behaving more and more like Aboriginals – were they being colonized and assimilated by the original inhabitants?

Modern white Australians painted their faces for sports events and parties, like Aboriginals do for corroborees. White Australians wore fewer and fewer clothes and sometimes went naked. Many often wore no shoes. (On a business class flight from Cairns, the man in the seat alongside me was barefoot.) They disappeared from work for days on a 'sickie', the white equivalent of an Aboriginal 'walkabout'. They took mind-bending drugs to enter 'dreamtime'. They cooked pieces of animals outdoors on an open fire. They ate kangaroos, emus and crocodiles. At meetings they sat around in a circle, like Aboriginals, instead of sitting in the traditional white man's classroom style. They began to worship their land and got angry when it was disturbed or when trees were cut down. They often took bush remedies instead of traditional white man's medicine. (Aboriginals had the most sophisticated medical pharmacopoeia on earth with some 900 medicinal plants.) They paid a lot of money in expensive restaurants to eat 'bush tucker' like wattle-seed bread, and drove around the inner-cities in the four-wheel-drive vehicles favoured by outback Aboriginals.

And they frequently held corroborees on beaches when various tribes from all over the area would paint their faces and compete with each other in trying to master the sea.

And when you come to think about it, even the relationship between white Australians and their cattle – letting them roam wild over thousands of square kilometres and then hunting them down and slaughtering them, was not that different from the relationship between Aboriginals and their kangaroos.

And finally, by 1999 it had even become fashionable in some circles in Australia to be able to show that you had a little Aboriginal blood – provided it happened a long while ago and you announced it first yourself.

This was different from what their critics called 'economic Aboriginals'. When land, money and other benefits were on offer it was to be expected that Australians who were not previously considered to be Aboriginals should decide that they were. In Tasmania, for instance, the number of people identifying themselves as Aboriginals increased from about 200 in 1971 to 13,873 in 1996.

In other states, as part of positive discrimination policies, certain jobs held by white Australians were designated as being solely for Aboriginals. Whereupon the apparently white Australians holding those jobs suddenly discovered that they had Aboriginal ancestors and were thus able to remain in the post. The issue of Aboriginal identity caused many a family dispute. Either from envy or genuine dismay, Aboriginals who had done well out of their Aboriginal identity were denounced by relatives who claimed that the family did not have Aboriginal blood at all. One said that they were descended from a black American whaler last century, another from a black American soldier posted to Australia in the Second World War. Suddenly it seemed that everyone in Australia wanted to be an Aboriginal. White Australians – and some Aboriginals – referred to those who had only recently discovered their Aboriginality as 'paper blacks', 'newcomers', or 'pop-up blacks'.

So the courts were asked to decide this delicate issue. Of course they failed. Justice Ron Merkel of the Federal Court heard a case in Hobart in 1998 about elections to an Aboriginal council. The plaintiffs, Aboriginals themselves, claimed that nine of the candidates were not really Aboriginals. The judge said that the question of racial identity in this day and age was not determined genetically but, socially, and it was for Aboriginals themselves to decide who was an Aboriginal. This was logical and interesting but not very helpful.

Tribal elders would probably claim that initiation was the crucial deciding factor. Unless a young man had been initiated, then neither he nor his family were real Aboriginals. But since initiation usually involved crude genital mutilation – circumcising the penis, then slicing open its underside to the urethra – modern young Aboriginals understandably wondered whether it was worth it.

* * *

After the Mabo judgment, the government had expressed its view that under common law, past valid freehold and leasehold grants extinguished native title. In other words, even if Aboriginals could show an association with a particular area of land, if someone else held the freehold or lease of that land, even if they were absentee landlords, the Aboriginals could not have it.

Two Aboriginal tribes, the Wik and the Thayorre, who come from remote northern Queensland up near the Gulf of Carpentaria, challenged this. They laid claim to two leasehold properties, Mitchellton and the Holroyd River holding. The claims ended in the High Court and there the second act of the drama that changed Australia was written, the one known as the Wik case, or, like Mabo, simply as Wik.

Both properties were in a bad way, which was probably a factor in choosing them for the legal challenge. One of the judges, Justice Michael Kirby, pointed out that the first lease on Mitchellton had lasted only three years before it was repossessed for non-payment of rent. The second lease was surrendered in an even shorter period. There were no boundary fences, no buildings and only a few feral cattle. All the while, the Thayorre tribe lived on their land in their traditional way without the faintest idea that white Australians held a lease on it. 'It would require a very strong legal doctrine to deprive [the Thayorre] of their native title,' Kirby said. At the Holroyd River Holding there were a few stockmen and wood gatherers, but the area was so large and so remote that the Wik tribe went about their traditional lives with no more than an occasional, brief contact with the leaseholder or his employees.

The court found by a bare majority that leaseholds did not necessarily extinguish native title rights. Each case would have to be determined on its own merits. There was no reason why the rights of native title holders and pastoral leaseholders could not co-exist. At first this seemed a victory for common sense and natural justice. Paul Sheehan, a Sydney journalist and author, wrote, 'The success of Wik and Thayorre hurt no one. The dispute gave rights to real people – Chocolate Thomas, Sheba Dignari, Button Jones, Dodger Carlton, Paddy Carlton, Ben Ward and many others – in favour of absentee landlords with nothing at stake. The judgment protected access to land used by local indigenous people for many generations, at virtually no cost to anyone else.'

But the Wik case led to hundreds of other cases all over Australia. Big, successful working stations found themselves under claim, often from competing tribes – the Mindara station in Western Australia faced fourteen separate native title claims. Some leaseholders willing to negotiate over claims on their properties had great difficulty in discovering with whom they should deal. Some tribes claimed enormous areas – in south-west Queensland one claim affected 3,000 leases. No one had imagined that the laudable aim of handing back to Aboriginals the land that the first white settlers had stolen from them was going to prove quite so complicated.

The conservative government of John Howard now moved to try to end the confusion by introducing into Parliament the Wik Bill, a long, complicated piece of legislation that not only failed to provide a satisfactory solution to most of the problems raised by the Mabo and Wik judgments, but created a number of new ones of its own. The Mabo judgment suggested that for an Aboriginal land claim to succeed, the Aboriginals would have to show continuous association with the land in question. But as claim followed claim it became clear that this was being relaxed and widened because no one was certain how strictly to define 'continuous association'.

The Northern Territory Aboriginal Land Commissioner's report on a claim for the area around Lake Amadeus said that the Aboriginal who had lodged the claim based it on the fact that he, his parents and his grandparents had travelled through the *vicinity* of the claim area. Further, he had a history of employment on the claim area, leading to a knowledge of the country, affiliation to the area through a spouse, and 'skin group' membership, which required knowledge of the stories associated with the claim area. Sydney columnist Padraic P. McGuinness noted, 'To these may be added the recently elaborated principle of development of traditions, which means that claims can be based on traditions which are of recent origin and sometimes suggested by anthropologists.'

The Wik Bill tried to deal with some of these problems by limiting the number of claimants and fixing the time allowed for making claims. But it satisfied no one. Aboriginal leaders issued a statement saying, 'Australia is at the racial crossroads.' An elder of the Wik community commented, 'We won the High Court claim with your law, with European law, not our law. That's something that should be pointed out.' And Noel Pearson, Chairman of the Cape York Land Council, said, 'We are on the verge, no less, of legal apartheid. We will have a national law that says there's two titles sitting alongside each other and the Aboriginal title can be extinguished and the other one remain in place. That's discrimination.' But the farmers said that the Wik Bill did little to protect farmers' interests.

Many got angry at the suggestion that only Aboriginals had a special relationship with the land, that only black people had sacred sites. Novelist and journalist Margaret Simons wrote, 'Throughout rural Australia, predominant white communities have their histories and mythologies of place – stories of floods and drought, birth and death, hardship endured . . . A farmer who knows the pattern of bark on each tree, who remembers digging every fence-post hole, and who knows his soil so well that he can predict crop growth-rates from paddock to paddock or even from one end of a paddock to another, is insulted by suggestions that he is somehow not at home. It is hardly surprising that some feel threatened by laws stating that older relationships with the land persist to the present day and must be acknowledged and reconciled.'

It took the Group Geographer of the Western Mining Corporation, Dr Stephen Davis, to point out that after the Mabo decision it was expected that about 38 per cent of the total land area of Australia would be subject to native title claim. The Wik decision expanded the area to 78 per cent of the whole of Australia. The Wik Bill would reduce the claimable area to 70 per cent – a minor reduction. But there were other possibilites lurking in the wings.

McGuinness wrote: 'There has been a steady development by human rights lawyers of the notion of collective rights attaching to indigenous peoples which override individual rights . . . This is the basis of a movement in favour of Aboriginal sovereignty which would lead to the establishment of what are called "domestic independent nations", geographically located or dispersed racial enclaves with their own system of government, claiming mineral and resource rights and the right to exclude others – as is already the case with some Aboriginal areas.'

Many Australians, both white and Aboriginal, considered that such a development would be a disaster. The country could become a series of Balkan-like kingdoms into which others could not enter without permission. But it was difficult to find anyone to say this for fear of being labelled unsympathetic to the Aboriginal cause, even racist.

CHAPTER 22

BACKLASH

'I and most Australians want our immigration policy radically reviewed and that of multiculturalism abolished. I believe that we are in danger of being swamped by Asians.'

– Pauline Hanson, founder of the One Nation Party

Into this confused, complicated and emotional situation stepped Pauline Hanson, a local council politician from Ipswich, Queensland. For two years she and her political party, One Nation, were to dominate the political agenda. She was 'old' Anglo-Australian, in that three of her grandparents had migrated from Britain before the First World War and the fourth, her maternal grandmother, was Queensland-born of British stock. Twice married and divorced, the parent of four children, Hanson won a seat on the Ipswich Council as an independent in 1994.

She won Liberal Party preselection for the Federal seat of Oxley, only to see it stripped from her after she wrote a letter to the *Queensland Times* about Aboriginals. 'How can we expect this race to help themselves,' she wrote, 'when governments shower them with money, facilities and opportunities that only these people can obtain no matter how minute the indigenous blood that flows through their veins, and this is what is causing racism. Until governments wake up to themselves and start looking at equality not colour then we might start to work together as one.'

When Hanson stood as an independent, the sitting member, Labor's Les Scott, refused to appear on the same podium with her, and local Aboriginal elders called her 'ignorant as a pig in mud'. This only enhanced her cause and to everyone's amazement, except that of Pauline Hanson herself, she won the seat with the biggest swing in the nation, almost 20 per cent. In her maiden speech in Parliament, since everyone already knew her views on Aboriginals, she concentrated on the other major problem she believed faced Australia.

'Immigration and multiculturalism are issues that this government is trying to address, but for far too long ordinary Australians have been kept out of any debate by the major parties,' Hanson said. 'I and most Australians want our immigration policy radically reviewed and that of multiculturalism abolished. I believe that we are in danger of being swamped by Asians.'

The reaction was amazing. In no time at all Pauline Hanson, a woman who said she had developed her political ideas by listening to what her customers said in her Ipswich fish and chip shop, became first a national figure and then an international one. As she expounded her views on Asian immigration and Aboriginals, tapping into many an Australian's previously hidden resentment, reporters and television crews began following her everywhere she went. She was a good story, big news, even though she had none of the usual politician's skill in handling the media. Her voice grated on the ear, her grammar was poor, and she had difficulty in reading a speech. But large numbers of Australians apparently loved her. They loved the way she left politicians gasping for answers to her *faux naif* questions:

> 'Would the Prime Minister please explain how his government can justify making available to Indonesia what may be as much as one point seven billion dollars when we already give approximately ninety million dollars in foreign aid . . . ? Has the government considered that perhaps it is more appropriate for President Soharto to help bail out his own people with some of the billions he and his family have profited from during his time in office?'

They loved the way she got the better of smart-alec journalists and television interviewers who tried to put her down:

> Interviewer: You've missed thirty divisions in Parliament since you became a member. How do you explain that?
>
> Hanson: But I didn't agree with what either side was saying. Why should I vote for them?

They loved the way that critics who attacked her in a personal manner found that their very criticism brought her more admirers. Responding to accusations that she was 'a racist dumbbell unable to speak properly' she told broadcaster Alan Jones of Sydney's Radio 2UE:

> 'People come up to me and say, "Listen, don't worry about it, mate. Keep going. We're just like you." And I will keep going despite what academic snobs say about me. They say it's so un-Australian that I don't speak properly and they chip me about it all the time. I think that they're

so damn jealous of the fact that a woman from a fish and chip shop can get up and have a voice as I do. But that's their problem, not mine. If it's waking them all up and giving them a shake-up, wonderful.'

One Nation won eleven seats in the Queensland state elections in June 1998 and put up 23 candidates for the Senate and 139 for the House of Representatives in the October 1998 Federal election. Hanson stomped the country creating controversy and confusion. She was not against migrants, she said, just against immigration at its present level. 'I go to citizenship ceremonies and I say "Welcome, as long as you give this country your undivided loyalty. If you don't then I'll be the first one to take you to the airport, put you on a plane and wave you goodbye".' She was not against Aboriginals, she said, just against 'professional' Aboriginals who sucked money from the government system at the expense of 'seriously-disadvantaged Aboriginal Australians'.

But a book, *Pauline Hanson: the Truth*, sold to raise funds for One Nation, written by an anonymous supporter but with the copyright attributed to Pauline Hanson, said that Aboriginals were cannibals who sometimes ate family members. The party's national director, David Ettridge, justified this claim, saying, 'The suggestion that we should be feeling some concern for modern day Aborigines for suffering in the past is balanced a little bit by the alternative view of whether you can feel sympathy for people who eat their babies.'

Hanson and her party saw conspiracies everywhere. Multiculturalism, Aboriginalism and Asianization were being imposed upon Australia by a new elite class which had deliberately earmarked Anglo-Saxon Australia for destruction. At the behest of the secret forces dominating the new world order, the new elite had decided to open Australia to the surplus population of Asia, particularly of China and India. The Hanson book imagines Australia in 2025. It has a population of 1.8 billion. Its capital is called Vuo Wah, and its president is a lesbian called Poona Li Hung, who has been chosen by the World Government.

Everywhere Hanson went in Australia she was met with demonstrations, insults and often violence. She was spat upon, threatened and abused. She videotaped a message to the nation to be broadcast on television in the event that she was murdered. Fellow members of Parliament accused her of dividing the nation, of introducing racist politics, of damaging Australia's image abroad, especially in Asia. The Thai foreign minister, Dr Surin Pitswan, said Australia had been in the process of trying to convince Asia that it was a part of it, that it wanted to be integrated. The race debate sparked by Pauline Hanson 'brought back the impression that some countries might have had of Australia from previous experiences' and had triggered a period of suspense and anxiety.

But she also won support, often from surprising quarters. In the New South Wales town of Mudgee, several police officers donned Pauline Hanson T-shirts to show solidarity. Senator Bill O'Chee reported receiving a telephone call from a 78-year-old woman who said she would vote for One Nation because she had grown up in the days of the White Australia Policy and she believed that it was immoral that people who looked like O'Chee should be allowed into the country. O'Chee said that the conversation then went as follows:

O'Chee: What do I have to do to be an Australian, because my family has been in this country for a hundred and ten years.
Old Lady: It doesn't matter.
O'Chee: I've got to look English, have I?
Old Lady: Yes.
O'Chee: What about the Aborigines?
Old Lady: They're Australian, too.
O'Chee: Can I just get this down for the record – you can look Aboriginal and be an Australian, or you can look English and be an Australian, but you can't look Asian and be an Australian?
Old Lady: That's right.

No dinner party anywhere in Australia in the run-up to the election failed to discuss Pauline Hanson. She was the best-known woman in Australia. In London, Australians struggled to explain that although five former Prime Ministers (Whitlam, Fraser, Hawke, Keating and Gorton) had issued a statement condemning Hanson's race-based policies, the current Prime Minister, John Howard, had failed to do so. Writer Ashley Hay remembers: 'A friend flew in from Melbourne bemoaning the advent of Pauline Hanson and the deterioration of any sort of racial tolerance in Australia. She said Bill Clinton visited recently and announced that Australia was the only country in the world that had managed to get multiculturalism to work. It would be shattering to see everything that we've all worked so hard for just disintegrate.'

Political historian Paul Kelly called on Australians to face Hanson's challenge. He said she symbolized an alienation within part of the community caused by a conjunction of forces: globalization, economic restructuring and social changes. People needed scapegoats to explain their frustration. 'Australia has to live with this and handle it because Hanson-type policies affect the principles of our society, the nature of human relationships, and how the rest of the world relates to Australia . . . Hanson is a test of our future.'

Australia passed. When the votes were counted, Hanson and her party were wiped out. She lost her seat. Her chief strategist, David Oldfield, failed to gain a seat in the Senate from New South Wales and One Nation won only a single

Senate seat from Queensland. The electorate decided, when it came to the crunch, that Pauline Hanson and One Nation were not up to running the country. Hanson had had her say and those who had supported her felt better for it, but it was time to get back to the real problems – unemployment, health services, the breakdown of rural communities, violent crime and corruption.

It could even be argued that the rise and fall of Pauline Hanson had some positive influences. It drew attention to the role of the Australian media in distorting Hanson's importance and news value. Kelly wrote: 'It is inconceivable that any other nation would have responded with such obsession about an extreme right-wing populist occupying just one seat in the national Parliament.'

It reminded everyone what freedom of expression was all about – that however much one disagreed with what Hanson had to say, there could be no denying that she had a right to say it. It showed that it was unhealthy to suppress criticism of minorities – be they Asian or Aboriginal – through fear of being labelled racist or politically incorrect. But it also set the boundaries for that criticism, defined what was harsh but acceptable and what was racist and out of order. As one young Aboriginal woman, a journalist, put it to me, 'It is OK to say you resent your tax dollar going to Aboriginals who spend it all on grog. It's not OK to say that Aboriginals will never be able to do anything about their excessive drinking because they have smaller brains.'

In the middle of all the Hanson controversy, Neville Bonner, the first Aboriginal to be elected to Federal Parliament, died. An elder of the Jagera tribe, he had gone to Canberra in 1971 as a member of the Senate for the conservative Liberal Party. But once there he became too radical and in 1983 his party dumped him. Long before Mabo and Wik he put before Parliament a motion calling on it to recognize that Aboriginals were the prior owners of Australia. He tried to get the mainstream parties to adopt more Aboriginals as candidates, but they were slow to do so. Yet in the year of Bonner's death, Aden Ridgeway took his seat in the Senate, the second Aboriginal to make it into the Federal Parliament. He did it – and many savoured the moment – by beating a candidate from Pauline Hanson's party.

Hanson and One Nation struggled on, a spent force, and in August 1999, the Supreme Court of Queensland found that she had fraudulently registered One Nation as a political party when it did not comply with Queensland electoral law. The judgment said that Hanson and two other party officials had misrepresented a 500-strong Hanson support group as party members when they must have known that the party 'had no members except themselves' – an appropriate ending to a threat that had always been exaggerated.

As might be expected, the Hanson affair had encouraged some harsh comments about Asians and Aboriginals. Canberra journalist David Barnett, writing in the Australian *Financial Review*, said he wanted to hear no more about the Aboriginals being the natural guardians of the land. Australia used to be covered with huge forests containing numerous forest-dwelling animals – 'all gone thanks to these natural guardians'. Fossil remains showed that before Captain Cook ever set foot on Australian soil, sixty large animal species had already been hunted to extinction. 'Yet today we have given the Aboriginals sole guardianship over Australia's great national park, the Kakadu, along with full hunting rights.'

The result, Barnett said, was that almost all the dinner-providing animals – wallaroos, wallabies, bustards, and tortoises – had been eaten. While America's Yellowstone Park teemed with life, visitors to Kakadu had to make do with the rare sight of a crocodile or two. At the start of each dry season, the Aboriginals burned everything above the Kakadu basin, in pursuit of what politically-correct Australians fatuously referred to as 'firestick farming'. Barnett said that this further destroyed the fauna and flora, eroded the soil, and poisoned the air. 'Kakadu would be in better shape if it had a nuclear power station in its midst.'

But there was a lot of positive news as well. In Sydney in May 1998, with the Pauline Hanson controversy raging throughout the country, two thousand Australians packed St Mary's Cathedral to say goodbye to Shirley Perry, MBE, AM, an Aboriginal activist, better known as Mum Shirl. It was a great public occasion, honouring a great Aboriginal woman whose own dreadful start in life inspired her to become a house mother to the wretched and the poor of Redfern, an inner-city suburb with a large Aboriginal population.

Mum Shirl was born into a staunch Aboriginal resistance family at Erambie mission near Cowra in southern New South Wales. Kicked out of the mission because her father was 'difficult', the family had to live under a bridge. Her schooling was perfunctory because she was an epileptic and considered unlikely to be able to cope with regular classes, so she never learnt to read or write and in later life had to rely on her prodigious memory. She was abused by white children who chanted, 'Nigger, nigger, pull the trigger. Bang! Bang! Bang!'

In the late 1950s the family moved to Sydney. Her older brother was often in trouble with the police and Mum Shirl visited him in prison. This began a mission of prison visiting. She would seek out Aboriginal and white prisoners who had no other visitors. When prison officers chided her about keeping bad company, she retorted, 'If no one else wants to be crucified between two thieves, I do.'

She expanded her welfare work to include neglected and homeless children, alcoholics and down-and-outers. It was said that there was not a street in Redfern where Mum Shirl had not at some time rented a house and packed it with troubled kids. 'There are always children, children being born, living, dying . . . Every one of these events leaves its mark on you.' She always left a window open so that anyone with a problem could crawl through and find help. 'She was like a nagging magpie,' said one of her nephews, taken in by Mum Shirl when he was a baby, 'But she knitted us all together.' Every Sydney policeman and magistrate knew her. 'I can't give this boy bail,' said the Newtown magistrate, 'Nobody knows anything about him.' The duty solicitor said, 'Mum Shirl'll vouch for him.' Magistrate: 'OK, that'll do me.'

She never discriminated between black and white and feared no one in authority. She invaded a dinner of Catholic social workers to which no Aboriginal had been invited and demanded to speak. When they tried to stop her she ripped off the tablecloths, shattering crockery as bishops fled in terror. When Neil Diamond first came to Australia, Mum Shirl went to the tour organizers and asked for some free tickets for disabled children in homes. 'Sorry,' the organizers said. 'All right,' Mum Shirl said, 'I'll get the Redfern All Blacks to throw a demo outside the concert and then no one will get in.'

The organizers not only changed their minds, but invited Mum Shirl and the children backstage to meet Neil Diamond who then gave her $65,000 dollars to fund her work and invited her to appear on his show in the United States. There he introduced her: 'I've never had a mother, but when I was in Australia I found one and here she is – Mum Shirl from Redfern, Sydney, Australia.' Her growing fame never changed or fazed her. When the Queen presented her with her CBE at Government House in Sydney in 1975, the royal courtiers listened in amazement as Mum Shirl said, 'Well, I'm pleased to meet you, Your Majesty, and I appreciate this what-yer-may-call it you've given me. But I can't stay here and yarn all day. I've got to get out to Long Bay. There's a lot of prisoners out there waiting to see me.' And the Queen not only agreed that Mum Shirl had indeed other things to do, but arranged for royal protocol to be waived so that Shirl could leave the gathering before the Queen did.

She went to Randwick Racecourse to see Pope John Paul II when he visited Sydney in 1966, but did not realize that she needed a ticket to get into the enclosure near the Pope. When she was stopped by attendants, one of the children with her said, 'But this is Mum Shirl', and she was immediately allowed through. As she walked up the track in front of the main grandstand someone recognized her and called her name. In seconds hundreds of others were shouting it.

Mum Shirl called herself an MRC – 'a Mad Roman Catholic' – but the priest who helped her with her work, Father Ted Kennedy, said she was a miracle worker and the greatest theologian he had ever known, never properly appreciated by the Federal government, which he called 'a government without a soul'. He said sculptor Bill Clements had created a life-sized image of Mum Shirl which would be cast in bronze and placed in Sydney as 'a permanent reminder to this city of overdogs of what she meant to the underdogs'.

But the most moving and yet disturbing tribute came from Paul Coe, one of her protégés who had become a barrister. He said he could not understand how Mum Shirl had loved white Australians as well as Aboriginals. 'When I was a boy my heart was filled with hate for white people,' he told the mourners. 'I still have this hate. I will die with this hate. I loved this woman but I do not understand her.'

Perhaps the most profound sociological change in Australia in the century since Federation was the role of women in Australian life and their relationship with Australian men. The Australian approach to empowering women was to concentrate on what went on in the workplace. 'It is no good for the state to try to liberate women in the home because what goes on there is not the state's business,' Susan Walpole, the Sex Discrimination Commissioner with the Human Rights and Equal Opportunity Commission, told me in 1994. 'But in the workplace you can compel people to behave in a different way.'

So let us begin with what happened in one workplace in Port Kembla, an iron and steel town about one hundred kilometres south of Sydney. The principal employer there was BHP, one of Australia's largest companies. BHP paid high wages for the heavy, dirty work in the steel mills, and the lure of this money brought large numbers of migrants, mostly from the former Yugoslavia, to Port Kembla. The men worked on the mill floor while Yugoslav women found light jobs elsewhere in the works. In 1980, to everyone's amazement, some of these women applied for jobs as steelworkers.

BHP rejected them, pointing out that health and safety regulations prohibited the employment of women in jobs where they were required to lift weights greater than 14 kilograms. The women quietly gathered evidence that when the work required the lifting of heavy weights, the men themselves used mechanical devices. When the Sex Discrimination Act was passed in 1984 the women lodged a complaint against BHP and won. BHP were ordered to employ the women who had applied to fill vacancies for steelworkers.

But the battle was by no means over. In 1988, pleading a downturn in the economy, BHP reduced its workforce. By agreement with the unions, they applied the 'last ones hired, first ones sacked' rule that was common in Australian industrial relations. Since the Yugoslav women were the most recent steelworkers hired, they were the first ones to be sacked. The women brooded for a while and then, advised by a public interest law centre, they brought another case against BHP.

Now they argued that they were the victims of indirect discrimination, in essence that if they had not been denied jobs as steelworkers in the first place – because they were women – then they would not have been the last ones to be hired and, therefore, not the first ones to be sacked. BHP fought the case to the highest court in the land. They lost and the women were awarded $1 million in compensation for lost earnings.

The Port Kembla case became a landmark in the struggle by all women for equal opportunities and their ranks in the workforce steadily grew, so that by the end of the 1990s they had broken through to the top in increasing numbers. It was only in 1962 that Australia had its first woman Queen's Counsel, but by 1991, the Law Institute had elected the first female President in its 132-year history. Women sat on the boards of major corporations – and managed motherhood at the same time. A male director of a leading Australian bank told as an example of its liberal attitude how a woman director had breastfed her infant throughout a board meeting – and had accidentally squirted milk onto his tie. The 1990s saw women pilots in the RAAF, two women state Premiers, a woman as Governor of a state.

But there remained areas where their presence was not tolerated, workplaces that were considered suitable for men only. Veterinary students at Sydney University were required as part of their training to visit the killing floor of an abattoir. Dr Katrina Warren recalls: 'The moment one of the slaughtermen saw me he started this terrible moaning noise. It took me a moment to realize what it was. He was chanting, "woo-man, woo-man, woo-man". All the other men stopped work and took it up until the place was resounding with it. It was awful, very intimidating. I left as soon as I could.'

At the same time as women were achieving equality in the workplace, their relationship with men underwent revolutionary changes. To begin with, the historic trend, common in pioneering societies, of men outnumbering women, was finally reversed in Australia, and by the 1990s women outnumbered men (8.7 million women to 8.6 million men in the 1991 census) with the trend expected to continue. Simultaneously, Australian men, for reasons difficult to determine but probably because they were unable to relate to the new, liberated Australian woman, married Chinese, Thai and Filipina women in greater numbers than ever before – between 1981 and 1991 such marriages increased fourfold. More and more Australian women lived alone – up 20 per cent between 1986 and 1992, and by the end of the century they outnumbered men living alone. The length of marriages grew progressively shorter as divorce became easier after the Family Law Act of 1975 and in a ten-year period dropped from a median of 14 years to only 10 years. The combination of more women than men, an increase in the number of gay men, and shorter marriages meant that eligible men were in short supply at the end of the Millennium in Australia.

Visitors were quick to notice the change. 'Dateless and desperate,' wrote English journalist Kathy Gyngell, about her Australian cousins, 'how Australia's women are shattering the myths of the nation's macho males.' Gyngell wanted to know why so many dignified, smart Australian women who had rewarding careers were prepared to be interviewed about their love life in newspapers and magazines and even admit the last time they had sex. 'The fact is that they are, to a woman, single and most of them haven't had a decent relationship in years.'

Perhaps to a greater extent than in any other industrialized country, a role reversal began to take place in Australia. In education, for example, in every year since 1984, more women than men started university courses. By 1996, in the age group from 20 to 24, 13 per cent of women held university degrees compared with just 8 per cent of men. A high proportion of these university-educated women would never marry and of those who did, half would still not have had children by their early thirties. The age-old pattern of women marrying up in age and status was in the process of being reversed. Dr Bob Birrell, of the Centre for Population and Urban Research, said in 1998, 'These women are the wave of the future.'

To know how far Australian women have come since Federation you have only to see the number of power-dressed young women with their mobile telephones in the streets of Australia's main cities, or sit next to a group of them dining out and behaving badly after work on a Friday night. Behaving badly? Another sign of reversed roles was 'Ladies Only' functions at which professional male strippers perform. These are common all over Australia in wealthy, professional suburbs as well as working-class ones. Sometimes they become rowdy. Geoffrey Martin, a nineteen-year-old 'bottomless waiter', was serving drinks at a Brisbane Telecom 'ladies only' party wearing only underpants under a long shirt, when some of the women ripped his underpants off.

Police Inspector Kowle Dwyer of the licensing branch, who was at the party to check that the liquor laws were being observed, arrested Martin and charged him with wilful exposure. Martin pleaded not guilty. 'When the ladies get a couple of drinks into them they start getting stroppy,' he said. 'The other two waiters had their undies on but the ladies ripped mine off me. Normally you can get them back after a while but they have very funny ways of hiding them. If you see them passing something under the table, that's probably them. But they could be in a handbag or down someone's cleavage. Nothing really bad happens, it's just a lot of fun. If the sheilas can do it, why not blokes?'

It is Friday night at a singles bar in North Sydney. The clink of glasses and the buzz of voices compete with the irritating trill of mobile phones. The place is full of well-paid young men and women in their late twenties and early thirties, what one writer has described as the 'expensive, high-maintenance crowd'. The men are saying things that would make the old, traditional Australian male cringe with shame. 'It's all too hard,' says David Smith, a 27-year-old who has not had a relationship for three years. As far as he is concerned, the effort it takes to find out if he is compatible with a woman is not worth it.

His friend James Cunningham, who is 28, looks at groups of young women but says he is not interested in joining them. 'It's just too hard to go up and approach a girl. There are too many risks. You might offend or you might feel embarrassed.' Anyway, he is in a new relationship, with a colleague. Yes, the woman had made the first move – 'She got a friend of hers to send me an e-mail.'

'All changed,' says an older man who is just sightseeing. 'Everything is being shaken up. No one knows where they're at. The old Aussie ideal of a lifelong marriage has gone and no one's yet worked out what to replace it with.'

CHAPTER 23

THE COMING OF AGE

'Australia is the most successful high immigration society and the most successful, tolerant multicultural society.'

– Padraic P. McGuinness, political commentator

No one ever said that coming to terms with what white Australia did to Aboriginals would be easy. It will take years, but Australia has made a start. The changes compared with say, 25 years ago, are enormous and the will for reconcilation is there – among the people, if not yet the government. There was a national Sorry Day in May 1998 to give white Australians a chance to say sorry for having stolen Aboriginal children from their parents. Hundreds of thousands of Australians took part, including the former Governor-General, Sir Zelman Cowen, state Governors, former Prime Ministers, leading churchmen, writers and film-makers. 'Sorry Books' travelled around the country to enable more than one million white Australians to record their apologies.

There were other heartening moves. More mining and land development companies showed a willingness to negotiate with Aboriginal communities, not only over mining and development rights but access to culturally sensitive areas and on long-term benefits for their communities. 'These are hard-headed commercial decisions,' a West Australian resources executive told me in 1999. 'In the old days we'd go to law and waste years and millions of dollars on lawyers. Now we negotiate. In general we've found that what Aboriginals want is not just lump sums but something for the next generation. We've got a company cadetship scheme now for young Aboriginals, with a promise of a genuine job worth up to a eighty thousand a year when they graduate. That's the sort of deal that interests them.'

And Aboriginals who have been unable to reach agreement with mining companies, like the Mirrar people at Kakadu national park, where Energy Resources of Australia wants to open the Jabiluka uranium mine, have found widespread support for their struggle from white Australians, particularly young ones. 'You can no longer slide on by,' says Penny Hood of the National Union of Students. 'You have to be aware of what is going on around you.'

In the process of adjustment and reconciliation and support many Australians have learnt a lot about their country and the fellow Australians they share it with. Ethel Munn is an Aboriginal woman of about seventy who was born near Bollon, a small inland town in southern Queensland. A member of the Gungarri tribe, she spent most of her adult life living in a tin shack on the river bank. Not far away, on a 22,000 acre farm called North Yancho, lived Camilla Cowley, a farmer who grew up a privileged white Australian. She went to boarding school, trained as a teacher and in the 1990s was running sheep on North Yancho with her husband, Kerry.

One day in September 1996, Mrs Cowley opened a letter from a solicitor acting for the Gungarri tribe, advising her that the Gungarri were making a native title claim over 3,000 leases in the area, including North Yancho. Mrs Cowley explained in 1999 what her reaction had been: 'I had never heard of the Gungarri. And I had never imagined that any Aboriginals would want to claim our land. I consulted some neighbours. One said he was glad he hadn't surrendered all his guns. He'd need them to shoot the first Gungarris who came on to his place.'

Soon afterwards the local farmers called a protest meeting in the nearby town of Mitchell. Mrs Cowley said that the atmosphere was almost hysterical, and when the chairman called for a vote on a motion to extinguish native title, the Cowleys put up their hands along with everyone else. The vote was apparently unanimous. Then the chairman noticed two people at the back of the hall, the only Aboriginals who had dared to attend. They were Gordon Munn and his wife Ethel. Ethel stood up and said, 'We vote against. I've worked all my life and paid my taxes like you. The only difference between you lot and me is the colour of my skin.'

Mrs Cowley still cannot explain what decided her next move. She left the meeting and walked across the street to the local Aboriginal Land Office to see what she could learn about the Gungarri and their history. Ethel Munn saw her and followed her inside. The two women gingerly struck up a conversation and by 1998 had become close friends. Mrs Cowley recalls, 'Ethel told me that all the Gungarri wanted was recognition and acknowledgement, and the right to come back to their country to visit. They didn't want to take anything from us. When I realized what native title was all about, I could only applaud. I feel ashamed now over how ignorant I was over my own country's situation.'

Ethel Munn and her husband now own a 1,000 acre cattle farm near Roma – a small venture by Australian standards – but both women continued to work for better understanding of native title. Mrs Cowley organized meetings between Aboriginal leaders and farmers and Mrs Munn concentrated on bringing Aboriginal and white women together to talk about their hopes and problems. Both felt that this was something that women could do better than men. 'It's the men who've been standing up and shouting misinformation,' Mrs Munn said. 'The women aren't stupid. We've got a better chance of getting it right.' Both condemned a National Farmers' Federation television advertisement which showed a black child and a white child fighting over land.

Just before Christmas 1998, the Cowleys held a formal ceremony with the Gungarri at North Yancho at which both sides acknowledged sharing the land for their different purposes – the Cowleys for raising sheep, the Munns for traditional ceremonies. Ethel Munn said, 'For us it just means being able to go there if we want, so future generations can say, "That's my mother's and my grandmother's country and those people at North Yancho don't mind if we go there."'

Where does the push for reconciliation go now? Most Australians agreed that a preamble to the Australian constitution must contain some recognition that Aboriginals were in Australia before the white man. In 1999 the argument was whether they were occupants, custodians or owners of the land. There were many small gestures: the Aboriginal flag of black, yellow and red flew over Perth's Central Police Station – an event that could well have caused riots even ten years ago – but as the century drew to a close there was a feeling that everyone in Australia was waiting for something.

Germaine Greer, opening an exhibition of Aboriginal paintings in London, suggested what it might be – a belated treaty giving Aboriginals the right to their land. 'It is a symbolic gesture, but Aboriginal people understand symbols. Australia belongs to Aboriginals and people of European descent and others are there on sufferance.'

If not a treaty, then what is undoubtedly needed is a grandiloquent, magnanimous, expansive, sensational, symbolic gesture – one that will stand out from history's pages. Australians are an inventive and imaginative people. They will come up with something.

If anyone had told my parents that on 18 April 1997, the *Sydney Morning Herald* would report that Sydney's Muslim community had the previous day celebrated Eid-Ul-Adha, the feast of Abraham's sacrifice, in the mosque at Lakemba, they would have laughed in disbelief. A mosque in Lakemba! Thousands of 'Australian Muslims!' I can hear my mother asking, 'What's an Australian Muslim?' And even if all this did happen, why is it being reported in the *Herald*, the bible of old Anglo-Australia?

What would she, or anyone of her generation, have made of the official photograph for class 1B at the state school in Bellevue Hill, an inner eastern suburb? Three rows of smiling boys and girls, some blonde and blue-eyed, some dark and brown-eyed, all typical six-year-olds, healthy, vigorous Australian kids. But look at the names: George Ergemlidze, Nathan Geller, Liron Israeli, Meir Azulay, Brent Wolfson, Alex Ocias, Sasha Davyskib, Kenji Hashimoto, Joshua Hoilliard, Richie Likht, Daniel Naoumenko, Natasha Day, Shelley Knowles Britten, Marina Reznikov, Ellen Juan, Alex Beavis, Scott Mitchell, Bukhari Yusof, Natasha Tcistoganoff, Keshia Levine, Alisa Sannikova, Bredgette Whalan, Sarah Zizer, Morgan Batista-Bath, Fallon Zofrin, Jackie Shvarts and Shelly Cezana. Twenty-seven Sydney children and hardly an Anglo Australian among them.

Well mother, that's modern, multicultural Australia. It sort of snuck up on us, but now that it is here, most Australians seem to like it. Some commentators even say that since most societies are multicultural it was inevitable. Padraic P. McGuinness points out that the First Fleet was multicultural and so was Aboriginal society, that 'Australia is the most successful high immigration society and the most successful, tolerant multicultural society.'

Despite the paranoia of Pauline Hanson's One Nation Party and its fear of being 'swamped' by immigrants, throughout the 1990s the proportion of Australian residents born overseas increased only slightly and in June 1998 was 4.4 million, about 23 per cent of the population. The United Kingdom continued to be the largest birthplace group, making up 6 per cent of the total population. About 5 per cent of the Australian population were Asian-born, a figure that will rise to about 7½ per cent by 2041.

True, if you add Australian-born children of Asian-born parents to Asian-born residents, the combined percentage would be much higher. But the National Multicultural Advisory Council objects to this. 'It would label Australian-born children with their parents' origin, rather than the country of their own birth, which is Australia. This runs the risk of classifying Australians according to their appearance, which would be both offensive and dangerous.' The heartening news is that the fastest growing classification in Australia in 2000 is one where people describe themselves as being of 'ethnic mix'.

But what about all those ghettos? What about historian Geoffrey Blainey's picture in *All for Australia* of 'old Australians' living in 'frontline suburbs' confronting 'the new Chinese'? The facts simply do not bear out the accusations. Dr James Jupp, director of the Centre for Immigration and Multicultural Studies, at the Australian National University, Canberra, pointed out in 1995 that in all societies there is a tendency for immigrant and ethnic minority groups to concentrate in particular areas.

'These are often wrongly termed "ghettos", suggesting isolation from the majority, poor and even criminal characteristics and undesirability. However, there are few recent instances of such deprived concentrations in Australia, compared, for example, with the situation in the United States or the United Kingdom.'

Professor Nancy Viviani, who taught international relations at Griffith University, analysed the 1996 census statistics to examine two accusations – that Asian Australians form ghettos and fail to mix. On the basis of earlier immigration and settlement patterns, she established that a concentration of about 10 per cent of the local population of any area for any ethnic group was about the norm for Australian cities.

In Sydney and Melbourne only two groups were over this figure. One was the Vietnamese. The other was a surprise – those born in the United Kingdom and Ireland. In Sydney they formed just under 11 per cent of the population of Manly and in Melbourne just over 10 per cent of the population in areas around Mornington and Frankston. Technically, these suburbs in Australia's two major cities were *British* ghettos. Further, the British and the Irish are the only two European groups to form ghettos. Other Europeans are diversified throughout the suburbs of Australia's largest cities (except for Italians in some areas).

In Fairfield in Sydney's West, 13 per cent of the population were born in Vietnam. In Melbourne's Footscray, 13.7 per cent, and in Springvale 12.8 per cent. But other Asians – Chinese, Indians and Sri Lankans – are not heavily concentrated in any particular area.

Gerard Henderson, the executive director of the Sydney Institute, summed up the findings: 'Asian immigrants to Australia behave much like their European counterparts. From time to time population concentrations, in the vicinity of 10 per cent, may occur. But they stabilize and diversify. Ghettos remain a sad fact of life in some US cities. But they have never really existed in Australia, certainly not since the slum clearance programmes of before and after World War II. Viviani's research of the 1966 census data show that in Australia right now, the word ghetto is a political expression, not a term describing housing.'

In multicultural Australia the animosity between Protestants and Roman Catholics appeared to have finally ended, even though the 1990s saw the replacement of Anglicans by Catholics as the largest single denomination. The unwritten tradition in the New South Wales Police Service that, to avoid sectarian enmity, the position of Police Commissioner would alternate between a Masonic officer and a Catholic officer, also vanished virtually overnight.

* * *

Just because Australia is a successful multicultural society without the old Protestant-Catholic divide, this does not preclude one group from receiving a caning from the others from time to time. When France was exploding its nuclear bombs in the South Pacific in 1995, the French community in Australia had to bear the brunt of Australian anger. '*Pourquoi les français sont des connards?*' said a headline in the *Sydney Morning Herald*. The article went on, only partly tongue in cheek, to accuse the French of all sorts of shortcomings – cowardice ('When invaded by the Germans in World Wars I and II, they rolled belly-up and waited for the Australians to save them'), incompetence ('How many Renaults do you see on the road these days?'), cheating ('Alain Prost won grands prix only by shouldering Ayrton Senna off the road'), cultural wankery ('French culture has been on the slide since Victor Hugo's aptly named *Les Misérables*, which only became a successful musical when translated into English'), and selfishness ('Single-handedly, the French obstructed the Uruguay Round of GATT because their smelly farmers wanted to continue driving limousines and living in *châteaux*'). The newspaper was flooded with letters of complaint but most Australians loved it.

And there are always the English migrants, the Poms, to turn on when no one else is readily available. 'Pom bashing', as it is called, ranges from inoffensive throw-away remarks about Britain's poor showing in sports, to criticism of the English pox, class. Warning his readers not to crow too much about a series of Australian sporting victories over the English, columnist Frank Devine, wrote, 'The Pommy bastards may be up to something and just leading us on by losing all the time.'

Broadcaster Mike Carlton, who describes himself as an 'Anglophile republican Australian', remembers dinner parties in his twelfth-century Surrey village: 'Confronted by an Australian accent at dinner, the English cannot find a class to put it in. It is sadly apparent that you are not Charterhouse and All Souls, rather more happily obvious that you never done Grimethorpe Comprehensive and Wormwood Scrubs, guv'nor. But where? And what? Are we dealing with Merv Hughes or a Booker Prize novelist, they wonder, baring their interesting teeth in a thin and nervous smile. It is important to put your hosts at ease by nonchalantly wielding the correct cutlery for the oleaginous Sainsbury's salmon roulade and by saying "pudding" instead of "sweets".'

Art critic Giles Auty, an Englishman who has lived in Australia since 1995, sees both sides. 'Contemporary Australia is among the safest, easiest and potentially most prosperous lands in human history. It is also among the most adolescent, sentimental and solipsist. A major part of this condition is manifested in adolescent contempt for the erstwhile parent, Britain, which in the eyes of almost any fashionable Australian academic, historian or writer, never did anything right.'

Sometimes the British in Australia must begin to think that they cannot win. In 1997 a report from the New South Wales Anti-Discrimination Board revealed that of all the ethnic communities in the state, the Anglo-Celtic group made more official complaints about racial vilification than any other, including Asians and Aboriginals. The Board said complaints tended to increase during English cricket tours of Australia, when people objected to being called 'whingeing Pommy bastards'. The Australian press then accused the Poms of being whingeing Pommy bastards for complaining that Australians called them whingeing Pommy bastards.

Part of the trouble is that the British and the Australians are prey to stereotypes of each other that no longer exist. Carlton recalled 'carving up' a Tory MP in a radio interview in London in a manner unremarkable in the Australian media, only for the MP to shout down the telephone when the interview ended, 'Why don't you go back to the convicts and the rabbits?'

Observer columnist Simon Hoggart reported that he flew into Australia to learn that a Melbourne brewery had just marketed a beer called 'piss' and that its advertising slogan was 'Taking the piss'. So at first he thought that Australia had not changed at all, but then he realized that it was very different. 'The notorious cultural cringe is going fast. In five days, I've been asked only once if I liked Australia (the only acceptable answer was always an enthusiastic, unqualified "yes"). Now they know that they've got a terrific country, and they don't require your approval, thank you very much. Instead there's a polite but unmistakable sense that you are to be pitied for living in England, where it is always cold.'

Hoggart did not discuss modern Australians' lifestyle, probably because so much has been written about it. French and British chefs trip over each other to praise Australian cuisine. Leading British restaurant-owner and designer, Terence Conran, said he preferred Australian chefs to French or British and employed about fifty of them. Food writer Robert Carrier claimed that 'the whole world is going Australian and I have been a prophet of it'. Even an American chef, Jeremiah Tower, from San Francisco, said in 1997, 'There are twenty amazing chefs in Australia and twenty in the States. There are two hundred million people in America and eighteen million in Australia. Something's definitely going on in Australia.'

Australian winemakers went to France in 1994 to teach the French how to make cheap, easy-drinking wines. Other winemaking countries got to hear of what they were doing and soon more than a hundred Australian winemakers were at work in Spain, Portugal, Germany, Italy, Hungary, Czechoslovakia, Romania, Moldova, Brazil, Chile, Argentina and Uruguay. In 1999, Jacob's Creek became Britain's wine of the year, while at the top end of the market quality Australian wines began to worry the best European winemakers.

* * *

Even the English were prepared to admit that there might be something they could learn from Australia. In January 1999, a group of British MPs, the Commons Culture Committee, travelled all the way out to Australia to find out why Australians were so much better at sport than the British. At the time of writing, Australia (with a population of eighteen million, remember) are world cricket champions, world Rugby Union champions, world tennis champions, world netball champions, world women's hockey champions, and world men and women surfing champions. Further, on a *per capita* basis, Australia was by far the most outstanding performer at the 1996 Olympic Games. The MPs' trip to discover why all this was so gave the *Observer* a chance to air a double stereotype: 'It's because that wonderful weather keeps children outdoors all day doing sporty things. Our weather keeps us indoors. That's why their cultural achievements are on a par with our sporting triumphs – few and prized as a consequence. Perhaps their cultural committee will fly over here to investigate why the finest flower of Australian TV is *Neighbours*.'

Not true. There is an Australian TV series called *Halifax FP* which ran on Britain's Channel 5 in 1997. Halifax is a woman, a forensic pathologist, and in the final episode she is confronted by two Sydney policemen, one called Derrida, the other Foucault. Watching the episode, a rural Australian shouted from his lounge room to his wife, 'Hey, Ethyl, leave the dishes. You won't bloody believe it. They've got a pair of cops on TV named after the grand old man of French deconstructionism and the para-Marxist mastermind of *les évènements* of sixty-eight.'

All right, no one can vouch for the second part of the story. But there *was* a panel game on Australia's Channel 9 called 'Spot the Emblematic Construct', and at the University of Queensland, the Rupe School of Semiotics and Media Discourse examined subjects like 'The Hidden Signifiers of *Neighbours*' and the soap's 'Chaucerian Analogues'.

The *New York Times*, no less, reviewing *Homesickness*, by Murray Bail, in 1999 said he ranks as 'one of three indisputably world-class Australian novelists currently practising'. The *Times* names the others as Peter Carey, winner of Britain's premier award for fiction, the Booker Prize (in 1988 for *Oscar and Lucinda*), and Thomas Keneally, who also won the Booker (for *Schindler's Ark* in 1982). Three world-class novelists from a population of eighteen million would put Australia ahead of Britain on a per capita basis in the cultural stakes (novelist's division) if one were so crass as to make such comparisons.

And, to take another field altogether, the New England Conservatorium of Music, at a provincial university in Armidale, New South Wales, working with the internationally acclaimed pianist Roger Woodward, and the German piano makers, Steinway and Sons, have started the world's first degree course in piano technology – all because Steinway told Woodward that there was a worldwide dearth of good piano tuners. Cultural desert?

There are other big changes. The Countess of Harewood (formerly Patricia Tuckwell, violinist, and then Bambi Shmith, model) said in 1994 that she shared the nightmare of all 'ex-pats': that she would wake up one morning and find herself trapped in Australia forever – in the 1950s. But the past twenty years have seen a new cultural confidence in Australia and the terminal decline of Australian expatriatism to Britain, especially the kind of compulsory expatriatism as represented by Germaine Greer, Clive James, Barry Humphries and Robert Hughes.

This confirms the prediction of Melbourne historian Jim Davidson who wrote in 1971 that a new generation of Australians would soon come to take Australian culture as their primary point of reference and regard their native land as a fully autonomous society and their metropolis.

In *Duty Free*, a study of Australian women abroad, author Ros Pesman tells of her own daughter who lives and works in Paris but also commutes between Europe and Australia: '[For her] the problem of whether she is or will become an expatriate is . . . not an issue.' Pesman says that sort of debate is now confined to 'an ageing, if still voluble cohort, the leftovers of the 1950s and early 1960s.'

As a 'leftover' myself I wonder whether England and Australia are in the process of swapping places, that it is now Britain that is on the margin of all the interesting, exciting things that are happening. When *The Times* sporting writer Simon Barnes asked the former Australian Rugby coach, Alan Jones, why England played boring Rugby and Australia played exciting Rugby, Jones replied: 'Because we're an exciting nation and England is a boring one.' Barry Humphries was exaggerating but had a point when he said, 'I've suddenly discovered that England is really a province of Australia.' Even accents are swapping places – a letter in the *Spectator* suggested that Tony Blair had stopped talking 'posh' and had gone in for 'the Australian sounds of the post-*Neighbours* generation.'

In any case, the Britain that proved so attractive to Australians in the 1950s has long since gone, subjected, according to author Richard West, to a 'reverse takeover bid' by her former colonial subjects – 'There are more people working in Britain's curry houses, not to mention Chinese restaurants, than in its coal, steel and shipbuilding industries put together.'

There are still many ties between the two countries. You cannot ignore the influence of a common language, culture, legal and parliamentary systems and – sometimes underrated – sport. In *Hope and Glory: Britain, 1900–1990*, Peter Clarke tells of Joe Chamberlain's efforts with protectionist tariffs to bind the ties of Empire. But, Clarke says, 'In the end cricket proved to be a more lasting legacy for Commonwealth unity than anything Joe Chamberlain achieved.'

And, for want of a better description, when the going gets tough there are always the emotional ties of blood. Actor Stephen Fry told of finding himself after a long day dealing with Hollywood producers, alone, tired and depressed in the Polo Lounge of the Beverly Hills Hotel. Then he heard the sound of a familiar voice. 'A healing wave of homesickness swept over me like a moist mountain wind. I forgot America and its billion dollar entertainment industry and suddenly I knew I was English and could never be anything else.' The familiar voice was that of Rolf Harris. 'When you're alone in the Polo Lounge, the fluting tones of Australia's greatest son beckon you home like a lighthouse.'

Economically, Australia at the end of the twentieth century was a very different place from at the beginning. It produced different things and did different things. In 1900, wool was king. A hundred years later, Australia made about six times more money from tourism than from wool. So successful was its manufacturing export business that despite removing most tariff protection it had a trade surplus for most years of the last decade of the century.

It ranked sixteenth in the list of the world's top industrial nations. It survived the Asian economic crisis intact. On average its economy was growing faster than the United States or most of the economies of Western Europe. Its savings rate was higher than that of Britain, the US or Canada and only slightly lower than France or Germany. Its tax rate was lower than Europe and only a little higher than the US. Government debt was lower than the US, Japan and most of Western Europe and its low inflation rate was the best of the world's advanced economies. Leading market economist John Edwards adds, 'Compared with the US, Japan, France, Germany (or New Zealand), the place is actually quite well run, with a high level of consensus and reasonably good decision-making. The hardest story to tell is of Australia's successes.'

Success was hard to quantify, however, in the century-old battle between bosses and workers. There was a series of bitter fights in the mining industry as new owners managed to win 'the right to manage', get rid of unions and impose individual contracts on employees. In 1989, with the support of a Labor government, the unions' traditional ally, all the unionized pilots in the domestic airlines were forced to resign in an effort to bring down the high costs of internal flights. In 1998, the stevedoring industry attempted to crush the Maritime Union of Australia, the waterside workers' union, by suddenly sacking all union members and replacing them with secretly trained non-union labour.

But many Australians were appalled at televised scenes of violent clashes between workers, 'scabs', police and company security guards. It recalled an era they thought had passed. The dispute got bogged down in the courts and ended in stalemate, but few doubted that the employers would probably try again. But, in general, the old days of strike after strike as demonstrations of union power – the Meat Industry Employees Union once called the cattle dogs out on strike at the Williamstown Abattoirs – had gone. Bosses now found they were dealing with a new sort of union leader – better educated, legally literate, and with a strong sense of union history.

Peter Harris, an elected organizer with the Construction, Forestry, Mining and Energy Union, said in 1998: 'I look at all those crooks in the *Business Review Weekly* rich list and I look at ordinary working people and I have no doubt who contributed more to this country. But if the workers can't get their act together, if they can't organize effectively, then there's nothing to prevent a wage decline like we've seen in the United States in recent times. Then, just like the US, we'll end up with a working poor and all the social problems it brings with it.'

But as Fred Argy, a leading Australian economist and one of the architects of financial deregulation in Australia, said in 1998, that whereas inequalities widened dramatically in Britain and the United States, this had not happened to anything like the same extent in Australia. 'There is something very ugly about a society that gets meaner as it gets richer. Elements of such a society have long been evident in the United States and began to appear in countries such as Britain and New Zealand during the 1980s and 1990s. It is a society in which many Australians would not want to live.'

Australians fought to protect their welfare state and as a result, as a proportion of output, government spending on welfare is little different from a decade ago. In the late 1990s, Australia still measured up to Ryszard Kapuściński's definition of a civilized society – it took care of its young, old, sick and poor, no matter what their means.

True, the outback suffered. As the rural population shrank, farms became vacant, services dwindled and cost-cutting governments paid less and less attention, life in the bush got harder. While Sydney was never more buoyant and prosperous, some workers in outback Australia were earning as little as $6 (£2.50) an hour and 40 per cent of outback children in New South Wales were living in dole-dependent families.

This rural collapse marked the last days of the legend of the bushman as the Australian hero. The tough, self-reliant, straight-talking Aussie from the outback had given way to the urban entrepreneur as a symbol of country. Martin Woollacott, a foreign correspondent for the *Guardian* and an experienced observer of Australia, wrote, 'Australia has long been a society more urban than rural, but it left its heart in that hard back country, far from the ports which became its capital cities. Those who have championed the two great shifts in modern Australian society – its multicultural reorientation and its economic deregulation, may now feel some guilt, and some fear, at having left this older, saddened Australia behind.'

Donald Horne agreed. 'Sadly . . . the bushman-Digger myth is now mainly in the hands of the souvenir and entertainment industries . . . In the mood of the times, the bush has become a statistical category called the rural sector.' Australian writer Rob Darroch put it more strongly: 'There has been the conscious, deliberate ethnic cleansing of that distinctive character, the Australian. A Chips Rafferty is a comic turn nowdays. Our accent is disappearing, our bush culture is being suppressed and our history denigrated and ignored. There is now almost no such thing as "an Australian" except in the older generation.'

If things were relatively so good in Australia in the late 1990s, why were so many Australians depressed? One in five were suffering clinical depression, one of the highest rates in the world. The country's youth suicide rate was high, but this was a worldwide problem – in the United States nearly half the country's adolescents were at 'high or moderate risk of seriously damaging their life chances'. In parts of rural Victoria, the suicide rate had risen 34-fold in 30 years. One Australian psychiatrist quoted in the *Sunday Times* said, 'We're possibly a bit spoilt out here; we've had it too good for too long.' Vitaly Vitaliev, a Russian writer, migrated to Australia in the early 1990s but lasted only two years. He went back to Europe complaining that life in Australia was too good and and he was suffering from 'spiritual heartburn'. As his fellow Russian, Andrei Sinyavsky, observed, 'If they are to have a soul, things must be ancient.' Or, perhaps, the soul must have its seasons – even the birds know that.

Others said it was fear of unemployment. Economic rationalism had finished off the old Australian idea of a career and a job for life if you wanted it; people were bewildered by the fact that a century that had been the most expansionist and technologically innovative period in world history had ended by giving people the same sense of foreboding, the same sense of job insecurity that they had at the end of the nineteenth century. In many country areas there were no jobs at all, especially for young people.

Still others said it was guilt because Australia had mortgaged and sold off the heritage of its children and grandchildren – of all the major industrialized countries, Australia had the highest level of foreign investment. People became depressed when they realized that so much of Australia's mineral resources belonged to foreign companies and so many Australian commercial and industrial icons were no longer Australian, among them Vegemite (US), Arnott's Biscuits (US), Aeroplane Jelly (US), Bundaberg Rum (UK), Billy Tea

(UK), Minties (Switzerland), Peter's Ice Cream (Switzerland), XXXX beer (New Zealand), Jacobs Creek wine (France) and Driza-bone clothing (UK).

Or perhaps they were confused about whether they wanted to take the final step in the long goodbye from Britain and become a republic. On 6 November 1999, after a long and acrimonious campaign, Australians were asked to vote yes or no to the following proposal: 'To alter the Constitution to establish the Commonwealth of Australia as a republic with the Queen and the Governor-General being replaced by a President appointed by a two-thirds majority of the members of the Commonwealth Parliament.'

When the votes were counted, 55 per cent had voted no to such a change. This was interpreted in the rest of the world as 'Aussies vote for the Queen'. They had not. An overwhelming majority wanted a republic but not one with a President chosen by Parliament. They wanted a say themselves in who their President would be, so they voted against the republic being offered to them. Prime Minister John Howard, a staunch monarchist, said that the issue was now dead, but Labor leader Kim Beazley pledged that when Labor was next returned to office it would hold another referendum, where Australians would be asked to vote for a republic with a President elected directly by the people.

The referendum result exposed not only a flawed campaign by the republicans, but the anti-authoritarian streak in all Australians that makes it very difficult to achieve a yes vote in any referendum. The monarchists succeeded in portraying the republicans as Chardonnay-drinking Catholics living in posh seaside suburbs of Sydney who were not interested in the views of beer-drinking members of the Returned Servicemen's clubs in the working class Western suburbs. The power of the anti-authoritarian sentiment was typified by an Australian who wrote to the *Guardian* in London to explain why he had voted no. 'I am an Australian republican, and have been for some 40 years. While I did not vote for the republic this time, it was not because the model was flawed, it was simply because I cannot trust those who pushed for a republic. [They] pushed too hard and when they push I push back.' So a longtime republican refused to vote for a republic because the republicans tried too hard to get him to do so. How to cope with this very Australian attitude, and an equally negative one – 'If it ain't broke, don't fix it' – will tax the republicans next time round.

As for the argument that Australians would have difficulty in agreeing on someone sufficiently worthy to be a popularly-elected president – 'We'd end up with a sports star or a radio announcer' – this is easily refuted. On the first day of the last year of the century Eric Rudd, one of Australia's most remarkable explorers, an old-fashioned Australian, died in Adelaide. Rudd was convinced, despite almost universal scepticism, that Australia was rich in oil and mineral resources and that the discovery of these was vital to the country's future. In the 1930s, as a field geologist with the exploration company Oil Search, he would spend eight or nine months traipsing around Western Australia without a break, mapping as he went.

'A tent fly was all we had for cover each night. We were living mostly on kangaroo and bush turkeys. I can remember one time the cook struggled with the slaughter of an emu. Eventually he managed to kill it and cooked the legs. They were like bundles of barbed wire around plastic.'

As early as 1951, long before any commercial strikes had been made, Rudd predicted the three prime areas for discovery of oil and gas would be north-west Western Australia, Gippsland and the Roma area of Queensland. But he could not interest anyone, even the Americans, who felt that the blocks of exploration land Australia was offering could not have any value. 'There were two Venezuelan colonels hawking blocks of a few acres in South America at that time for a million dollars an acre,' Rudd recalled. 'It's ironic that the US companies took far more notice of them than of me.'

He continued to press for Australia's leaders and the public to get behind the search and development of Australia's resources and he lived long enough to see this happen. He could have furthered his career by moving to the United States – he was Hearst Professor at the University of California at Berkeley, where he lectured on the mineral potential of Australia. But he remained an Australian nationalist, and was still searching for oil and gas in the Frome Embayment area of South Australia in the last years of his life, the sort of man of vision who would have made a fine President of an Australian republic.

You could argue that Australians were worried, if not depressed, at the beginning of this new century because they were alone in the world, a fact brought brutally home to them by the civil war in East Timor, the island just 300 miles to the north of Australia. Geographically part of the Indonesian archipelago, this old Portuguese colony was occupied by the Indonesian army in 1975 and ruled by force for the next 24 years. In a United Nations-supervised referendum in 1999, a majority of East Timorese voted for independence from Indonesia. Fighting immediately broke out between the pro-Indonesian militias, supported by the Indonesian army, and pro-independence groups. A United Nations force, headed by 4,500 Australian troops, had to go in to restore order and prevent further massacres.

There was always the possibility that the East Timor situation might blow up. But Australia had remained confident that it could rely on Indonesia to maintain order. After all, Australia had carefully cultivated its giant northern neighbour and had gone to great lengths to win over its senior Indonesian army officers, inviting them to take part in joint operations, training some of them and getting to know and trust them.

When it became clear that far from maintaining order, Indonesia was encouraging the excesses of the militia in East Timor, Australia turned to the United States, calling on the 'special relationship' it believed existed between the two countries. After all, Australia had sent troops to Korea and Vietnam and backed American policy in the Persian Gulf, so surely the United States would reciprocate by pledging American soldiers to patrol East Timor? Washington said no.

It appeared that Australia had forgotten the Nixon Doctrine (or the Guam Doctrine), set out by President Nixon in an impromptu news conference on Guam during his 1969 trip to welcome home the Apollo 11 astronauts from the moon. In essence, the doctrine said that the United States would no longer send troops to fight for Asian nations, but would confine its help to providing sophisticated American weapons. 'Asian hands must shape the Asian future,' Nixon said. The idea is to avoid ground fighting altogether or to get someone else to do it – anything to prevent American casualties.

'Australia has discovered the unhappy truth,' wrote Doug Bandow in the *Japan Times*. 'The world is full of countries claiming to have special ties to America . . . Washington rightly believes that East Timor is largely Canberra's problem, which demonstrates that the alliance with America is much less than many Australians apparently believed . . . Washington's unwillingness to jump into the Indonesian imbroglio reflects a long overdue sense of realism . . . It's time Canberra and other US allies began assuming responsibility for their own security.'

Since the United States navy, 'The Great White Fleet', visited Australia in 1908 (drawing bigger and more enthusiastic crowds than the Federation ceremonies), since Curtin's historic turn to the United States in the Second World War, since Harold Holt's 'all the way with LBJ' in the Vietnam War, since the Australian government's agreement to allow American bases on Australian soil during the Cold War, since the ANZUS treaty of 1951 (which Canberra thought bound America to support Australia unconditionally, but apparently does not), and despite the Nixon Doctrine of 1969, Australians had thought that, come the crunch, the United States would be there for them. East Timor marked Australia's coming of age. From then on, whether an Asian nation or a Western nation in Asia, as far as her defence was concerned, Australia was a nation on her own.

Did a different character develop in the Great South Land? At the end of Australia's first hundred years as a nation, is there an Australian distinctly different from his or her forebears, not an Aussiefied Pom, a second-hand

European, or an Asian with an Aussie accent, but a new breed, *Homo Australiensis*?

The answer is yes, but defining what it means to be an Australian is not easy. Ask Australians themselves and you get a variety of platitudes. Migrants have thought more about it than native Australians, but both agree that mateship figures as an important element, so important in fact that the Prime Minister, John Howard, worked it into his draft for the preamble to the Australian Constitution – 'We value excellence as well as fairness, independence as dearly as mateship.' Yet mateship is indefinable, but a topic that is guaranteed to break up any Australian dinner party. The simple question put to men and women around the table: 'Who's your best mate?' results in most women naming their husband and most men naming a male friend (see Chapter 2).

A deeply ingrained sense of fairness, 'a fair go . . . a fair crack of the whip . . . a fair suck of the sav(eloy)' is certainly part of the Australian character. Jesus Arroyo, a Spanish doctor who lived in Switzerland, came to Australia on a two-year study posting and put his children into Australian schools. Soon after his return to Switzerland he was summoned by the headmistress and told his ten-year-old daughter was being suspended for 'disruptive behaviour' and 'a wrong attitude she has obviously picked up in Australia'. It turned out that the teacher of his daughter's class had been unjustly picking on another pupil over a long period. Finally Arroyo's daughter stood up in class and said, 'That's not fair.' The headmistress pointed out to Dr Arroyo that the authority of the teacher and discipline in the school was at stake in this case, and asked him what sort of schools did they have in Australia, where pupils were apparently taught that fairness was more important than authority?

At the Stockholm Olympics in 1912, Cecil Healy, who was to die on the Somme a few years later, qualified for the finals of the 100 metres freestyle swimming event. Then he learnt that all the American swimmers, including the great Hawaiian swimmer, Duke Kahanamoku, the favourite to win the event, had failed to turn up because of a managerial misunderstanding about the starting time. Healy went to the international jury and said the semi-finals would have to be swum again, otherwise it would not be fair. The jury granted his request, and Kahanamoku went on to win the title, pushing Healy into second place.

A love of adventure and a willingness to eschew the safe way in favour of the risky is an Australian trait they do not advertise. Robert Haupt, an Australian journalist who lived for many years in Moscow, wrote *Last Boat to Astrakhan* about some of his travels in the old Soviet Union. There is a photograph in it captioned 'Robert Haupt (in red) and photographer Jack Picone in eastern Russia'. Haupt is wrestling with the handlebars of a giant blue motorcycle and sidecar. Picone is slumped in the sidecar under several blankets. The motorcycle and sidecar are leaning periously to one side as it ploughs through what appears to be thick, black sand. There is not a road, a house, a tree, a shrub, a person or, indeed any sign of life anywhere in sight. Where are they? Where have they come from? Where are they going?

This sense of adventure makes Australians good people to have with you in a tight corner. In May 1999, the *Sun Vista*, a luxury cruise ship, caught fire and sank off the coast of Malaysia. Passenger Thomas Bonnard, from Cleveland in Britain, recalled after his rescue, 'All the lights went out and the next thing we knew we were being herded into lifeboats. The crew were more panicky than the passengers. Nobody seemed to want to take charge. Luckily, there was an Australian in our lifeboat who seemed to know what he was doing. He got us lashed to two other boats and we could stand off the ship. The last we saw of her was a patch of hull just above the waterline that was glowing from the heat.'

What about that streak of the larrikin that is said to be part of the Australian character? The Canadian newspaper proprietor, Conrad Black, owned the *Sydney Morning Herald* for a few years in the 1990s, but sold it after 'an uphill battle to build a permanent relationship with Australia'. Black's parting shots at Australian leaders included a summary of Paul Keating as being 'the king of all larrikins'.

But the Australian larrikin's bad reputation is historically unfounded. Larrikins were members of gangs or 'pushes' among the Sydney and Melbourne working classes in the late nineteenth century, but their danger was exaggerated, and Mark Twain, visiting Australia at that time, said they were better behaved than their English equivalents. The lyrical poet, C. J. Dennis, made them famous and their mixture of irreverence, disdain for authority, humour, sentimentality and charm came to represent part of the national identity.

Visitors to today's Australia who arrive expecting to find bad behaviour of the legendary larrikin type – at least in sporting teams – often come away surprised. Patrick Collins, the *Mail on Sunday* sports columnist, wrote in 1998: 'This week I travelled across this improbably vast continent with the Australian cricket team. They wore ties which they neither loosened nor removed during the five-hour flight. They mingled affably with their fellow passengers, signing autographs with a smile and answering questions with courtesy . . . Nobody climbed into the overhead lockers, nobody groped a stewardess, nobody sang or swore, nobody even raised his voice. They weren't remotely miserable and, being Australians, they certainly weren't conformist. In their civilized maturity, they were the most impressive travelling sports team I have ever seen.'

That was because Australia expected its sporting teams to be proud ambassadors for a new, mature Australia. British sports writers noticed this – and its significance – before their Australian colleagues. Frank Keating of the *Guardian* wrote as early as 1997, 'Australians, as part of their collective nature emerge from the womb fully bonded with street, village, town and nation. They do not need lessons at it. Their cricketers kiss their cap badges when they score centuries. Fair enough but, if a well-adjusted Englishman kept doing that, we would see him as a bit of a berk.'

We will return to Frank Keating's reference to Australians' 'collective nature' and love of country, but to show that it is not only in cricket that it exists, let us look at the Bledisloe Cup Rugby match between Australia and New Zealand in August 1999. A world-record crowd of 107,000 people watched Australia beat the mighty New Zealand All Blacks 28–7 in Sydney's new Olympic Stadium. Rugby writers were lyrical about the match – 'Can it get any better than this? . . . This is what Rugby supporters want to see on the other side of the Pearly Gates.' They agreed that what spurred the Australians on to an unexpected victory was the sound of more than 100,000 fellow Australians singing *Waltzing Matilda* in full-throated unison. 'After that, we were out-passioned,' said one of the New Zealand players.

A few days after Tony Blair had become Britain's new Prime Minister in 1997, Martin Kettle, a leading British political commentator, received a call from an Australian friend of Blair's, offering advice on what made Blair tick. 'You've got to realize that Tony's not British at all,' he said. 'He's Australian.'

Kettle thought about this, running a few Australian stereotypes through his mind – Blair the boozy ocker, Blair the crafty larrikin. None of them seemed to fit. Then he realized what his Australian friend had meant. 'Blair is not in awe of the past. He is not intimidated by class. He is a meritocrat, a doer and a practical, problem-solving politician. He is not particular about where he gets his ideas from. He is simply happy making his own history. He is not inhibited by history or deference from changing what needs to be changed.

'These are qualities with which the Blair generation of public figures and politicians feel comfortable. Blair, who has spent formative time in Australia, is at ease with them too. If this makes him an Australian, then it sounds like a pretty good compliment.'

The references to class are enlightening. There are – and always have been – classes in Australia, although not the infinite divisions that exist in Britain. George Orwell is said to have pondered long and hard over whether he was upper-lower middle class or lower-upper middle class. Most Australians, if forced to do so, would place themselves in one of three classes – upper, middle or working class. But again, unlike Britain, these classes are defined not so much by what school you went to, what accent you have, or what work you do, but by how much you earn. The important point is that Australians seldom think about class and when they do, they are not intimidated by it. They take people as they find them.

Simon Hoggart of the *Guardian* summed it up: 'Unlike us, Australians don't make an instant assessment of everyone they meet, based on speech, dress, accent and general appearance. It takes us a few seconds and we do it instinctively. By contrast, most Aussies like you until you prove them wrong, in which case they will be greatly displeased. ("You know what you can do with the rough end of a pineapple.") The British, having been more suspicious from the start, are less likely to be disappointed.'

On 15 September 2000, the opening night of the 27th Olympic Games, the new, modern, confident Australia had an opportunity to show the world what it could do. The omens had not been good. In London the *Guardian* newspaper noted that preparations for the Games had been plagued with disaster and 'Sydney looks like being an Olympic horror story'. Australians responded to concern by saying, 'No worries'. But the *New York Times* concluded that the use of this expression showed that 'Australians are incredibly insecure'. This was why Australia, about to honour its founders, 'can't remember their names' and 'know more about George Washington than Edmund Barton'.

Australians were indeed nervous. Sydney weather in September is uncertain – it might well rain throughout the Games. The programme for the opening ceremony, which would set the whole tone, was rumoured to be too ambitious and accident-prone. What role, if any, would Aboriginals play? Would some of them disrupt the ceremony? Would Sydney's public transport be able to cope? For one hundred days the Olympic torch had been carried through one thousand Australian towns but who would actually light the Olympic flame remained a secret.

The night before the opening ceremony one million Australians said, 'Stuff it', gave up worrying, and took to the streets of Sydney for an all night party. Most clustered around the Harbour foreshores where they could see the five huge Olympic rings on the Harbour Bridge and shells of the Opera House floodlit in different colours.

The weather the next night was perfect, Australian Spring at its best. More than 110,000 spectators packed the Olympic Stadium while four billion people around the world watched on television. The president of the International Olympic Committee, Juan Antonio Samaranch, welcomed everyone to Sydney, 'the most beautiful city in the world', and the first of 12,000 performers began a pageant designed by the director of ceremonies Ric Birch and the artistic director David Atkins to celebrate Australia's history and peoples.

The question about Aboriginal participation was quickly answered. Nine hundred of them staged a corroboree, *Awakening*, with dreamtime spirits represented by Djakapurra Munyarryn and hundreds of clan members. This most haunting part of the ceremony was interspersed with thirteen-year-old Nikki Webster floating high over the arena and then wandering among the Aboriginals hand in hand with an elder. The Aboriginal actor and entertainer Ernie Dingo, who was one of the commentators, grasped the significance and expressed it in his own inimitable way as symbolising reconciliation: 'So we can all be as one mob – the ancient culture and the youth of today.'

The finale was a 2,000 piece marching band which introduced the 12,000 athletes from 200 countries, the largest representation of any Olympic Games. North and South Korea marched together for the first time in nearly a century and by special dispensation there were individual representatives from war-torn East Timor.

Herb Elliott, the 1,500m gold medalist at Rome carried the torch into the stadium and then handed it over to six women to honour one hundred years of women's participation in the Olympics – Betty Cuthbert (whose wheelchair was pushed by Raelene Boyle), Dawn Fraser, Shirley Strickland-de la Hunty, Shane Gould and Debby Flintoff-King. Then the Aboriginal runner Cathy Freeman lit the cauldron to a background of flame and water and a sky exploding with fireworks that could be seen from miles away.

After such an inspiring beginning, fate would have been cruel indeed if the rest of the Games had not gone like clockwork. The weather was wonderful, the harbour sparkled, the transport worked and thousands of smiling volunteers ran around everywhere asking bemused visitors how they could help them. Australians suddenly realised that they were staging the best Olympics ever, what the staid London weekly, the *Spectator*, described as 'the greatest seventeen days of sport in the history of the planet'.

But it was more than that. Australians re-affirmed their communal solidarity, their warmth for their fellow citizens, their sense of fun, of shared joy. My sister told of waiting at Central Station for the first train to Parramatta early in the morning after the fireworks display that marked the end of the Games. Someone began to sing 'Show Me The Way to Go Home, I'm Tired and I Want to Go to Bed'. The whole platform joined in, then people waiting on other platforms for other trains, until the entire railway station was singing.

And the rest of the world looked beyond the Games and saw the new Australia. Simon Barnes of *The Times* wrote: 'My salute to the Games is also a salute to Australia . . . this most aspirational of cultures, seeking to discover itself so eagerly without the baggage of history, and to the envy of Australia that invariably stirs in the breast of the visiting English.

'We have too much history, the nightmare from which we cannot awake. Australia has sport instead of history, optimism instead of tradition . . . It has been going through the standard adolescent phases: rejection of parent country (us), too fervent admiration of the hero country (US) and now, with the triumph of the Games, the nation is emerging as its own person, filled with a dashing enthusiasm, and yet with its sense of humour intact.'

The party spirit continued almost without interruption into the new Millennium and on 1 January 2001 Australia marked the centenary of its federation. To be a nation its citizens must arrive at their own sense of identity. As all over the country Australians celebrated, there was no doubt that they knew who they were. Out of that unpromising contrast of jailers and their prisoners, augmented later by waves of migrants seeking a new life, gold rush adventurers, war-shattered Europeans, and then – in a makeover the speed of which surprised the world – new settlers from all over Asia, has grown a multicultural society that calls itself Australian.

From this long and educational journey, writing about the land in which I was born, if I were asked to name one quality about Australians that stood out, I would nominate their sense of collectivism. Although it has faltered at times, the predominant Australian characteristic is their feeling that whatever it may be, come the crunch, they are in it together, one for all and all for one. Since they excel at sport – 'Australians rule the world', said the *Guardian*; 'Australia is the greatest [sporting] nation on earth,' said the *Spectator* – I turn to cricket to illustrate this point.

When Mark Taylor played his 50th Test match as captain of his country, Stephen Fay, a writer more English than whom it would be hard to imagine, wrote a long newspaper profile of Taylor. Most of it was about cricket, one of Fay's passions, and when he had exhausted this topic he switched off his tape recorder and they both relaxed. Then suddenly Taylor began to muse about the Australian character and Fay switched on his recorder again.

'Competition is a part of life, and it's about losing, too. I'm trying to teach my six-year-old this. He loves winning, but the bottom line is that there are always times when you'll have a bad run and you'll need to call on your inner strength and become a better person. That's what I like so much about team sport.

'In England last year and before that I hadn't played any worse throughout my career, and yet I was still a winner. I was a loser personally, but the team was winning. What better lesson can you get for living than that? That although you're not doing well yourself, if you can just hang in and play your part, you can still be a winner.'

And when he had reported this Australian philosophy of life, the Englishman Stephen Fay added his own comment: 'The Aussies are lucky to have Mark Taylor.

'But then, they made him.'

'Just the cicadas
spinning their metallic yarns
into the empty street
the air full of the scent
of frangipani.'

– Antigone Kefala

Acknowledgements

I could fill a small book with the names of those who have helped me learn about my own country. It was a Polish writer, Ryszard Kapuściński who first planted the idea, and then Rachel Calder, of the Tessa Sayle agency, and Dan Franklin and Tom Maschler of Random House who made it possible to get on with it. Dozens of friends and hundreds of Australians – some acquaintances, many strangers – then made their contributions. All I had to do was ask the key question – 'What is an Australian?' – to spark off prolonged and often heated debate. In particular I am grateful to:

Neil McDonald (a tireless researcher and critic), Gil Appleton, Maurice Wilmott, Andrew Clark, Robert Darroch and Sandra Jobson, Max Suich, Jan Mayman, John Shaw, Judith Mitchell, Bill Broes, Robert Treborlang, Tony Healey, Murray Sayle (for his expertise on General Monash), Caroline Gathorne-Hardy, Richard Victor Hall (for his all-encompassing general knowledge of Australia and Australians), Paddy Pearl, Tony Griffiths, Gordon and Mary Ellen Barton, Darcy Farrell and the government of Western Australia which showed me the richness of the state's resources, Mark Lopez (multiculturalism), Robert Milliken, Pera Wells, Margaret Simpson and all the members of Cheltenham women's tennis club, Sydney (for delving into their memories of times past), Kevin Walsh of Apple (for rescuing me at a crucial time), Val and Jeffrey Meyers, Tony Delano, and all the people who gave me interviews (mentioned by name in the book). I hope that they enjoy the reading as much as I enjoyed the writing. Sadly, Moi Moi Treborlang, who appears in Chapter One, died from cancer in 2001.

The photograph by Bill Bachman on the dustjacket, a brilliant design by Suzanne Dean, is of West Australian semi-precious stones found at Meekatharra, an old goldmining town 1000 km north-east of Perth.

I am grateful to Antigone Kefala for permission to use her poem about cicadas. This is from her collection *Sydney Afternoon Poems*, part two. Laurie Duggan kindly gave permission to use his poem quoted on page 43. The hit song 'I'm Forever Blowing Bubbles' is by Jaan Kenbrovin and John William Kellette and was originally published by B. Feldman and Co. Copyright is now divided among several companies in different countries. It proved impossible to locate them all in time for publication but the author will be happy to acknowledge them in later editions.

SELECT BIBLIOGRAPHY

Aitken, Jonathan. *Land of Fortune*, Secker & Warburg (London) 1971.

Alcorta, F. X. *Darwin Rebellion*, N.T. University (Darwin) 1984.

Alomes, Stephen and Provis, Michael. *French Worlds Pacific Worlds*, Two Rivers Press (Melbourne) 1998.

Australian Institute of Political Science. *Australia and the Migrant*, Angus & Robertson (Sydney) 1953.

Baker, A. J. *Social Pluralism*, National Library of Australia (Canberra) 1997.

Ball, Desmond and David Horner. *Breaking the Codes*, Allen & Unwin (Sydney) 1998.

Bardon, Geoffrey. *Papunya Tula*, McPhee Gribble (Victoria) 1991.

Bates, Daisy. *The Passing of the Aborigines*, Murray (London) 1938.

Batstone, Kay. *Our Century*, Viking (Sydney) 1999.

Baume, Michael. *The Sydney Opera House Affair*, Nelson (Melbourne) 1967.

Beams, Adler, Grey, Moore and Harris. *The Canberra Coup*, Worker's News (Sydney) 1976.

Birch, Charles. *Confronting the Future*, Penguin (Victoria) 1975.

Bird, Carmel (ed.) *The Stolen Children*, Random House (Sydney) 1998.

Blainey, Geoffrey. *The Tyranny of Distance*, Sun Books (Melbourne) 1971.

Braddon, Russell. *Images of Australia*, Collins (Sydney) 1998.

Broes, Bill. *That Was Happiness*, privately published.

Carey, Gabrielle. *Australian Story*, ABC Books (Sydney) 1997.

Cheers, Gordon (ed.) *Australia Through Time*, Random House (Sydney) 1997.

Clark, Manning. *A Discovery of Australia*, Ambassador Press (Sydney) 1976.

Manning Clark's History of Australia, Ab. by M. Cathcart, Chatto & Windus (Melbourne) 1993.

Clarke, F. G. *Will-o-the-Wisp* Oxford (Melbourne) 1983.

Cochrane, Peter. *Simpson and the Donkey*, Melbourne University Press (Melbourne) 1992.

Colebrook, Joan. *A House of Trees*, Farrar, Straus & Giroux (New York) 1987.

Crisp, L. F. *Ben Chifley*, Longmans Green & Co. (Melbourne) 1961.

Cronin, Leonard. *A Key Guide to Australian Mammals*, Reed (Melbourne) 1991.

Curthoys, Martin and Tim Rowse. (eds) *Australians from 1939*, Fairfax (Sydney) 1987.

Davis, Mucke, Narogin and Shoemaker (eds) *Paperbark*, University of Queensland, 1990.

Davison, Hirst, MacIntyre (eds) *The Oxford Companion to Australian History*, Oxford University Press (Melbourne) 1998.

Day, David. *The Great Betrayal*, Angus & Robertson (Sydney) 1998.

Deen, Hanifa. *Caravanserai*, Allen & Unwin (Sydney) 1995.

Edmund, Mabel. *Hello, Johnny!*, Centenary Queensland University Press (Rockhampton) 1986.

Edwards, John. *Keating*, Penguin (Melbourne) 1996.

Elliott, Laurence W. *World War II*, Reader's Digest (Sydney) 1970.

Flannery, Tim. *The Future Eaters*, Reed (Melbourne) 1994.

Foster, David. *Self Portraits*, National Library of Australia (Canberra) 1991.

Fraser, Rosalie. *Shadow Child*, Hale & Iremonger (Victoria) 1998.

Goodall, Heather. *Invasion to Embassy*, Allen & Unwin (Sydney) 1996.

Grey, Anthony. *The Prime Minister was a Spy*, Weidenfeld & Nicolson (London) 1983.

Griffith, Paddy. *Battle Tactics of the Western Front*, Yale (London) 1994.

Griffiths, Tony. *Contemporary Australia*, St Martins (New York) 1977.

Harding, R. W. *Police Killings in Australia*, Penguin (Victoria) 1970.

Hardjono, Ratih. *White Tribe of Asia*, Hyland House (Victoria) 1992.

Hardy, Frank. *The Unlucky Australians*, Pan (London) 1978.

Haste, Cate. *Keep the Home Fires Burning*, Penguin (London) 1977.

Horne, Donald, *The Lucky Country*, Penguin (Melbourne) 1964.

The Coming Republic, Sun (Sydney) 1992.
The Little Digger, Macmillan (Melbourne) 1983.

Horner, David. *Blamey*, Allen & Unwin (Sydney) 1998.

Horton, W. M. introduction *The Hazel de Berg Recordings*. National Library of Australia (Canberra) 1989.

Hughes, Robert. *The Fatal Shore*, Collins (London) 1987.

Irving, Winton. *The Pictures Tell the Story*, Angus & Robertson (Sydney) 1995.

Jones, Cecil. *To Enjoy the Interval*, National Library of Australia (Canberra) 1986.

Keegan, John. *The First World War*, Hutchinson (London) 1998.

Keneally, Thomas. *The Great Shame*, Chatto & Windus (London) 1998.
Memoirs from a Young Republic, Heinemann (London) 1993.
Hombush Boy, Hodder & Stoughton (London) 1995.

Laffin, John. *Digger*, Transworld (London) 1960.

Lambden Owen, W. L. *Cossack Gold*, Hesperian Press (W. A.) 1984.

Lang, J. T. *I Remember*, Invincible Press (Sydney) undated.
The Great Bust, Angus & Robertson (Sydney) 1962.

Langford, Ruby. *Don't Take Your Love to Town*, Penguin (Victoria) 1988.

Learmonth, A. T. and A. M. *Encyclopedia of Australia*, Warner (London) 1968.

McDonald, Neil and Peter Brune. *200 Shots*, Allen & Unwin (Sydney) 1998.

McDonald, Neil. *War Cameraman*, Lothian (Melbourne) 1994.

MacDonnell, Freda. *The Glebe: Portraits and Places*, Ure Smith (Sydney) 1975.

McMullin, Ross. *The Light on the Hill*, Oxford (Melbourne) 1991.

McQueen, Humphrey. *Suspect History*, Wakefield Press (South Australia) 1997.

Manne, Robert. *The Way We Live Now*, Text (Melbourne) 1998.

Mayer, Henry. *The Press in Australia*, Lansdowne (Melbourne) 1964.

Mercer, Derrik. *Chronicle of the Twentieth Century*, Chronicle (London) 1988.

Milliken, Robert. *No Conceivable Injury*, Penguin (Victoria) 1986.

Mitchell, Susan. *Icons, Saints and Divas*, Harper Collins (London) 1997.

Montgomery, Michael. *Who Sank the Sydney?* Leo Cooper (London) 1981.

Moore, Andrew. *The Secret Army and the Premier*, NSW University Press (Sydney) 1989.

Moorhouse, Geoffrey. *Sydney*, Weidenfeld & Nicolson (London) 1999.

Morris, Jan. *Sydney*, Viking Press (London) 1992.

Nicholson, Margaret. *Aussie Fact Book*, Penguin (Victoria) 1997.

O'Connor, Mark. *This Tired Brown Land*, Duffy & Snellgrove (Sydney) 1988.

Park, Ruth. *The Companion Guide to Sydney*, Collins (Sydney) 1973.

Phelan, Nancy. *Some Came Early, Some Came Late*, Macmillan (Melbourne) 1970.

Pilger, John. *A Secret Country*, Cape (London) 1989.

Pring, Adele (ed.) *Women of the Centre*, Pascoe (Victoria) 1990.

Pybus, Cassandra. *The Devil and James McAuley*, University of Queensland 1999.

Read, Peter. *A Rape of the Soul So Profound*, Allen & Unwin (Sydney) 1999.

Reynolds, Henry. *This Whispering in Our Hearts*, Allen & Unwin (Sydney) 1998.

Rowe, John. *Vietnam, the Australian Experience*, Time-Life (Sydney) 1987.

Rowley, C. D. *The Remote Aborigines*, Pelican (Sydney) 1972.

Rusbridger, James and Eric Nave. *Betrayal at Pearl Harbour*, Michael O'Mara (London) 1991.

Salter, Elizabeth. *Daisy Bates*, Angus & Robertson (Sydney) 1971.

Sekuless, Peter. *A Handful of Hacks*, Allen & Unwin (Sydney) 1999.

Sheehan, Paul. *Among the Barbarians*, Random House (Sydney) 1998.

Smolan, Rick and Andy Park (eds) *A Day in the Life of Australia*, National Library of Australia (Canberra) 1981.

Souter, Gavin. *Lion and Kangaroo*, Fontana (Sydney) 1976.

Tacey, David J. *Edge of the Sacred*, Harper Collins (Melbourne) 1995.

Tench, Watkin. *1788*. (ed. Tim Flanner) Text (Melbourne) 1996.

The Bulletin, 1880–1980, Australian Consolidated Press (Sydney) 1980.

Turbet, Peter. *The Aborigines of the Sydney District Before 1788*, Kangaroo Press (NSW) 1989.

War Letters of General Monash, Angus and Robertson (Sydney) 1934.

Ward, Russell. *The Australian Legend*, Oxford University Press (Melbourne) 1958.

Wilkinson, Anne. *Upcountry: Portraits of Rural Australia*, Cassell (Sydney) 1981.

INDEX